Microfluidics for Biotechnology

For a listing of recent titles in the *Artech House Microelectromechanical Systems Series*, turn to the back of this book.

Microfluidics for Biotechnology

Jean Berthier
Pascal Silberzan

ARTECH
HOUSE

BOSTON | LONDON
artechhouse.com

Library of Congress Cataloging-in-Publication Data
Berthier, Jean, 1952–
  Microfluidics for biotechnology/Jean Berthier, Pascal Silberzan.
    p. cm.—(Artech House microelectromechanical systems (MEMS) series)
  Includes bibliographical references and index.
  ISBN 1-58053-961-0 (alk. paper)
    1. Biotechnology. 2. Microfluidics. I. Silberzan, Pascal. II. Title.
  III. Microelectromechanical systems series.

  TP248.2.B3746 2005
  660.6—dc22   2005051315

British Library Cataloguing in Publication Data
Berthier, Jean
  Microfluidics for biotechnology.—(Artech House microelectromechanical systems series)
  1. Microfluidics 2. Biotechnology 3. Microelectromechanical systems
  I. Title II. Silberzan, Pascal
  660.6
  ISBN-10: 1-58053-961-0

Cover design by Yekaterina Ratner

© 2006 ARTECH HOUSE, INC.
685 Canton Street
Norwood, MA 02062

International Standard Book Number: 1-58053-961-0

10 9 8 7 6 5 4 3 2 1

To Erwin, Linda, and Rosanne
—Jean Berthier

To David, Lena, and Flora
—Pascal Silberzan

Contents

Preface

Since the concept of the first DNA biochip, biotechnologies have soared, deeply changing the world of biology. In fact, they have already had direct implications on each of us. Since the very beginning of this science in the 1980s, spectacular advances have been made, such as the analysis of the human DNA genome sequence, while dramatic changes have broken out in the field of proteomics, and more are still expected in the domain of cellular analysis.

The field of investigations of biotechnology is constantly increasing, from the first biochips built to analyze sequences of DNA and investigate its mutations, to protein analysis and the study of the role of proteins in human life, and toward the comprehension of the complex mechanisms that take place inside the cells.

Biotechnology is a science that is not only dedicated to assisting biologists in their desire to understand the complexity of life, but also has very practical applications, especially in bioanalysis and biodetection. For example, progress in the rapidity of detection of viruses has been spectacular, and it is expected that soon direct analysis of viruses may be performed in a few minutes at the doctor's office.

Biotechnology is not only dedicated to *in vitro* analysis, it is transforming *in vivo* medical treatments. Concerning the in vivo domain, the impact of the new technologies is manifold. First, monitoring of the correct functioning of vital organs in patients at risk is going to be possible. Second, miniaturization techniques—brought about by biotechnology—will greatly reduce the invasiveness of external interventions inside human body. Third, new biotechnological devices—such as functional micro- and nanoparticles—will help internal drug guidance to find their targets inside the human body. All of these advances will likely results in drastic changes in medical treatment of such deadly diseases as cancer and diabetes, and will transform our everyday lives in the years to come with the emergence of automated medical help and monitoring right at home.

Biotechnological microsystems are known by different names that have more or less the same physical meaning, like biochips, or bio-MEMS (for *micro-electromagnetic-systems*, or *lab-on-a-chip*—meaning that many of the different operations performed in a lab are done on a single microdevice) and sometimes μTAS (for *micro-total-analysis-systems*).

It is surprising how the concept and development of the first DNA biochip opened the way to a whole new domain of technology. Very soon, it appeared that many other concepts could be imagined and that miniaturization had many advantages in biochemical science. The advantages of biotechnological microsystems concern industry as well as research.

A first advantage resides in the automation and streamlining of biological processes, as it was shown with the DNA biochip: using microchips with thousands of wells—each one testing a specific DNA sequence—reduces considerably the time needed for the recognition process. Another example is that of the proteomic reactors breaking proteins into peptides by enzymatic catalyst inside microchannels; the peptides are then transported by a buffer fluid to a spray injector and to a mass spectrometer, where the peptides are identified. Such a protein chip realizes many operations in sequence. Otherwise many different manipulations and a lot of time would have needed to obtain the same result.

Another advantage of biotechnological microdevices is the reduction of costs of biological analysis due not only to the streamlining and parallelization of the operations, but also to the reduction of the quantities of reactants. Because the reactants needed to perform the sequence of biological reactions are usually quite expensive, it is important that they are used in very limited quantities. Of course, biochips may still be somewhat expensive, especially if etched silicon is used, but the use of cheaper materials like glass or plastic is increasing. Overall the price of the biochip is more than compensated by the gain in the mass of reactants.

It has been also found that the danger of working with toxic, dangerous bacteria, or even explosive substances—in chemistry—is greatly reduced by miniaturization of the reaction scale. Explosive substances are no longer dangerous at very low concentrations, and dangerous viruses and toxic bacteria can be more easily confined in microsystems.

It is also expected that biochips or bio-MEMS can provide higher sensitivity than usual macroscopic systems. For example, some diseases caused by a virus can be detected earlier, at a number of viruses much smaller than that required by the usual diagnostics, leading to more effective treatment and a reduction of the contagion possibility. It has been also shown that chemical reactions are more complete in microsystems.

On the research point of view, there are also many advantages brought by biotechnological microsystems. For example, in vivo interventions are facilitated by the small size of the new biotechnological devices, reducing the invasiveness of the drug delivery system. Another example is that of the technology of encapsulated active microparticles targeted specifically in the human body. It is also expected that biochips will contribute to the discovery of new drugs by testing many new molecules at the same time on living cells isolated in labs-on-a-chip for cells. In that sense, biotechnology is more and more considered a very useful complement to biology itself, as it may contribute to the discovery of new drugs by automatic testing of many molecules at the same time.

As we just have mentioned the complementarity between biotechnology and biology, we point out here that the central theme of biotechnology is the control, displacement, and guidance of the different microsized objects that are present in biologic buffer liquids.

In reality, there are two types of biological objects: "natural" objects like DNA, proteins, antibodies, antigens, peptides, cells, bacteria, red and white blood cells, and so forth, which constitute most of the time the biological targets. Their sizes range from about 20 nm (short strands of DNA) to 100 μm (for the larger cells). The other type of particle is constituted by micro- and nanoparticles, which we may consider "artificial" and which are used as tools to perform specific tasks. In this category, we can list magnetic beads, different fluorophores (CY3, CY5, FITC), quantum dots, gold microparticles, polypyrolles, carbon nanowires, surfactants, and the like. These objects are smaller than the previous ones, ranging from 10 nm to 2 μm. Some natural biological objects can be also used as tools, especially for biorecognition processes: a DNA strand can be considered a tool to immobilize a complementary DNA strand. We will refer to all of these objects under the names of micro- and nanoparticles and macromolecules. A simplified statement is that biologists study the functional behavior of biological objects, whereas biotechnologists focus on the mechanical behavior of these objects.

In this book, we present and detail the mechanical behavior of the different micro- and nanoparticles and macromolecules that are used in biotechnological microsystems.

The two first chapters are dedicated to the microfluidics aspects of the buffer fluid flowing in biochip microchannels. In order to predict correctly the behavior of particles, the physical behavior of the buffer (carrier) fluid must be first determined. Chapter 1 treats continuous single-phase microflows and two-phase plug microflows of immiscible liquids. The second chapter is dedicated to digital microfluidics, sometimes called drop microfluidics, which is less classical but appears to be a promising way of transporting biological objects. In such a scheme, microdrops of less than 500 μm of radius are moved individually step by step on a flat surface.

Because the micro- and nanoparticles and macromolecules we are interested in are much larger than the fluid molecules, their behavior differs from that of the fluid. So, after the first two chapters concerning the carrier fluid flow, the next two chapters focus on the mechanical behavior of the particles themselves, under the action of diffusion (Chapter 3) and transport by advection (Chapter 4). Different numerical approaches are presented, and the important case of confined volumes is examined.

All of the studies on the buffer (carrier) fluid flow and the behavior of the particles in this flow are aimed at controlling the motion of the particles of interest to have them reacting at some specific location. Chapter 5 is then dedicated to the study of biochemical reactions. First, the principle of biorecognition is presented, and the different biochemical reactions to recognize DNA sequences and antibodies are studied. Next, to take into account the transport of the reactants by the buffer fluid, a coupled approach including diffusion/advection of reactants and the biochemical reaction itself is examined.

At this stage, it appears useful to present in more detail the most interesting micro- and nanoparticles and the macromolecules that are used in biotechnology and to give to the reader some concrete insight into the usual experimental techniques used to manipulate these particles. This is the object of Chapter 6.

Because it was found that transport by the buffer fluid is often not specific enough, complementary methods have been developed. In Chapter 7, we present the principle of labeled magnetic microbeads and show how these beads are used to bind with the targeted biological objects and to transport them into specifically designated areas.

Another typical way of controlling the motion of microparticles is based on the use of electric fields. In Chapter 8, we present the different ways electric fields act on the particles, including electrophoresis, dielectrophoresis, and electro-osmosis.

Acknowledgments

We thank our colleagues, particularly N. Sarrut and F. Rivera, for their contribution with photographs and pictures, and A. Buguin, who has been kind enough to review and comment on some chapters. Our discussions with A. Ajdari, R. Austin, D. Chatenay, J.-F. Joanny, P. Massé, F. Perraut, J. Prost, and L. Talini have fueled many parts of this book, particularly Chapters 6, 7, and 8. We would also like to thank J. Chabbal, the director of the Biotechnology Department at LETI, and P. Le Ber, head of the Biochip Laboratory at LETI, for their support for this project. We would also like to thank the editing team of Artech House for their help.

Finally, we wish to express all our gratitude to our spouses, Susanne and Isabelle, for their patience and their support during the long hours of writing of this book.

CHAPTER 1

Microflows

1.1 Introduction

1.1.1 On the Importance of Microfluidics in Biotechnology

Biotechnology is closely linked to microfluidics. Biological targets are nearly always transported by a buffer fluid or carrier fluid, as well in vitro and in vivo. In the human body, any bio-MEMS has to deal with body fluids. In *in vitro* microsystems, the target molecules/particles are nearly always transported by a buffer fluid for many reasons: first, the target molecules/particles are most of the time extracted from a liquid (e.g., DNA and cells); second, the biochemical reactions on these targets are performed in an aqueous environment; and third, confinement of the targets is easier in a liquid than in a gas. Very few examples of biotechnological microsystems exist that do not require the use of microfluidics. One counterexample might be the "electronic nose," where detection of target molecules transported by ambient air is done directly on a dry contact surface by mass spectrometry. Even for bacteria carried by air, like legionella, it is now thought to extract and concentrate them in a water base.

In this chapter, we deal with liquid microflows, and we will not consider gas flows. For the reader whose concern is gas microflows, useful information can be found in [1].

If liquids are the most frequent carrier of microparticles, the flow pattern can be very different. In the next section, we present the different patterns of microflows currently used in biotechnology.

1.1.2 From Single Continuous Flow to Droplets

In biotechnology, microfluidics is present under different forms depending on the different applications. The most usual form is single-phase microflow in channels and capillary tubes (Figure 1.1). Typically, this is the general case of buffer liquids carrying biologic targets and circulating in an assembly of microchannels, where different biologic processes successively take place. Different reactions, biochemical analysis, and detection can be done in microchambers placed alongside the flow. Usually, the fluid is moved under the effect of pressure (from a syringe or a micropump), sometimes by electric forces (electro-osmotic flow).

In order to accelerate the speed of biorecognition—for high throughput screening, for example—samples of different buffer solutions have to be treated simultaneously and continuously. In such a case, the solution is to convey successive buffer fluid plugs in capillary tubes; the plugs are separated by a nonmiscible,

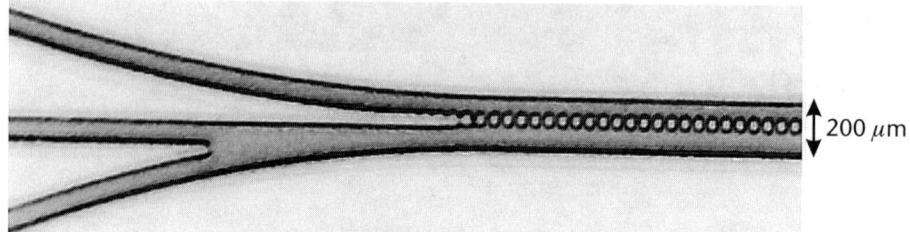

Figure 1.1 Continuous flow microfluidics: microchannels designed for liquid-liquid extraction. Two immiscible liquids are moving in parallel, separated by pillars aligned in the middle. Micro- and nanoparticles or macromolecules carried by one of the liquids migrate through the interface into the other liquid. If the flow rate of this last liquid is small, the particles are automatically concentrated. (Courtesy of N. Sarrut, CEA/LETI.)

biocompatible liquid (Figure 1.2). Such types of flows are commonly called multiphase flows or, if there are only two liquids, two-phase flows. One of the fluids may be a gas (e.g., alternate plugs of liquid separated by air bubbles are used to create microemulsions or sprays).

A similar, but slightly different category of flow results from the break up of one phase into droplets (Figure 1.3). It is still a multiphase flow, but the discontinuous phase is more or less dispersed in the continuous phase. This way of proceeding may be interesting to concentrate and isolate some biological targets in very small volumes of buffer fluid.

Finally, microdrops can be considered separate entities and manipulated individually (Figure 1.4). This type of fluid motion is often called *digital microfluidics*. The aim here is to manipulate with precision extremely small quantities of fluid—of the order of tens of nanoliters—containing the biological target (e.g., DNA, cells).

In this chapter, we deal with continuous and two-phase microflows. Digital microfluidics will be treated in Chapter 2.

1.2 Single-Phase Microflows

As we saw in the preceding section, the first category of microflows is the single-phase continuous microflow. It is the most widely used in biotechnology for transporting biological targets and detection probes. In biochips, the fluid is flowing under the effect of a driving pressure (imposed by a micropump), under the action of a piston (syringe or deflected membrane), or under the action of an electric field,

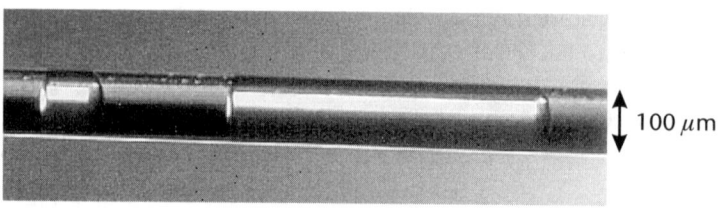

Figure 1.2 Liquid plugs in a capillary tube. Alternate plugs of buffer solution—separated by immiscible silicon oil plugs—circulate inside a capillary tube.

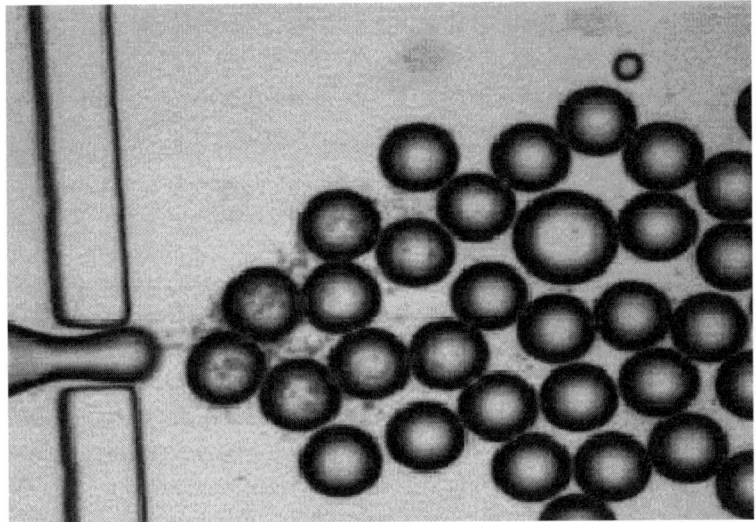

Figure 1.3 Water droplets in oil. Transition from a continuous microflow (left part of the photograph) to microdrops (right part of the photograph) by shearing through an aperture in a solid wall. (Courtesy of A. Shen, Washington University, St. Louis, Missouri.)

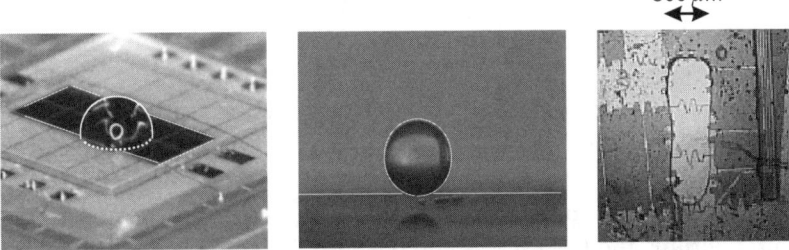

Figure 1.4 Digital microfluidics: photographs of a water microdrops on an electrode (left), a water microdrop in silicon oil submitted to a pulsed electric field (middle), and a water drop in silicon oil in an EWOD microdevice. (Courtesy of C. Peponnet, CEA/LETI.)

inducing an electro-osmotic effect [1]. Different apparatuses are design to monitor and actuate the flow as valves, compliance chambers, passive or active micromixers, and such. One of the major difficulties is to integrate many different functions in a miniaturized system; this is due to limitations brought by the fabrication process, the different connections between the fluidic elements, and taking into account the handling and packaging of the components.

In the following, we will not go into the details of the many microfluidics apparatuses, but we present the equations that govern fluid flows, and we focus on the physics and particularities of microflows.

1.2.1 Navier-Stokes Equations

The macroscopic approach for the calculation of the velocities and pressure in a fluid is based on the continuum hypothesis (i.e., in every elementary volume of the fluid, there are enough molecules to define statistical properties like velocity and pressure). The continuum hypothesis works well at a microscopic scale for liquids; for gases, the hypothesis breaks down at the nanoscopic scale where the characteristic Knudsen number (Kn) becomes of the order of 1.

$$Kn = \frac{\lambda}{L} \tag{1.1}$$

where λ is the mean free path of the molecules and L is the characteristic dimension of the channel. For gases, the limit for L is about 1 μm. In liquids, the mean free path is much smaller, and the continuum hypothesis is applicable to most microsystems.

On the most general point of view, fluid flows are determined by the knowledge of velocities $U = \{u_i, i = 1,3\}$, pressure P, density ρ, viscosity μ, specific heat C_p, and temperature T. For each fluid, density, viscosity, and specific heat are related to pressure and temperature (or enthalpy) via characteristic equations of state (EOS):

$$\rho = f(P, T)$$
$$\mu = g(P, T) \tag{1.2}$$
$$C_p = h(P, T)$$

Pressure and temperature characterize the number and the state of the molecules that are present in a given volume. Equations of state are generally complicated, but they can be approximated by analytical functions if the domain of variation of the parameters (P and T) is not too large. Thus, we are left with five unknowns: u_x, u_y, u_z, P, and T. These unknowns are related by a system of three equations: (1) a scalar equation for the mass conservation, (2) a vector equation for the conservation of momentum, and (3) a scalar equation for the conservation of energy. In biotechnology, fluid flows are often isothermal or variation of temperature is negligible. Note that this is not the case for microchemistry, where chemical reactions are seldom isothermal. If temperature is constant or nearly constant, we have to deal with four unknowns u_x, u_y, u_z, and P, with the help of the mass conservation equation and the conservation of momentum equation, plus the EOS $\rho = f(P), \mu = g(P)$. Some authors give the name Navier-Stokes (NS) equations to the whole system; others confine this name to the second equation (momentum).

1.2.1.1 General Case: Governing Equations

The first equation is the mass conservation (or continuity) equation. For simplicity, we demonstrate here only the two-dimensional form of this equation. Assume a velocity field (u, v) and an element of volume $(\Delta x, \Delta y)$ as sketched in Figure 1.5.

The mass conservation equation requires

$$\frac{\partial (\rho \Delta x \Delta y)}{\partial t} = \rho u \Delta y + \rho v \Delta x - \left[\rho u + \frac{\partial (\rho u)}{\partial x}\right]\Delta y - \left[\rho v + \frac{\partial (\rho v)}{\partial y}\right]\Delta x$$

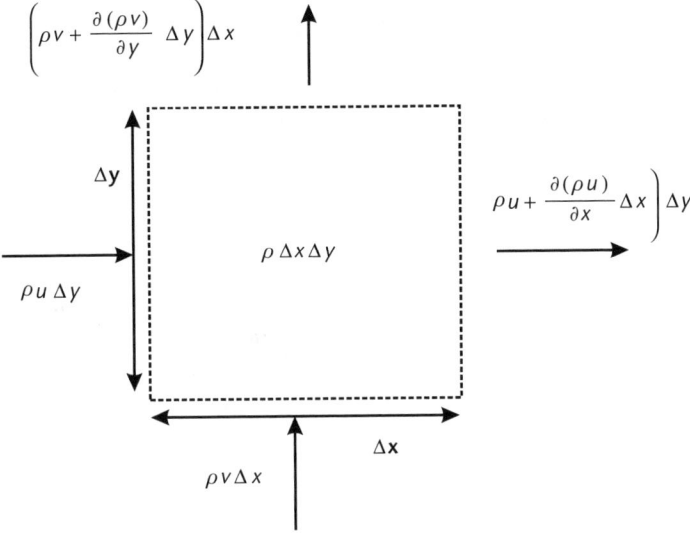

Figure 1.5 Two-dimensional conservation of mass.

or, dividing by $\Delta x\,\Delta y$,

$$\frac{\partial \rho}{\partial t} + \frac{\partial (\rho u)}{\partial x} + \frac{\partial (\rho v)}{\partial y} = 0 \tag{1.3}$$

This equation may be written as

$$\frac{\partial \rho}{\partial t} + u\frac{\partial \rho}{\partial x} + v\frac{\partial \rho}{\partial y} + \rho\left[\frac{\partial u}{\partial x} + \frac{\partial v}{\partial y}\right] = 0 \tag{1.4}$$

and, under a vector form

$$\frac{D\rho}{Dt} + \rho\nabla \cdot \vec{V} = 0 \tag{1.5}$$

where the operator D/Dt is

$$\frac{D}{Dt} = \frac{\partial}{\partial t} + u\frac{\partial}{\partial x} + v\frac{\partial}{\partial y} + w\frac{\partial}{\partial z} \tag{1.6}$$

in a three-dimensional Cartesian coordinate system. Liquids may generally be considered incompressible, and the mass conservation equation is then reduced to

$$\nabla \cdot \vec{V} = 0 \tag{1.7}$$

In Cartesian coordinates, we have

$$\frac{\partial u}{\partial x} + \frac{\partial v}{\partial y} + \frac{\partial w}{\partial z} = 0$$

and in cylindrical coordinates, the axisymmetric form of (1.7) is

$$\frac{\partial v_r}{\partial r} + \frac{v_r}{r} + \frac{\partial v_z}{\partial z} = 0$$

The second equation is the momentum conservation equation (or NS equation). The change of momentum in a fluid element is equal to the balance between inlet momentum, outlet momentum, and exerted forces [2]. Figure 1.6 shows the inlet and outlet momentum in the x-direction, and Figure 1.7 represents the forces on the same fluid element.

Projecting all of these forces on the x-direction, and using the mass conservation equation, we obtain

$$\rho \frac{Du}{Dt} = -\frac{\partial \sigma_x}{\partial x} + \frac{\partial \tau_{xy}}{\partial y} + F_x \tag{1.8}$$

Normal stress and tangential stress are for most fluids (called *Newtonian fluids*, see the following discussion) given by the constitutive relations

$$\sigma_x = P - 2\mu \frac{\partial u}{\partial x} + \frac{2}{3\mu}\left(\frac{\partial u}{\partial x} + \frac{\partial v}{\partial y}\right)$$

$$\tau_{xy} = \mu\left(\frac{\partial u}{\partial y} + \frac{\partial v}{\partial x}\right) \tag{1.9}$$

where μ is the dynamic viscosity. Combining (1.8) and (1.9), and extending the formulation to the three-dimensional case, yields the NS equation

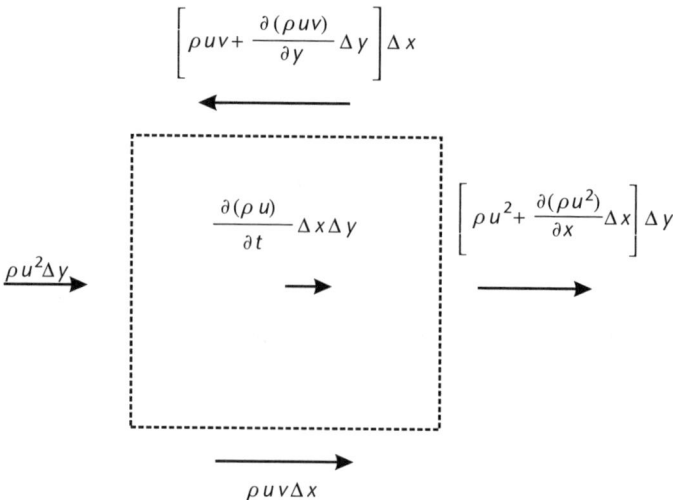

Figure 1.6 Momentum in the x-direction.

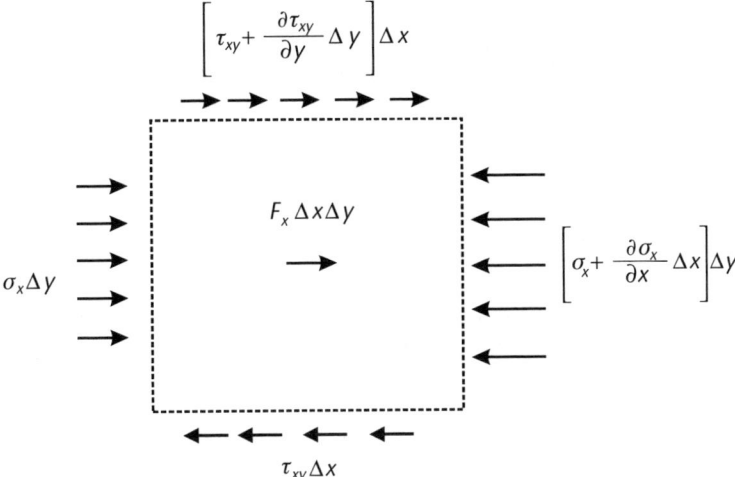

Figure 1.7 Force balance in the x direction. Normal stress σ_x, tangential stress τ_{xy}, and body forces per unit volume F_x on a $\Delta x\,\Delta y$ element.

$$\rho\left(\frac{\partial u}{\partial t}+u\frac{\partial u}{\partial x}+v\frac{\partial u}{\partial y}+w\frac{\partial u}{\partial z}\right)=-\frac{\partial P}{\partial x}+\mu\left(\frac{\partial^2 u}{\partial x^2}+\frac{\partial^2 u}{\partial y^2}+\frac{\partial^2 u}{\partial z^2}\right)+F_x$$

$$\rho\left(\frac{\partial v}{\partial t}+u\frac{\partial v}{\partial x}+v\frac{\partial v}{\partial y}+w\frac{\partial v}{\partial z}\right)=-\frac{\partial P}{\partial y}+\mu\left(\frac{\partial^2 v}{\partial x^2}+\frac{\partial^2 v}{\partial y^2}+\frac{\partial^2 v}{\partial z^2}\right)+F_y \qquad (1.10)$$

$$\rho\left(\frac{\partial w}{\partial t}+u\frac{\partial w}{\partial x}+v\frac{\partial w}{\partial y}+w\frac{\partial w}{\partial z}\right)=-\frac{\partial P}{\partial z}+\mu\left(\frac{\partial^2 w}{\partial x^2}+\frac{\partial^2 w}{\partial y^2}+\frac{\partial^2 w}{\partial z^2}\right)+F_z$$

The vectorial notation is

$$\rho\frac{D\vec{V}}{Dt}=-\nabla P+\mu\Delta\vec{V}+\vec{F} \qquad (1.11)$$

where $\vec{V}$ is the velocity vector (u,v,w) and F is the body force per unit volume.

1.2.1.2 Axisymmetric Formulation of the NS Equations for Incompressible Liquids

It is useful to have the NS equation written in a cylindrical axisymmetric coordinates system

$$\rho\left(\frac{\partial v_r}{\partial t}+v_r\frac{\partial v_r}{\partial r}+v_z\frac{\partial v_r}{\partial z}\right)=-\frac{\partial P}{\partial r}+\mu\left(\frac{\partial^2 v_r}{\partial r^2}+\frac{1}{r}\frac{\partial v_r}{\partial r}-\frac{v_r}{r^2}+\frac{\partial^2 v_r}{\partial z^2}\right)+F_r$$

$$\rho\left(\frac{\partial v_z}{\partial t}+v_r\frac{\partial v_z}{\partial r}+v_z\frac{\partial v_z}{\partial z}\right)=-\frac{\partial P}{\partial z}+\mu\left(\frac{\partial^2 v_z}{\partial r^2}+\frac{1}{r}\frac{\partial v_z}{\partial r}+\frac{\partial^2 v_z}{\partial z^2}\right)+F_z \qquad (1.12)$$

1.2.1.3   Energy Equation

In case there is a change in the liquid temperature, an energy conservation equation is added to complete the preceding system. This energy equation is, in Cartesian coordinates,

$$\rho C_p \left(\frac{\partial T}{\partial t} + u \frac{\partial T}{\partial x} + v \frac{\partial T}{\partial y} + w \frac{\partial T}{\partial z} \right) = \frac{\partial}{\partial x} \left(k \frac{\partial T}{\partial x} \right) + \frac{\partial}{\partial y} \left(k \frac{\partial T}{\partial y} \right) + \frac{\partial}{\partial z} \left(k \frac{\partial T}{\partial z} \right) + q \quad (1.13)$$

where C_p is the specific heat (in J/kg/K), k is the conduction coefficient (in W/m/K), and q is a source or sink term (in W/m^3).

1.2.1.4   Newtonian and Non-Newtonian Rheology

In the preceding sections, we have used constitutive relations for the normal and tangential stresses. A short discussion about these relations might be appropriate here because it appears that some buffer fluids transporting a large concentration in polymers may show a different behavior than that of (1.8).

Assume a fluid flowing parallel to a solid wall with a velocity u. At the wall surface, the velocity is zero, and there is a velocity profile $u(y)$ along the direction y normal to the wall, as sketched in Figure 1.8.

For a vast majority of fluids, the shear stress is linked to the shear rate by Newton's relation

$$\tau_x = -\mu \frac{\partial u}{\partial y} \quad (1.14)$$

where μ is the dynamic viscosity expressed in kg/m/sec (International System), and τ_x is the wall shear stress expressed in N/m^2. All of the fluids responding to such a relation are called Newtonian fluids or viscous fluids.

Not all fluids follow the relation (1.14); there is a category of fluids called viscoelastic—the opposite of viscous—where the stress-strain tensor is more complicated that the preceding Newtonian formulation. Specially, there are two different kinds of non-Newtonian fluids; the first ones have a shear stress/shear rate relation of the form

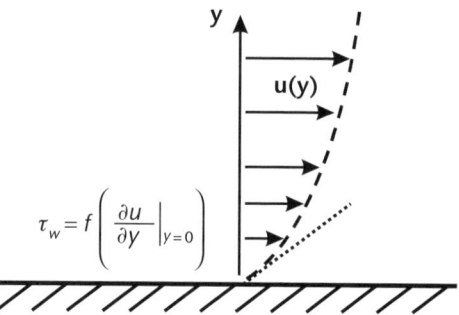

Figure 1.8 Friction on a solid wall is due to the shear stress τ, related to the shear rate by the relation $\tau = -\mu \frac{\partial u}{\partial y}$.

$$\tau_x = \tau_0 + K\left(\frac{\partial u}{\partial y}\right)^n \tag{1.15}$$

where K is a consistency index. Relation (1.15) is called the Herschel-Bulkley law [3]. The shear stress versus the shear rate has been plotted in Figure 1.9 for Newtonian and non-Newtonian fluids. By comparing (1.14) and (1.15), we may derive an apparent viscosity

$$\mu = -K\left(\frac{\partial u}{\partial y}\right)^{n-1}$$

There is a second category of non-Newtonian fluids—called time-dependent non-Newtonian—where the apparent viscosity depends on a relaxation time: For these liquids, there is a relaxation effect after any deformation of an element of liquid, as shown in Figure 1.10.

It appears that some biological liquids—like blood or buffer fluids carrying large amount of polymers—show some non-Newtonian behavior, especially under transient conditions. A very simplified explanation of the non-Newtonian rheology of these fluids lies in the morphology of the transported vesicles and polymers, as shown in Figure 1.11.

A similar explanation stands for blood, which is a slightly non-Newtnonian fluid due to the interactions and deformations of the vesicles like red blood cells (RBCs) or erhythrocytes and white blood cells (WBCs). The rheological behavior of blood is that of a Casson liquid [4].

$$\sqrt{\tau_x} = \sqrt{\tau_0} + K\sqrt{\frac{\partial u}{\partial y}}$$

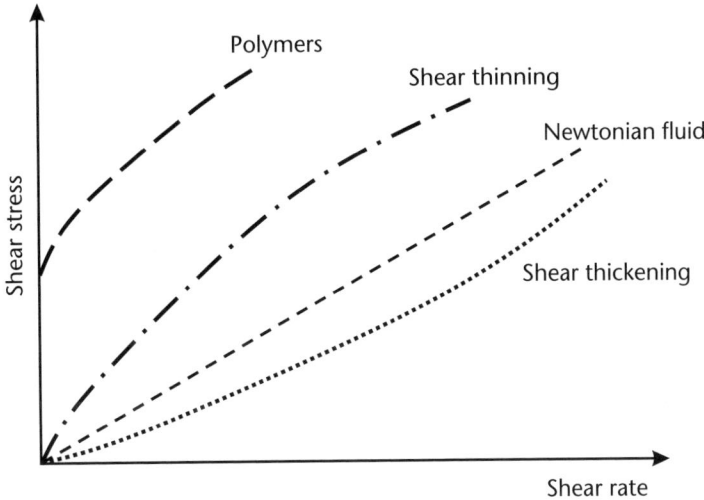

Figure 1.9 Shear stress versus shear rate for different fluids. For viscoelastic fluids, the apparent viscosity depends on the shear rate.

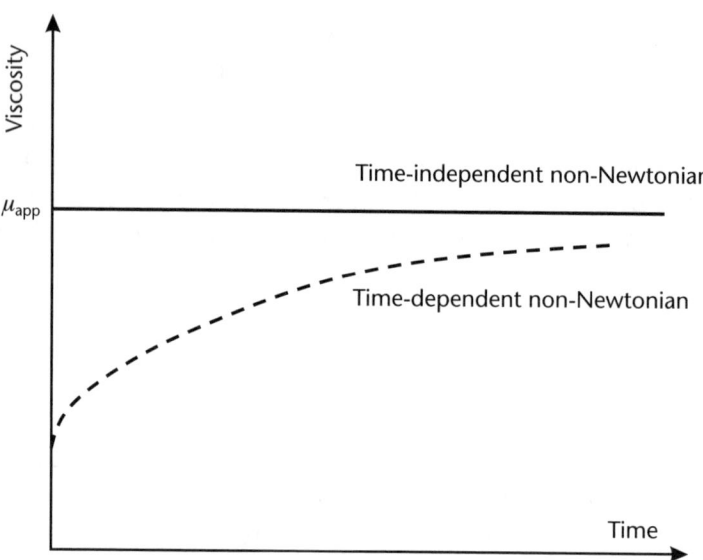

Figure 1.10 Time-dependent non-Newtonian behavior. The apparent viscosity depends also on the relaxation time.

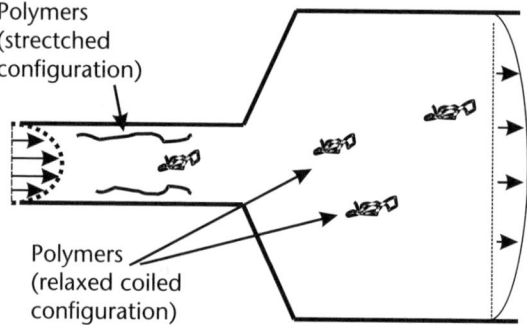

Figure 1.11 Non-Newtonian behavior of a buffer fluid transporting long polymer explained by the different configurations of the polymer chain: under a shear stress, the polymer is stretched, and when the shear stress is decreased, the polymer reconfigures to a coiled shape. This change of configuration dependence on the shear stress explains the time-dependent apparent viscosity if there is a sufficient concentration of polymers in the buffer fluid.

However, we shall deal here with small concentration of polymers or vesicles, and we may assume in the following a Newtonian behavior for the most of the buffer fluids. Blood is a special case, and, until recently, not much attention has been given to its rheological properties.

1.2.2 Laminarity of Microflows

Regardless of the size—macroscopic or microscopic—a fluid flow is said to be laminar when viscous forces dominate inertia. When this is the case, turbulences cannot develop, and the fluid flow lines are, at least locally, parallel. One can picture it by considering that the flow is locally laminated. On the other side, a turbulent flow presents fluctuating random vertices, even at very small scales. It is intuitive to think that microflows will predominantly be laminar, since there is a strong limitation for vertices development and randomness due to the proximity of the solid walls. One important point here is that flow recirculation (vortex) does not necessarily means that the flow is turbulent, as it is shown in Figure 1.12.

A nondimensional number—the Reynolds number—determines the ratio between inertia (convective forces) and viscous forces.

$$Re = \frac{UD}{\nu} \tag{1.16}$$

where U is the average fluid velocity, D is a characteristic dimension of the channel (or the obstacle), and $\nu = \dfrac{\mu}{\rho}$ is the kinematic viscosity (expressed in m^2/s). The Reynolds number naturally appears by performing a dimensional analysis of the NS equations. For simplicity we consider the two-dimensional Cartesian NS equation with no external forces:

$$\begin{aligned}
\frac{\partial u}{\partial t} + u\frac{\partial u}{\partial x} + v\frac{\partial u}{\partial y} &= -\frac{1}{\rho}\frac{\partial p}{\partial x} + \frac{\mu}{\rho}\left[\frac{\partial^2 u}{\partial x^2} + \frac{\partial^2 u}{\partial y^2}\right] \\
\frac{\partial v}{\partial t} + u\frac{\partial v}{\partial x} + v\frac{\partial v}{\partial y} &= -\frac{1}{\rho}\frac{\partial p}{\partial y} + \frac{\mu}{\rho}\left[\frac{\partial^2 v}{\partial x^2} + \frac{\partial^2 v}{\partial y^2}\right]
\end{aligned} \tag{1.17}$$

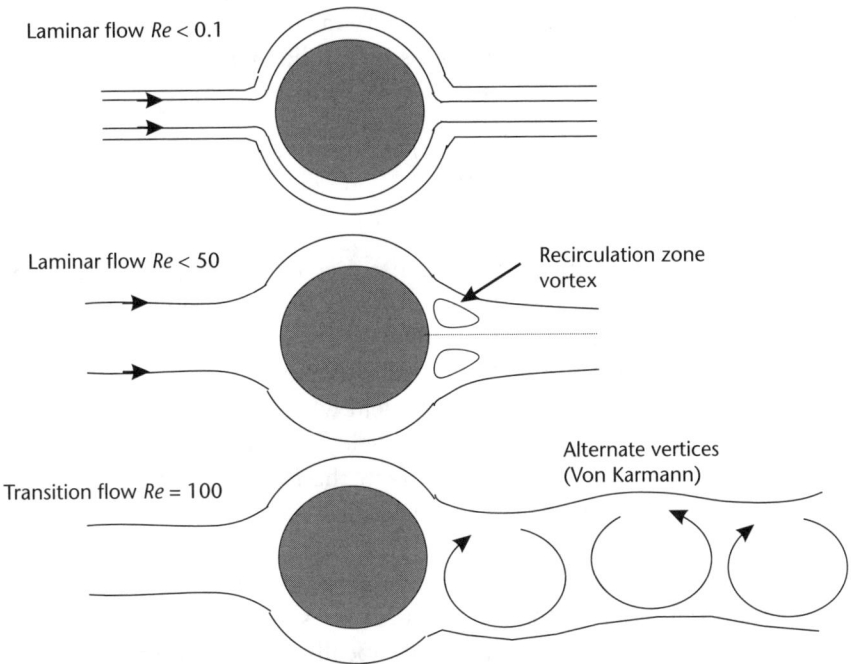

Laminar flow $Re < 0.1$

Laminar flow $Re < 50$

Recirculation zone vortex

Transition flow $Re = 100$

Alternate vertices (Von Karmann)

Figure 1.12   Different patterns of laminar and transitional flows behind a cylinder.

If U is a velocity reference (the average velocity), and D is a length reference (a characteristic dimension of the flow such as the diameter of the tube for a microflow in a capillary), we use the following scaling

$$u^* = \frac{u}{D}, \; v^* = \frac{v}{U}, \; x^* = \frac{x}{D}, \; y^* = \frac{y}{D}, \; t^* = \frac{t}{D/U}, \; p^* = \frac{p}{\rho U^2}$$

Note that ρU^2 has the same unit as the pressure. Then, the system (1.17) becomes

$$\frac{\partial u^*}{\partial t^*} + u \frac{\partial u^*}{\partial x^*} + v \frac{\partial u^*}{\partial y^*} = -\frac{\partial p^*}{\partial x^*} + \frac{\nu}{UD} \left[\frac{\partial^2 u^*}{\partial x^{*2}} + \frac{\partial^2 u^*}{\partial y^{*2}} \right]$$

$$\frac{\partial v^*}{\partial t^*} + u^* \frac{\partial v^*}{\partial x^*} + v^* \frac{\partial v^*}{\partial y^*} = -\frac{\partial p^*}{\partial y^*} + \frac{\nu}{UD} \left[\frac{\partial^2 v^*}{\partial x^{*2}} + \frac{\partial^2 v^*}{\partial y^{*2}} \right]$$

(1.18)

The system (1.18) is nondimensional with only one nondimensional parameter, the Reynolds number. Note that this conclusion agrees with Buckingham's theorem [5], which states that if a problem depends on N dimensional parameters containing M different units, the nondimensional form depends on $N - M$ nondimensional numbers. In the present case, there are $N = 4$ dimensional parameters ρ, μ, D, and U. The M units contained in these four parameters are kilos, meters, and seconds. From Buckingham's theorem, it results that there is $N - M = 1$ nondimensional number for the dimensionless system.

The characteristic scales D and U depend on the geometry of the problem. For a flow inside a tube, U is the average axial velocity and D is the tube diameter; for an obstacle in a fluid flow, U is the velocity far from the obstacle and D is the obstacle characteristic dimension (usually its hydraulic diameter).

The criterion for laminar flow has the form

$$Re = \frac{UD}{\nu} < Re_{trans}$$

(1.19)

Re_{trans} is the transition threshold between laminar and turbulent flow. For flow in tubes and pipes, Re_{trans} is of the order of 1,000–2,000 and for a flow past an obstacle, Re_{trans} is of the order of 64–100 [6].

The main difference between macroscopic and microscopic flow is that most of the time macroscopic flows are turbulent whereas microscopic flows are laminar. In biotechnology—or microchemistry—velocities are mainly small, and it is very seldom that the flow is turbulent. In fact, the *laminarity* of the flow is usually high. Typical fluid velocities are of the order of 1 mm/sec at the most in channels of cross dimensions of 1 mm at the most. As the cinematic viscosity of water is $\nu = 10^{-6}$ m^2/sec, the Reynolds number is—at the most—of the order of 1. More often, the Reynolds number is of the order of 0.1. Thus, the character of the flow is very laminar, meaning that the streamlines are locally parallel and that even obstacles in the flow are not going to induce any turbulence.

This high degree of laminarity implies that the streamlines are locally parallel. A simple picture one can retain is that the flow is composed of tubes similar to fibers that can be bent but still are parallel. Figure 1.13 illustrates this principle.

We will see in the next section that for very small Reynolds numbers, the NS equations may be simplified and are reduced to the Stokes approximation.

1.2.3 The Stokes Equation

For a stationary flow, at very low velocities, inertial forces become very small compared to the viscous forces. The Reynolds number is smaller than 1 and the inertia terms on the left of (1.11) may be neglected. In this regime, the NS equation reduces to the Stokes equation:

$$\frac{\partial p}{\partial x} = \nu \left[\frac{\partial^2 u}{\partial x^2} + \frac{\partial^2 u}{\partial y^2} \right] + \frac{F_x}{\rho}$$

$$\frac{\partial p}{\partial y} = \nu \left[\frac{\partial^2 v}{\partial x^2} + \frac{\partial^2 v}{\partial y^2} \right] + \frac{F_y}{\rho} \tag{1.20}$$

$$\frac{\partial p}{\partial z} = \nu \left[\frac{\partial^2 w}{\partial x^2} + \frac{\partial^2 w}{\partial y^2} \right] + \frac{F_z}{\rho}$$

In the case where the external force is just the gravity force, the simplification is considerable because the system (1.20) is now linear:

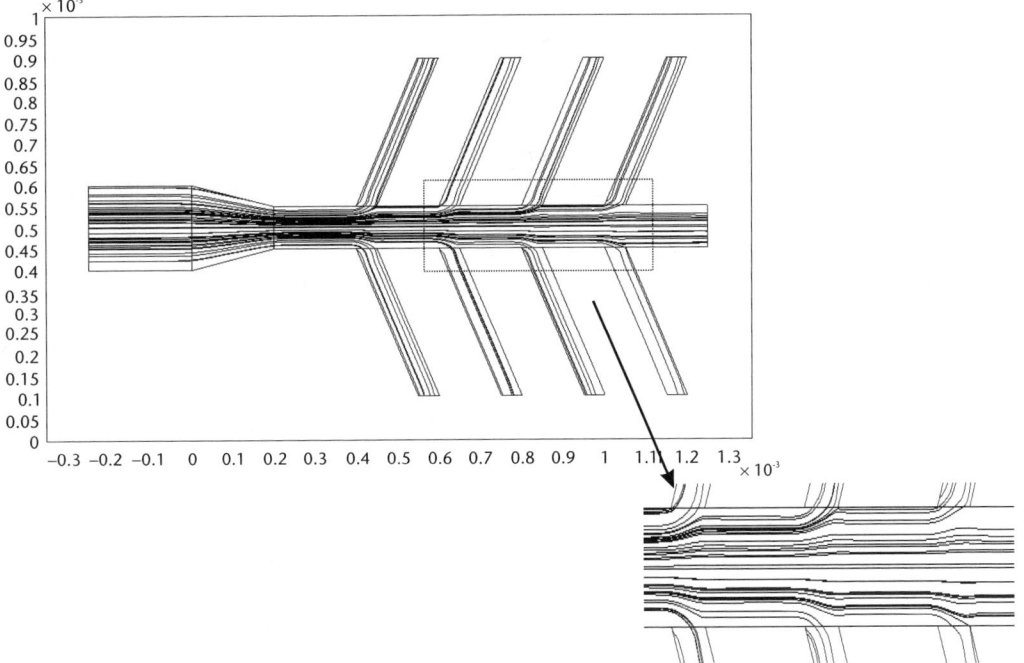

Figure 1.13 Microflow in a capillary with many exits. The flow is totally predictable (i.e., the *flow tube* exiting at a given outlet can be pinpointed at the entrance).

$$\nabla p = v\Delta \vec{V} + \frac{1}{2}\nabla z^2 \qquad (1.21)$$

where we have used the notation $\Delta = \nabla^2$ for the Laplacian operator. By taking the rotational of (1.21), and using the following mathematical relations

$$curl(grad\ P) = \nabla \times \nabla p = 0$$

$$\Delta(curl\ \vec{A}) = curl(\Delta\vec{A})$$

we obtain

$$\Delta(\nabla \times \vec{V}) = \Delta\vec{\omega} = 0 \qquad (1.22)$$

where ω is the vorticity of the flow. Thus, in the Stokes formulation, vorticity is a harmonic function [7] and the problem can be solved in the vorticity-streamline formulation as soon as the values of the vorticity on the boundaries are known.

Stokes formulation for creeping flows is very attractive because an apparently complex problem can be simplified to a *linear* formulation. Besides *linearity*, Stokes equation is *reversible* [8] (i.e., a change of the velocity u to its opposite $-u$ on the boundaries of the domain will result in a change of all the velocities to their opposite).

An example of this reversibility property can be done by considering a microdeflector in a microflow, as sketched in Figure 1.13. The calculation of the flow has been performed with the numerical software FEMLAB using the complete NS equations. The flow lines are shown in Figure 1.14, for an inlet velocity of 1 mm/sec from left to right. If the flow is reversed, the pattern of the flow lines is exactly the same (Figure 1.15). This is typically a case where the Stokes equations are sufficient to describe the flow.

The property of reversibility is very important because it shows that, at very low velocities, it is not possible to design a microfluidic *diode* where the pressure drop would be small in one direction and large in the opposite. If one wants to design such a diode, the flow velocity must be sufficiently large to be outside the Stokes hypothesis. For example, micropipes can give directionality to micropumping, based on a dissymmetrical design, as shown in Figure 1.16, if the fluid flow is sufficiently important.

A coarse simulation of such a design has been performed using the FEMLAB software (Figure 1.17). For a flow of 1 mm/sec from left to right, the pressure drop is 296 Pa; it is 324 Pa if the flow is reversed. For a flow of 100 μm/sec, the pressure drop is 29 or 31 Pa depending on the direction of the flow. Thus, such a design functions as a diode for flow velocities larger than approximately 1 mm/sec.

Besides linearity and reversibility, the Stokes equation has a *unique solution*, meaning that there cannot be bifurcations of the solution linked to the development of instabilities.

It is worth noticing that care must be taken when deciding to apply Stokes' simplification: the condition $Re \ll 1$ must be verified everywhere in the fluid domain; if there is only one location, however small, where this condition is not realized, then the simplification may not be valid and will bias the whole solution.

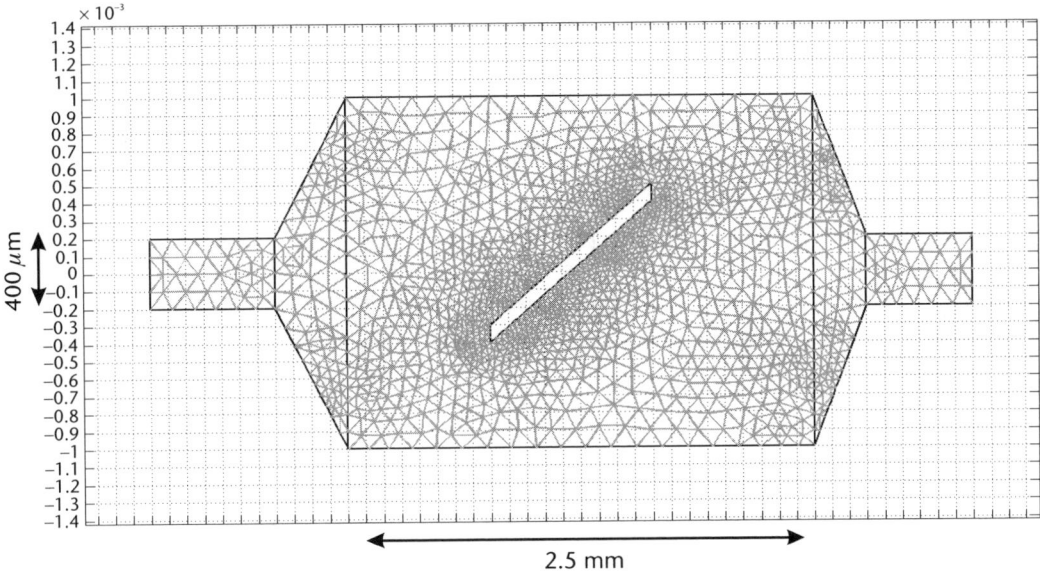

Figure 1.14   Computational domain and calculation grid of a 2D microchamber with a deflector. The maximum width is 2 mm and the length of the chamber is 2.6 mm. Note that the meshing depends on the shape of the solid walls.

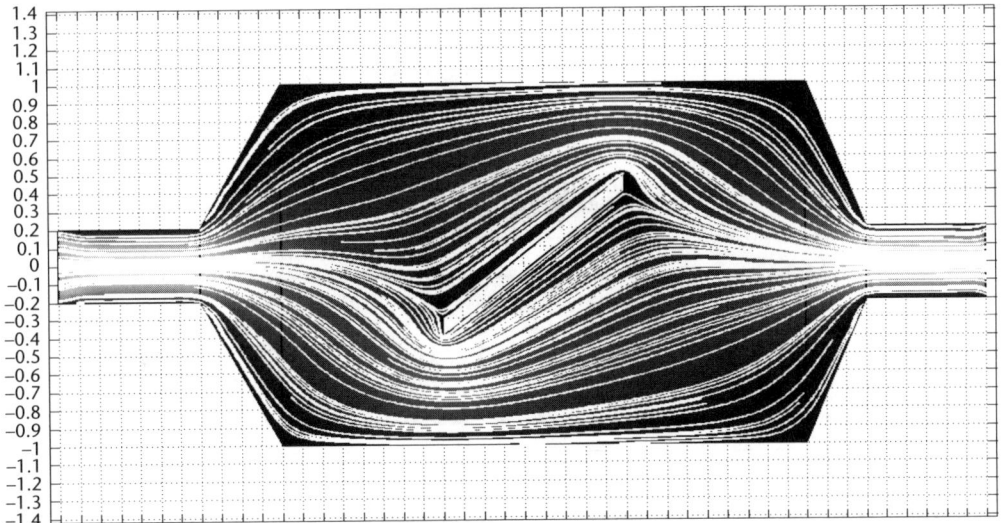

Figure 1.15   Flow lines in the microchamber. The deflector does not induce any recirculation.

1.2.4   Hydrodynamic Entrance Length—Establishment of the Flow

When a fluid enters a tube, at the entrance of a tube, there is a length where the flow is not yet established: the shape of the velocity profile evolves until it reaches the established profile (Hagen-Poiseuille profile for a laminar flow), as shown in Figure 1.18. The evolution of the profile is due to the progressive increase of the boundary

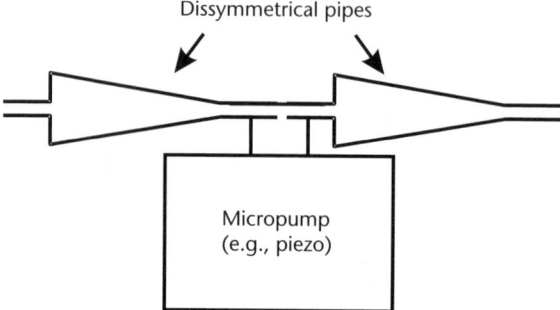

Figure 1.16 Directionality of pumped flow obtained by dissymmetrical pipes. Oscillations of the piezoelectrically actuated membrane triggers a directional flow only in the case of a sufficiently high Reynolds number.

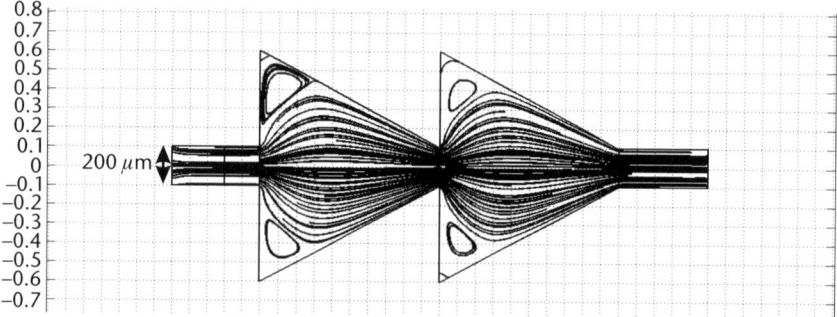

Figure 1.17 Modeling a fluidic diode: stream lines in dissymmetrical convergents (FEMLAB software). The flow is laminar but includes recirculating regions.

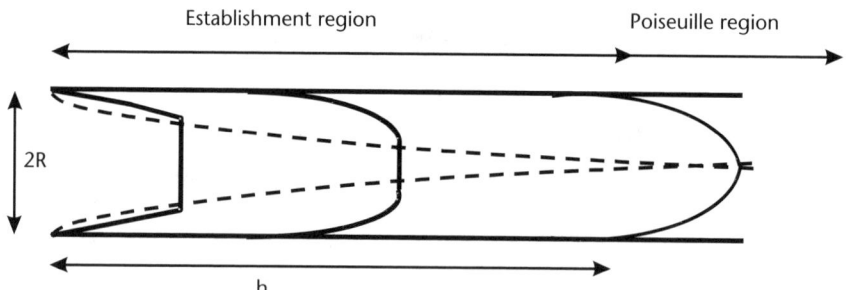

Figure 1.18 Evolution of the velocity profile at the entrance of a tube. Outside the boundary layer, the flow field presents no shearing.

layers at the wall. When the boundary layers merge at the center axis, the flow is established.

The length of the establishment region may be approximated by using the boundary layer theory. If we suppose that the entrance length corresponds to the length where the boundary layer grows until it reaches the value $R = D/2$, we can use Blasius' boundary layer correlation [2]:

$$\frac{\delta}{x} \cong 5Re_x^{-\frac{1}{2}} \tag{1.23}$$

where δ is the boundary layer thickness, and x is the axial distance. The entrance length h is the distance at which the two boundary layers merge; h is obtained from (1.23) by setting $\delta = R = D/2$:

$$\frac{h}{D} \cong 0.01 \, Re_D \tag{1.24}$$

Actually, this approximation does not take into account the acceleration of the fluid in the core region. A more accurate expression has been obtained by Schlichting [9]:

$$\frac{h}{D} \cong 0.04 \, Re_D \tag{1.25}$$

There is an important difference between macroscopic and microscopic flows. At a macroscopic scale, the establishment length may be long, whereas it is very short in microfluidics. An approximate entrance length for a 100-μm-radius tube is of the order of 8 μm for a Reynolds number of 1, which is quite small. It is often experimentally observed that during biochemical reactions in capillary tubes, there is an anomalous region at the entrance of the tube, and the reason observed is wrongly attributed to the flow pattern in the entrance region. It is much more likely that there is an anomalous concentration of immobilized chemical species—or tracers—in the entrance region, which results in anomalous reaction—or detection (see Chapter 5).

1.2.5 Hagen-Poiseuille Flow

In practice, it is common to deal with cylindrical and rectangular microchannels or those in a domain limited by two parallel plates (which is the case of a rectangular channel with a very small aspect ratio). In these cases, when the flow is laminar, there exists an analytical exact solution—for the cylindrical duct and the parallel plates—and an approximated solution—for the rectangular duct [2]. This solution is of interest because it simplifies considerably the partial differential equation (PDE) system governing the convection of particles (Chapter 5).

1.2.5.1 Cylindrical Tube

NS equations can be solved analytically in the particular case of a cylindrical duct. This solution is classical [2, 7] and we indicate here only the result. The velocity in the axial z-direction is given by

$$u(r) = 2U\left[1 - \left(\frac{r}{R}\right)^2\right] \tag{1.26}$$

where U is the average velocity. Equation (1.26) shows that the flow profile is parabolic and the same in any cross section. The mean velocity U is the averaged velocity in a cross section

$$U = \frac{1}{\pi R^2} \int_0^R u(r) 2\pi r dr$$

In the case of a tubular duct

$$U = \frac{u_{max}}{2}$$

Instead of using the mean velocity U to integrate the NS equations, we could have used the pressure difference between inlet and outlet, and we would find a relation between U and the pressure drop $\Delta P = P_{in} - P_{out}$

$$\Delta P = \frac{8\mu UL}{R^2} \tag{1.27}$$

where L is the length of the tube. This equation (1.27) is sometimes called Washburn's law [10].

1.2.5.2 Parallel Plates

The same reasoning may be done for a laminar flow limited by two parallel plates. If the distance between the plates is D and the mean velocity is U, the velocity field is

$$u(y) = \frac{3}{2} U \left[1 - \left(\frac{y}{D/2} \right)^2 \right] \tag{1.28}$$

where y is the transverse direction. Again, the profile does not depend on its location and is parabolic, with a maximum velocity of

$$u_{max} = \frac{3}{2} U$$

and the pressure difference between inlet and outlet is given by

$$\Delta P = \frac{12\mu UL}{D^2} \tag{1.29}$$

1.2.5.3 Rectangular Ducts

Generally in microtechnologies, capillaries of circular cross section are used to link a fluid reservoir to the microsystem. Due to the microtechniques of etching in silicon, glass, or plastic, capillaries in bio-MEMS are often rectangular [1]. An approximated, closed-form solution exists for laminar flows in rectangular channels. The real flow profile is given by a series expansion [11], which is not always practical for applications. An approximation to this expansion was given by Purday [12]. The flow velocity in the z direction inside a rectangular channel of dimensions $2a$ and $2b$ is approximated by

$$u(x, y) = u_{max}\left[1 - \left(\frac{x}{a}\right)^s\right]\left[1 - \left(\frac{y}{b}\right)^r\right] \tag{1.30}$$

where the exponents s and r depend on the aspect ratio $\alpha = b/a$ of the channel. A good approximation of these exponents is

$$s = 1.7 + 0.5\alpha^{-\frac{1}{4}}$$
$$r = \begin{cases} 2 & \text{for } \alpha \le 1/3 \\ 2 + 0.3(\alpha - 1/3) & \text{for } \alpha > 1/3 \end{cases} \tag{1.31}$$

Relation (1.31) shows that the values of r and s are generally close to 2. Integration of (1.30) over the duct cross section yields

$$\frac{u_{max}}{U} = \left[\frac{s+1}{s}\right]\left[\frac{r+1}{r}\right]$$

Taking into account (1.31), the value of the maximum velocity collapses to the value 3/2 U of the two parallel plates solution when one dimension of the rectangle (e.g., b) tends to infinity.

Using the exponents given by (1.31), (1.30) approximates the velocity within an accuracy of a few percent. Equation (1.30) shows that the velocity profile is nearly parabolic in the planes defined by the unit vectors (x,z) and (y,z) (Figure 1.19).

1.2.6 Pressure Drop and Friction Factor

Pressure is dimensionally an energy-per-unit volume. All along a flow, there is a redistribution of energy between pressure, inertia, and gravity. However, there is a

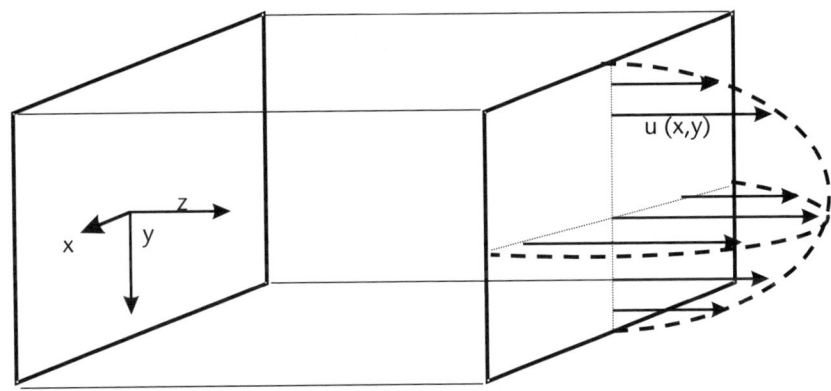

Figure 1.19 Velocity profiles in a rectangular duct.

loss of energy due to the friction at the wall. In the following, we give estimates of pressure drops in tubes of different shapes.

1.2.6.1   Friction Factor

Consider a duct of length L and cross section S. The momentum theorem in this duct yields

$$S\Delta P = \tau_w pL \tag{1.32}$$

where p is the perimeter of the cross section (wetted perimeter). We can replace the unknown wall shear stress by a dimensionless unknown, the friction factor f defined as

$$f = \frac{\tau_w}{\dfrac{1}{2}\rho U^2} \tag{1.33}$$

Combining (1.32) and (1.33) yields

$$\Delta P = f \frac{pL}{S}\left(\frac{1}{2}\rho U^2\right) \tag{1.34}$$

Equation (1.34) may be transformed into a standard form by introducing the notion of hydraulic diameter.

1.2.6.2   Hydraulic Diameter

Tubes and ducts are compared through their hydraulic diameter defined as

$$D_H = 4\frac{S}{p} \tag{1.35}$$

where S is the cross section and p the wetted perimeter. Some values of the hydraulic diameter are indicated in Figure 1.20.

1.2.6.3   Pressure Drop for Different Duct Shapes

Using the definition of the hydraulic diameter (1.35) and substituting it into (1.34) yields

$$\Delta P = f \frac{4L}{D_H}\left(\frac{1}{2}\rho U^2\right) \tag{1.36}$$

For a cylindrical duct, the comparison between (1.27) and (1.36) yields

$$f = \frac{16}{Re_{D_H}} \tag{1.37}$$

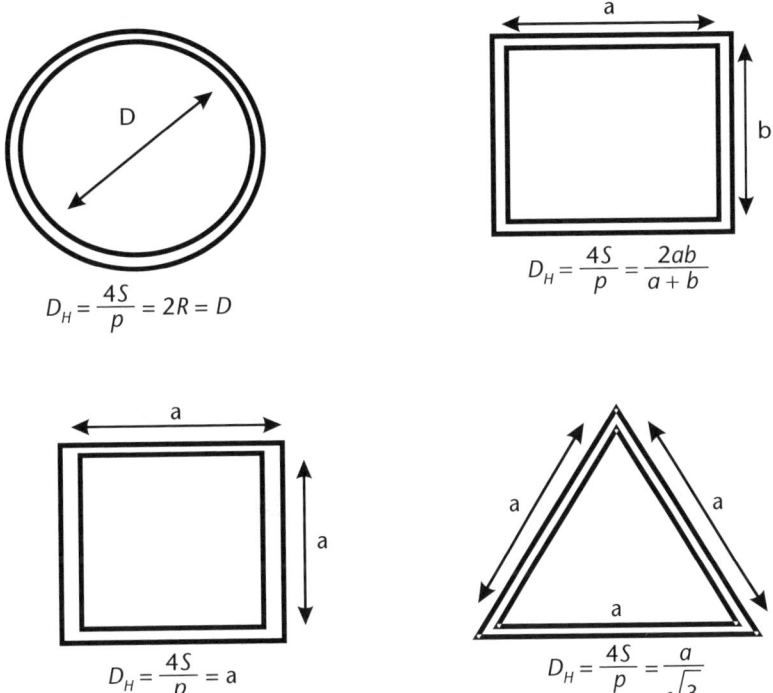

Figure 1.20   Hydraulic diameter of different tubes.

and for the two parallel plates configuration, from (1.29) and (1.36), we obtain

$$f = \frac{24}{Re_{D_H}}$$ (1.38)

Similar expressions are obtained for triangular and square ducts. In the case of an equilateral triangular duct

$$f = \frac{13.3}{Re_{D_H}}$$

and for a square duct

$$f = \frac{14.2}{Re_{D_H}}$$

In the literature, there exist catalogs of friction factors for very different shapes of ducts, with or without obstacles [13].

1.2.6.4   Rectangular Ducts

Friction in rectangular ducts of different aspect ratio $\lambda = b/a$ ($\lambda < 1$) has been the object of numerous investigations. The expression of Shah and London [12] is

$$f = \frac{24}{Re_{D_H}}\left(1 - 1.3553\lambda + 1.9467\lambda^2 - 1.7012\lambda^3 + 0.9564\lambda^4 - 0.2537\lambda^5\right) \qquad (1.39)$$

where Re_{D_H} is the Reynolds number based on the hydraulic diameter.

$$Re = \frac{UD_H}{\nu}$$

Another expression for the pressure drop in a rectangular capillary of cross dimensions a and b is [14]

$$\Delta P = \frac{8\eta LU}{a^2} g(\lambda) \qquad (1.40)$$

where the function $g(\lambda)$ is defined by using the Heaviside function H

$$g(\lambda) = \left(\frac{1+\lambda}{\lambda}\right)^2 H(4.45 - \lambda) + \frac{3}{2} H(\lambda - 4.45) \qquad (1.41)$$

1.2.7 Bernoulli's Approach

Bernoulli's work in hydraulics dates back to 1738. However, Bernoulli's equation is probably the most frequently used in engineering hydraulics today, and it has recently found interesting applications in microfluidics [15]. This equation relates the pressure, velocity, and height in a steady motion of an ideal fluid. But its energy form has extended its domain of application to viscous fluids. However, there is an important restriction to the applicability of Bernoulli's equation: the flow has to be one dimensional. This is not very restrictive in microfluidics, because microflows in capillary and microchannels are mostly one dimensional.

There are three forms of Bernoulli's equations. The first form derives directly from the NS equation under rather severe restrictions. First, suppose that the fluid is an ideal fluid so that the diffusion terms in the NS equation may be neglected (Euler's form of the NS equations)

$$\frac{\partial \vec{u}}{\partial t} + \vec{u}.\nabla\vec{u} = \vec{F} - \frac{1}{\rho}\nabla P \qquad (1.42)$$

Equation (1.42) is a vector equation having only one dimension, that of the one-dimensional fluid flow. Suppose also that the velocity field derives from a potential

$$\vec{u} = -\nabla\phi$$

If the external force field is conservative—that is, derive from a potential

$$\vec{F} = -\nabla\Omega$$

and if the fluid is supposed incompressible

$$\rho = const$$

then (1.42) may be cast under the form

$$\frac{\partial}{\partial t}\left(-\nabla\phi\right) + \vec{u}\cdot\nabla\vec{u} = -\nabla\Omega - \nabla\frac{P}{\rho}$$

Then the following gradient is identically zero

$$\nabla\left[-\frac{\partial\phi}{\partial t} + \frac{u^2}{2} + \Omega + \frac{P}{\rho}\right] = 0$$

Thus, the function must be constant

$$-\frac{\partial\phi}{\partial t} + \frac{u^2}{2} + \Omega + \frac{P}{\rho} = C$$

If we assume that the flow is stationary, we obtain the Bernoulli's equation

$$\frac{u^2}{2} + \Omega + \frac{P}{\rho} = C \tag{1.43}$$

This first approach requires quite severe conditions; other forms of Bernoulli's equations have been derived with a less restricting hypothesis.

The second form of Bernoulli's equation arises from the fact that in a steady flow, the particles of fluid move along streamlines, as on rails, and are accelerated or decelerated by the forces acting tangent to the streamlines (Figure 1.21).

This formulation does not require the irrotational hypothesis for the flow, and we are left with the equation

$$u\frac{du}{ds} = -\frac{d\Omega}{ds} - \frac{1}{\rho}\frac{dP}{ds} \tag{1.44}$$

where s is the distance along the streamline and u is the velocity directed along the streamline. The integration of (1.44) yields

$$\frac{u^2}{2} + \Omega + \frac{P}{\rho} = C$$

Figure 1.21 Bernoulli's equation along a streamline.

The third form of Bernoulli's equation is derived from the conservation of energy. Energy balance is often a powerful and elegant method in physics, and that was the initial Bernoulli's approach. In the case described in Figure 1.22, an element of fluid is transferred from one point to another in a duct with impermeable, rigid boundaries. We can look at it as though there were imaginary pistons moving with the speed of the fluid.

The energies per unit volume, made up of kinetic, potential, and pressure, are equated to obtain

$$U = \frac{\rho u_1^2}{2} + \rho g z_1 + P_1 = \frac{\rho u_2^2}{2} + P_2 \qquad (1.45)$$

so that by taking arbitrary points

$$U = \frac{\rho u^2}{2} + \rho g z + P = const \qquad (1.46)$$

The advantage of the energy approach is that it is general: First, it contains the streamline equation, just by taking imaginary pistons corresponding to the streamline as shown in Figure 1.23. Second, the fluid may be assumed compressible. Third, friction can be accounted for, by assuming a loss of energy due to friction on the rigid wall between any two cross sections. Streamlines do not account for wall friction since they are not bounded by walls, but energy conservation in a cross section does.

As we have seen before, the pressure drop caused by friction on the wall is usually expressed by

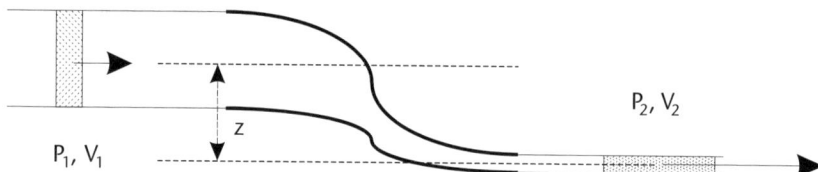

Figure 1.22 Sketch of the displacement of a fluid element in a duct.

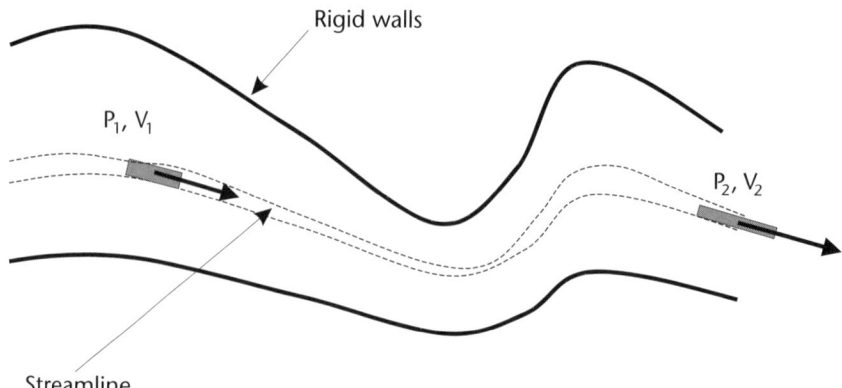

Figure 1.23 Bernoulli's law: energy conservation along a streamline. *P* is the pressure, and *V* is the velocity.

$$\Delta P = f\,\frac{4L}{D_H}\left(\frac{1}{2}\rho U^2\right)$$

where f is the friction factor (nondimensional). The value of f depends on the particular geometry of that part of the duct where the conservation equation is applied. In some cases, like for laminar flows in circular or rectangular ducts, the coefficient f can be calculated; most of the time it is given by algebraic expressions using one or more parameters derived from experiments.

Bernoulli's equation incorporating friction pressure drop is then

$$\frac{\rho u_1^2}{2} + \rho g z_1 + P_1 = \frac{\rho u_2^2}{2} + \rho g z_2 + P_2 + \Delta P_{1,2} \tag{1.47}$$

where $\Delta P_{1,2}$ is the pressure loss by friction on the walls between points 1 and 2. Written in such a form, Bernoulli's equation is a powerful tool for many applications in microfluidics, as we will see in the next sections.

1.2.8   Modeling—Lumped Parameters Model

Because it takes into account the complex geometry of the boundaries, finite element method is the preferred method of modeling microfluidic flows. We will not deal here with the finite element method and its application to microfluidics; this would be a book by itself and some aspects are already detailed in the literature [16]. However, we present the lumped parameters model, which is a simplified calculation method and gives very interesting and accessible results in some cases.

For complex hydraulic circuits, including many different parts having different functions, modeling with Finite Elements numerical software can quickly become impossible, because too many nodes are required and the capacity of the computer is exceeded. Besides, the description of the geometry can be quite difficult. In such a case, the circuit may be decomposed in connecting parts, each part—or branch—corresponding to a precise function, like connecting channels, microchambers, micropump, valves, and such. Such a model is called a lumped model (Figure 1.24) [17, 18].

The model requires us to define nodes $\{i = 1, N\}$ and branches $\{j = 1, M\}$; the nodes are the extremity of the different branches. As we deal with a flow field calculation, the unknowns are the average velocities U_j and the pressure at the nodes P_i. Thus, the vector of unknowns is of dimension $N + M$

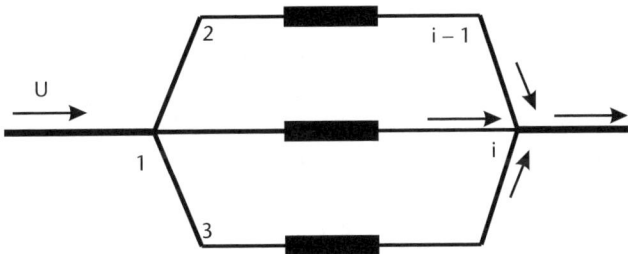

Figure 1.24   Schematic view of a microfluidic circuit.

$$A = \begin{Bmatrix} U_1 \\ \\ U_M \\ P_1 \\ \\ \\ P_N \end{Bmatrix}$$

A first set of equations is constituted by the mass conservation equations at each node i of the net. At a node i, the equation for the conservation of the flow rate is

$$\sum_{j_i} U_{j_i} S_{j_i} = 0 \tag{1.48}$$

where j_i is the index corresponding to all of the branches connected to node i. S_{j_i} and U_{j_i} are, respectively, the cross section of these branches and the average velocities. Note that the velocities U_{j_i} are signed.

The second set of equations is constituted pressure drop relations. For the branch $[i-1, i]$, this relation can be cast into the form

$$P_{i-1} - P_i = f\left(U_{i-1,i}, S_{i-1,i}\right) \tag{1.49}$$

If the branch $[i-1, i]$ is simply a channel of constant section, or any type of channel where Bernoulli's equation applies, the preceding relation collapses to

$$P_{i-1} - P_i = \frac{8\eta L_{i-1,i} U_{i-1,i}}{R_{H,i-1,i}^2} \tag{1.50}$$

Many different functions may be taken into account by the lumped model, like reduction or widening of cross section, Venturis, change of direction of the circuit, pumps, and the like.

The system of (1.48) and (1.49) constitutes a system of $N + M$ equations, because there is an equation of type (1.48) at each node and an equation of type (1.49) for each branch. With proper boundary conditions, such a system can be solved either by matrix inversion, if it is a very complex system, or by direct calculation for the simplest cases. In the following section, we give an example of the usefulness of lumped models.

1.2.9 Worked Example: Microfluidic Flow Inside a Microneedle

An example of emergence of new technologies in medical science is that of microneedles. For external uses on human skin, the classical needle with syringe has found a replacement with the patch of microneedles [19, 20]. For internal delivery, new concepts of needles are currently being developed [21, 22]. One of these new concepts is that of a needle of a smaller cross section—to be less invasive—and with many microscopic side channels in order to diffuse more efficiently the injected molecules.

Flow motion in this last type of microneedle is a good example of microflow and an example of the utility of lumped models. For this reason, we detail in this section the flow distribution in such a microsystem and the approach to the dimensioning of such a system.

1.2.9.1 Drug Delivery and Injection System

Diffusion in cell clusters and tumors will be presented in Chapter 4 concerning the mechanism of diffusion. We just recall the principle of drug injection (Figure 1.25): the active molecules transit through the needle tip into the cells' extracellular space (ECS), and finally inside the cells.

Let us focus now on the injection device. The rapidity of dispense and its efficiency can be increased by a microneedle that has many exits uniformly distributed on the sides (Figures 1.26 and 1.27). The advantage of this type of needle is that drug delivery starts from all side exits disposed along the needle. For the best possible efficiency of the microneedle, all side exits should have the same flow rate.

Eventually, the needle may be used as an electrode to increase the uptake rate. It has been observed that an electric field increases the uptake rate [24], so the needle can be electrically actuated [21]. However, we discuss here only the microfluidic aspects.

1.2.9.2 Lumped Model

What should be the dimensions of the different side exits if we want them to deliver the same flow rate? Intuitively, it is clear that the side exits close to the needle tip should have larger cross sections than that in the vicinity of the flow inlet. If all of the side exits were to have the same section, then the flow would exit in the first side branches and not reach the needle tip.

A first solution might be to assign initial cross sections to the side exits, run a calculation with a numerical software, memorize the values of the flow rates at each exit, calculate a "distance" to the expected uniform flow solution, find an optimization algorithm to change the dimensions of the side exits, and run the process iteratively until convergence. However, this way of proceeding is long, and, most

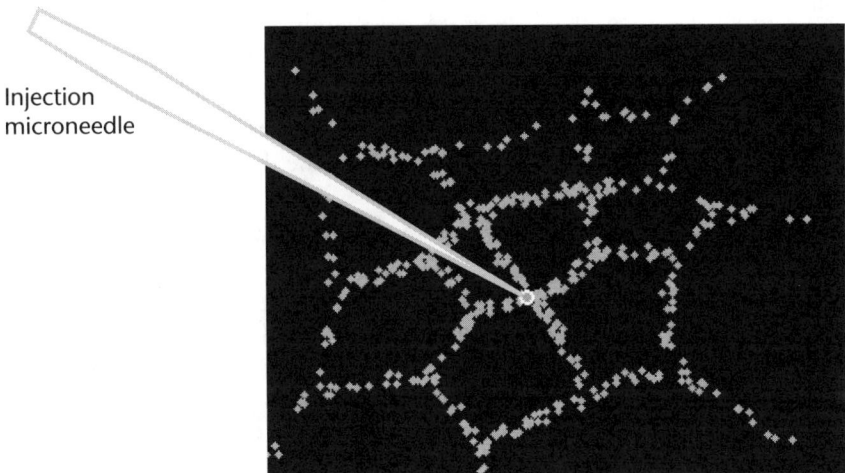

Injection
microneedle

Figure 1.25 View of drug injection in a cell cluster (obtained by calculation [23]). The ECS is delimited by the white dots. (*From:* [25]. © 2005 NSTI. Reprinted and revised with permission.)

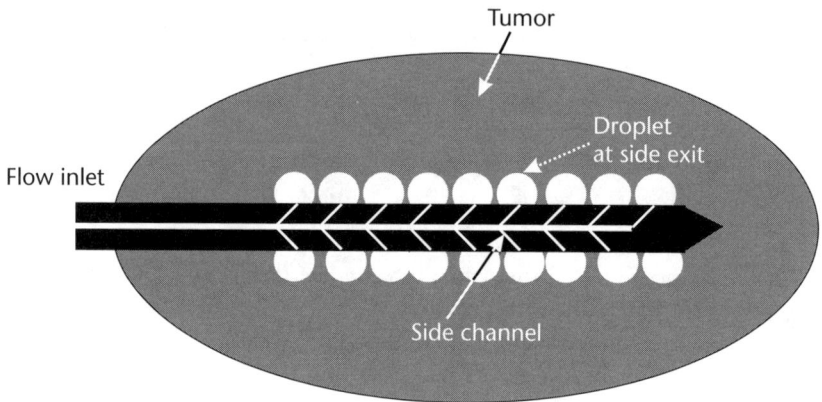

Figure 1.26 Schematic view of the drug-dispensing needle. (*From:* [25]. © 2005 NSTI. Reprinted and revised with permission.)

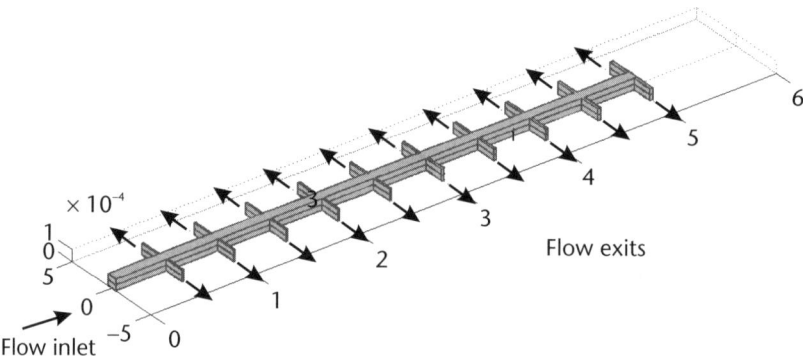

Figure 1.27 Schematic view of the central channel and the side exits: all side channels must deliver the same flow rate. (*From:* [25]. © 2005 NSTI. Reprinted and revised with permission.)

importantly, what should be the algorithm to perform an optimization on N parameters (the N widths of the $2N$ side channels $\{a_i, i = 1, N\}$)?

In such a case, it is preferable to search for an inverse algorithm based on a lumped model formulation. Using a lumped element model for taking care of all of the different microfluidic segments in the needle, and imposing the constraint that the flow rate at all outlets is the same, a recurrence relation for the pressure at the nodes can be derived and solved to obtain the desired channel widths [25].

1.2.9.3 Algorithm

The algorithm has three steps: (1) calculation of the velocities in the central channel by a recurrence relation starting from the needle tip, (2) establishment of a recurrence relation for pressure at the nodes starting from the needle tip, and (3) calculation of the pressure at each node using the two first steps.

Step 1: Velocities in the Axial (Central) Channel

Let the letters P, Q, V stand respectively for the pressure, flow rate, and fluid velocity. The density of the liquid is ρ. Because of the process of fabrication of the needle, the vertical dimension of the microchannels b is the same for all of the channels. For simplicity, it is assumed that the spacing L (axial distance) between the side channels is constant and that the width a_0 of the main channel is a given constant; as a consequence, the length of the side channels is also a constant L_s. A schematic view of the flow channels is given in Figure 1.28.

Because there are $2N$ exits, the total mass conservation equation can be written

$$Q_{in} = (2N)Q_{exit} \qquad (1.51)$$

By a recurrence approach, starting from the far end of the needle and progressing to the front end, we obtain

$$V_i = \frac{2Q_s}{\rho a_0 b}(N-i), \quad i = 0,\ldots,N-1 \qquad (1.52)$$

Step 2: Pressure at the Axial Nodes

We have already seen from (1.40) and (1.41) that Washburn's law for a rectangular capillary of cross dimensions a and b ($\lambda = b/a$) is

$$\Delta P = \frac{8\eta L V}{a^2}g(\lambda)$$

where the function $g(\lambda)$ is defined by using the Heaviside function H

$$g(\lambda) = \left(\frac{1+\lambda}{\lambda}\right)^2 H(4.45 - \lambda) + \frac{3}{2}H(\lambda - 4.45)$$

At an intersection, there is a distortion of the laminar flow lines. This problem is complex in a rectangular geometry and we simplify by

$$\Delta P_{inter} = \frac{8\eta(13a)V}{a^2}g(\lambda) \qquad (1.53)$$

and in a side branch of length L, the linear pressure drop is reduced to

$$\Delta P_{linear} = \frac{8\eta(L-4a)V}{a^2}g(\lambda) \qquad (1.54)$$

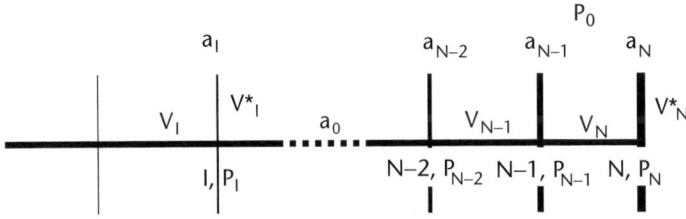

Figure 1.28   Schematic view of the main and secondary channels.

So that the total pressure drop of a side branch is

$$\Delta P = \Delta P_{linear} + \Delta P_{inter} = \frac{8\eta(L+9a)V}{a^2}g(\lambda) \tag{1.55}$$

Again, starting from the last node in the axial channel and applying (1.55), we obtain the pressure at the last node

$$P_N = P_o + \frac{8\eta(L_s+9a_N)V_N^*}{a^2}g(\lambda_N)$$

where V_N^* is the velocity in the Nth side branch, and P_o is the pressure at the outlets (atmospheric pressure). By replacing the flow rate by the flow velocity in the Nth side branch, the pressure can be cast into the form

$$P_N = P_o + \frac{8\eta(L_s+9a_N)Q_o}{ba_N^3}g(\lambda_N) \tag{1.56}$$

Now, we progress toward the front end of the axial channel and deduce the recurrence relation

$$P_i = P_o + \frac{8\eta(L_s+9a_N)Q_o}{ba_N^3}g(\lambda_N) + \frac{8\eta L Q_o}{ba_o^3}g(\lambda_o)(N-i)(N-i+1) \tag{1.57}$$

Equation (1.57) is a recurrence relation for the pressure at the nodes versus the widths of the side channels; this relation has been obtained by considering the velocities in the axial main channel only (to the exception of the last side channel). The pressures at the nodes are now directly calculated using the side channels.

Step 3: Pressure at the Nodes from Side Channels

The pressure at a node is directly related to the outside pressure by the relation

$$P_i = P_o + \frac{8\eta(L_s+9a_i)Q_o}{ba_i^3}g(\lambda_i) \tag{1.58}$$

and the solution is obtained by equating the values of the pressure at the nodes from (1.57) and (1.58)

$$\frac{(L_s+9a_i)}{a_i^3}g(\lambda_i) = \frac{(L_s+9a_N)}{a_N^3}g(\lambda_N) + \frac{L}{a_o^3}g(\lambda_o)(N-i)(N-i+1) \tag{1.59}$$

Relation (1.59) is a purely geometrical relation between the different geometric parameters. For any given value of the width of the last side channel a_N, the right-hand side of the previous relation is known, and we find the implicit relation for the a_i of the type

$$\frac{\left(L_s + 9a_i\right)}{a_i^3} g(\lambda_i) = B_i \tag{1.60}$$

where B_i is the value of the right-hand side (which depends only on the width of the last side channel). The inverse solution is then reduced to an implicit solution of an analytical function, and this is tractable. Depending on the value of the λ_i, one has to solve either one of these two third-order polynomials

For $\lambda_i > 4.45$,

$$\frac{2}{3} B_i a_i^3 - 9a_i - L_s = 0 \tag{1.61}$$

For $\lambda_i < 4.45$,

$$\left(B_i b^2 - 9\right)a_i^3 - \left(L_s + 18b\right)a_i^2 - b\left(2L_s + 9b\right)a_i = L_s b^2 \tag{1.62}$$

It can be shown that these two polynomials have one real root and two imaginary roots, so that the real root is the desired solution. It is interesting to note that the fluid viscosity does not appear in (1.60)—neither does the value of the total flow rate (Q_i)—so that the calculated dimensions of the microsystem will satisfy the constraint of delivering the same flow rate at the outlets for different fluids and different inlet flow rates. That shows a generality in the system. Another advantage of this method is that it is straightforward to change the angle between the main and side channels by changing the pressure drop expression at an intersection, and replacing (1.53) by

$$\Delta P = \frac{8\eta\left(L + 9a\sin^2\alpha\right)V}{a^2} g(\lambda) \tag{1.63}$$

For $\alpha = 90°$, (1.63) is the same as (1.53). Figure 1.29 shows that the distribution of the a_i in the case of 2×10 side channels, for a value of $a_{10} = 30\ \mu m$.

We now can use a direct model to verify the accuracy of the algorithm. Here, the direct model is the SABER code from the COVENTOR package [26]. Calculated widths are used as inputs in the SABER calculation. If the algorithm is correct, the results should agree. It is checked in Figure 1.30 that the SABER results agree well with the lumped model results.

Microneedles based on the results of the dimensioning algorithm have been realized in silicon using microtechnologies for silicon etching and assembling (Figure 1.31). Needles of different sizes (from $500\ \mu m$ to $300\ \mu m$) and with different side channel angles have been fabricated (Figures 1.32 and 1.33).

Using methyl blue-colored water and disposing the needle on a flat blotter, it is checked that all of the flow at all exits are identical (Figure 1.34).

1.2.10   Distributing a Uniform Flow into a Microchamber

Bio-MEMS and microfluidics immunoassays currently use microflow circuits where small capillary channels are linked to microchambers where the biologic functions

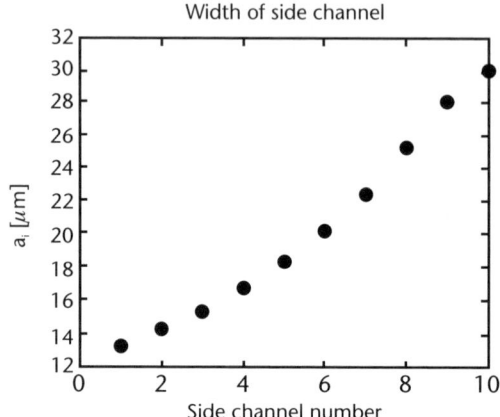

Figure 1.29 Calculated width of side channels as a function of the channel number. The width increases from the entrance to the tip of the needle.

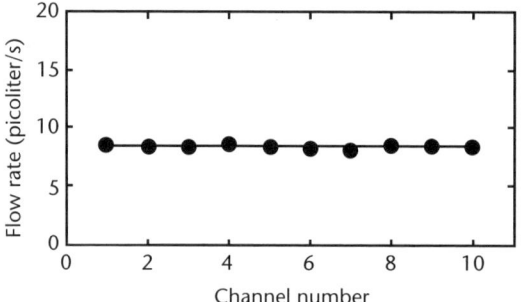

Figure 1.30 Comparison of side exit velocities—continuous line: present algorithm; dots: SABER results.

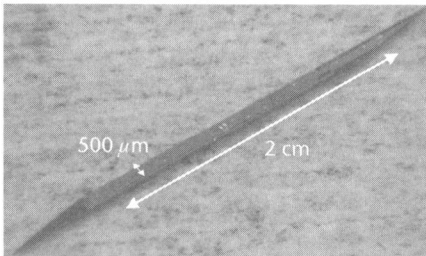

Figure 1.31 View of a microneedle. (*From:* [25]. © 2005 NSTI. Reprinted and revised with permission.)

are installed. Inside the capillary channels, the velocity profile is a Hagen-Poiseuille flow profile, parabolic for cylindrical and rectangular channels, showing an important transverse gradient. For the purposes of bioanalysis, labeled surfaces are installed inside the microchamber, and it is important that flow velocity should be as uniform as possible above all of the labeled surfaces. Labeled surfaces affected by a weak part of the main flow do not function correctly.

Figure 1.32   Detailed view of the microneedle: main channel and oblique side channels. (*From:* [25]. © 2005 NSTI. Reprinted and revised with permission.)

Figure 1.33   Microscope views of axial cut of two needles, the first one with oblique side channels (left), the other one with perpendicular side channels (right). (*From:* [25]. © 2005 NSTI. Reprinted and revised with permission.)

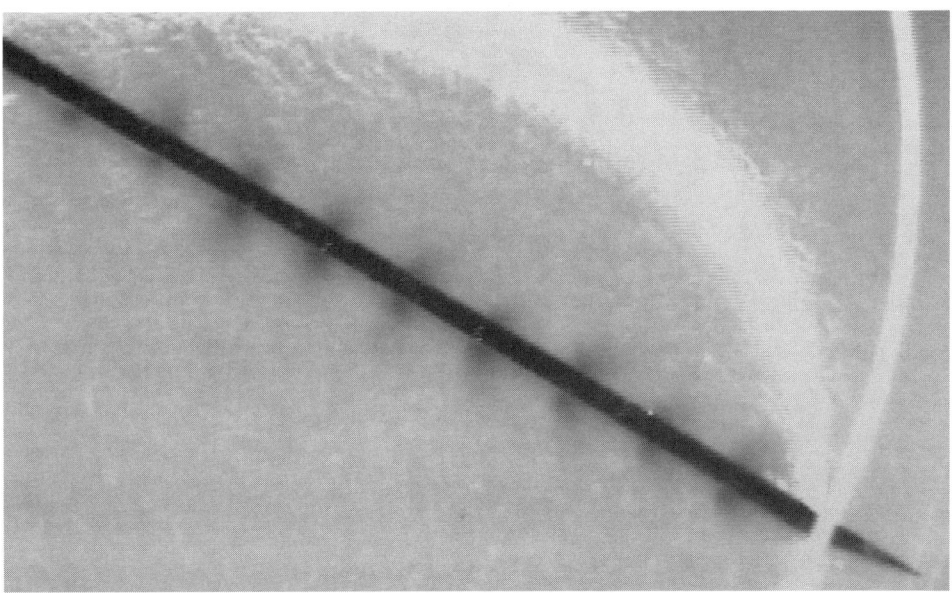

Figure 1.34   Visualization of equal flow rates at all exits. (*From:* [25]. © 2005 NSTI. Reprinted and revised with permission.)

A simple design of a divergent cone between the capillary channel and the microchamber cannot be satisfactory because it would result in a very nonuniform velocity flow profile in the microchamber. We have used the FEMLAB software

[27] to solve the incompressible NS equation. Figure 1.35 shows that the velocity profile in a cross section of the microchamber is nearly parabolic. The conditions for correct functioning of the device are then not met.

If the divergent cone is separated in subchannels, as shown in Figure 1.36, the velocity profile is very much improved, at least in the middle part of the cross section of the microchamber. Taking into account the requirements of lithography techniques, we obtain the classical solution schematically represented in Figure 1.37. From left to right, each branch is divided into two subbranches so that the velocities are the same in all of the branches of the same size. The velocity profile is very flat in the middle part of the microchamber cross section. This principle is applied to design proteomic reactors (Figure 1.38), which will be described in the following section.

We have shown here that by diverting the flow in many secondary flows, a satisfactory widening can be achieved. However, the price to pay is an increased friction, and the pressure drop is much higher in the solution of Figure 1.37 than that of Figure 1.35.

1.2.11 The Example of a Protein Reactor

In a proteomic reactor, proteins are broken into peptides (or peptidic segments) by the action of enzymes. The peptides are then separated according to their size inside

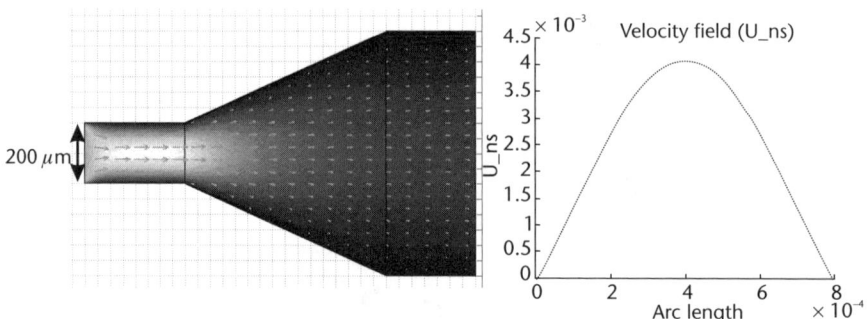

Figure 1.35 Left: widening from a microfluidic channel to a microchamber; right: nearly parabolic velocity profile in a cross section, showing that the microchamber will not be fed by a uniform fluid flow.

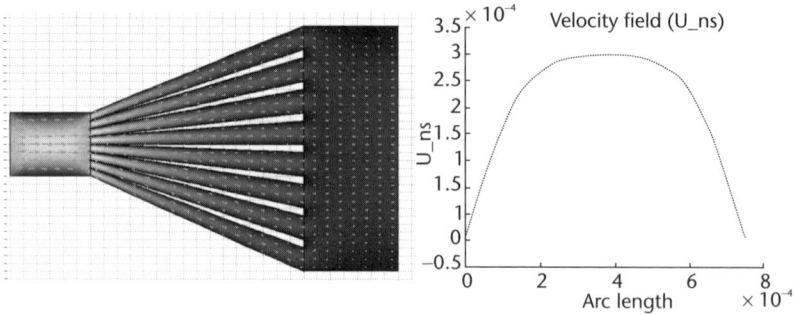

Figure 1.36 Left: widening from a microfluidic channel to a microchamber using subchannels; right: velocity profile in a cross section of the microchamber. The profile is improved compared to the preceding figure.

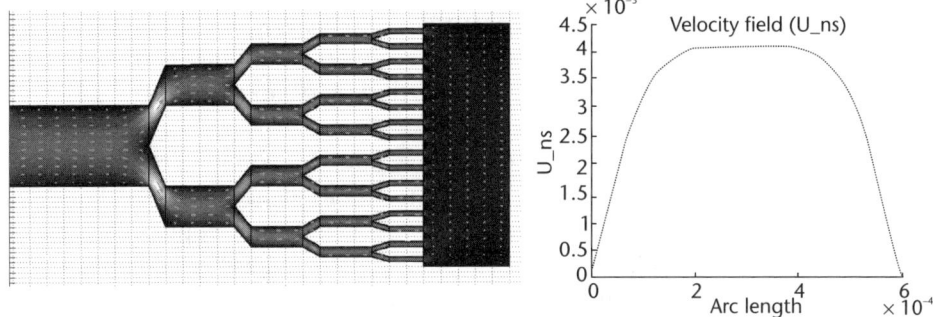

Figure 1.37 Left: widening from a microfluidic channel to a microchamber using lithography technology; right: velocity profile in a cross section of the microchamber.

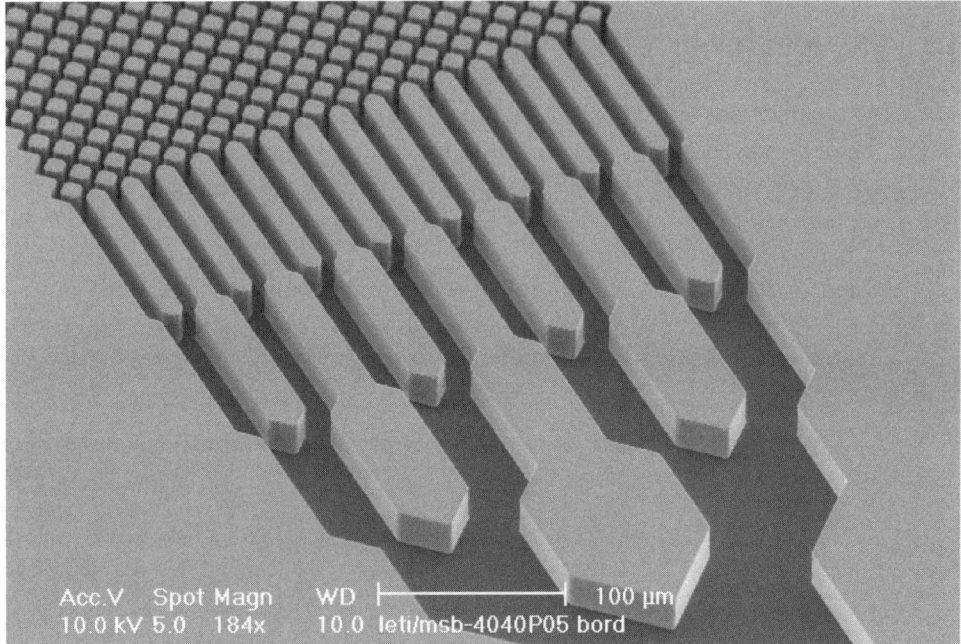

Figure 1.38 Inlet channels of a microfabricated proteomic reactor. The flow rate is divided in two at each step. (Courtesy of O. Constantin, CEA/LETI.)

a chromatography column and finally expulsed into a mass spectrometer. The composition of the proteins is reconstructed after analysis of the identified peptides by reconstruction algorithms.

The part of the microreactor where the proteins are "digested" by the enzymes must be carefully designed in order to have a complete reaction. The proteins are transported by the flow, and the enzymes have been previously immobilized on the walls of the reactor. In order to have a complete reaction, a maximum surface area for the contact between proteins and enzymes must be available, as shown in the upper part of Figure 1.38 [28].

This is typically a microfluidic problem coupled with a technological challenge. The microfluidic problem resembles, at another scale, the problem of the heat exchanger, where obstacles are introduced to increase the contact surface. However, the technological constraints are much more demanding at the microscopic level, with the fabrication of micropillars or the use of textured surface. Typical feasible pillars are shown in Figures 1.39 and 1.40.

The use of computer simulation is essential to compare the efficiency of the different designs proposed by the etching technology. In Figures 1.41 and 1.42, a comparison of the flow in two different arrangements of micropillars has been made (the size of the pillars is less than $10\,\mu$m). The flow field has been computed with the help of the finite element numerical software FEMLAB. The differences between the two arrangements are obvious: there are poorly irrigated channels in Figure 1.41, but not in the case of Figure 1.42. After computation of different motives, the configuration of Figure 1.42 has been adopted for the reactor (Figure 1.38).

1.3   Multiphase Microflows and Liquid Plugs

Microflows are not always constituted by a single liquid. Frequently, we deal with small quantities of liquid (buffer liquid containing the molecules of interest) under the form of liquid plugs moving inside capillary tubes and separated by inert, biocompatible, nonmiscible plugs of another liquid (Figure 1.43).

Two-phase microflows are often linked to the parallelization of biological operations, such as high-throughput screening and biodiagnostics. For example, Figure 1.44 shows simultaneous plugs flowing in parallel capillary tubes and convey to separate biodiagnostic microchambers.

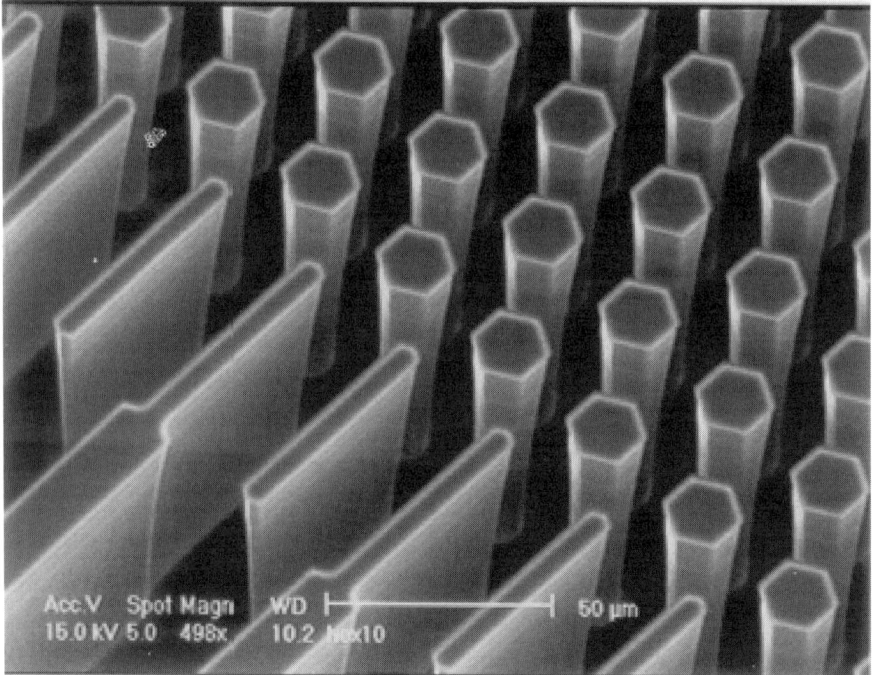

Figure 1.39   Detail of the fluid inlet in a proteomic reactor. Protein digestion takes place on the walls of the hexagonal pillars. (Courtesy of N. Sarrut, CEA/LETI.)

Figure 1.40   Nanostructuring of a proteomic reactor. (Courtesy of N. Sarrut, CEA/LETI.)

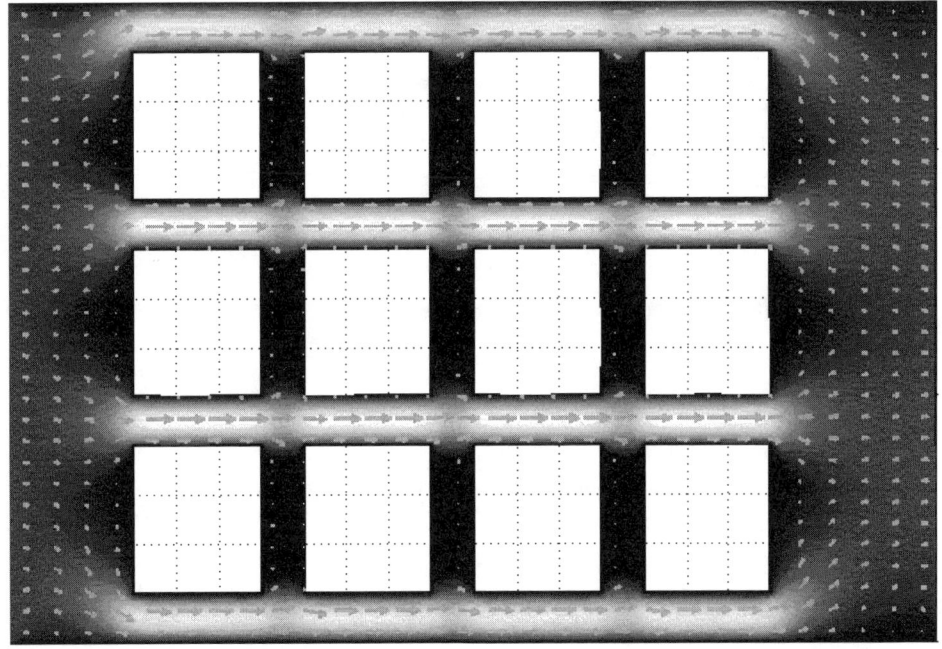

Figure 1.41   Simulation results obtained with the numerical software FEMLAB showing the poorly disposed pillars. The transverse channels have very low fluid velocity and do not contribute to the biochemical reaction.

Conceptually, liquid plugs may be seen as a transitional state between microflows and digital microfluidics (digital microfluidics or microdrops will be treated in Chapter 2).

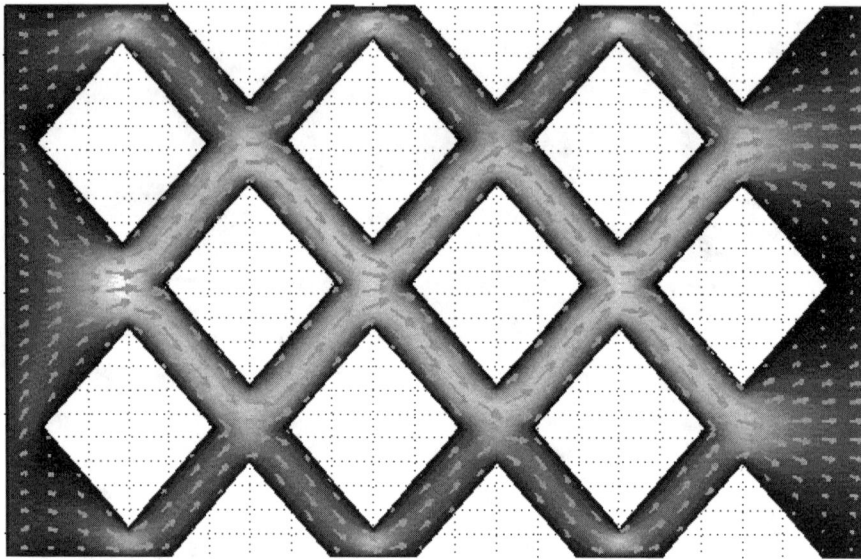

Figure 1.42 Correctly disposed channels: the flow affects all channels, and a maximum of surface of reaction is obtained.

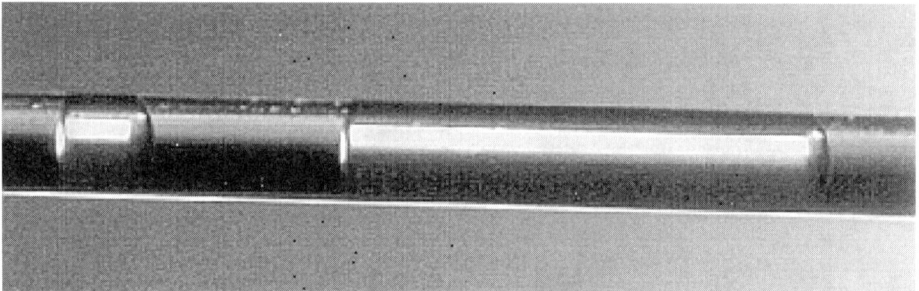

Figure 1.43 View of buffer fluid plugs separated by silicon oil plugs.

1.3.1 Interface and Meniscus

The shape of liquid plug in a capillary tube depends on the capillary forces. Capillary forces will be detailed in the next chapter. A liquid plug moving inside a capillary tube (or between two parallel plates) is limited by two meniscus, one corresponding to the advancing front (index a), the other one corresponding to the receding front (index r), as shown in Figures 1.44 and 1.45. For microcapillaries, menisci have spherical shapes. Note that receding, advancing, and static contact angles are not identical.

1.3.2 Dynamic Contact Angle

The contact angle formed between a flowing liquid front (advancing or receding) and a solid surface is not constant but reflects the balance between capillary forces and viscous forces. The relative importance of these forces is often expressed by the nondimensional capillary number Ca defined by

Figure 1.44   Synchronization of buffer fluid plugs in parallel capillaries.

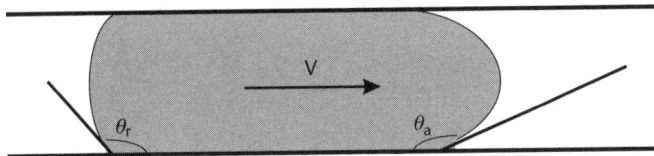

Figure 1.45   Schematic view of a liquid plug in a capillary tube with the advancing and receding contact angles.

$$Ca = \frac{\eta U}{\gamma} \qquad (1.64)$$

where η is the dynamic viscosity of the moving fluid (unit kg/m/s), U is its velocity (m/s), and γ is its surface tension (N/m). The capillary number is a scale of the ratio between the drag force of the flow on a plug and the capillary forces. In a cylindrical tube, from (1.27) we deduce that the order of magnitude of the drag force is

$$F_{drag} \approx \eta l U \qquad (1.65)$$

where l is the axial dimension of the plug. The capillary/wetting force is given by

$$F_{cap} \approx \gamma l \qquad (1.66)$$

We deduce from (1.65) and (1.66)

$$\frac{F_{drag}}{F_{cap}} \approx \frac{\eta U}{\gamma} = Ca \qquad (1.67)$$

Hoffman first proposed an expression for the dynamic contact angle based on experimental observations [29]; however, this correlation is rather complicated and Voinov and Tanner [30] have established the more workable correlation

$$\theta_d^3 - \theta_s^3 = ACa \tag{1.68}$$

where θ_d and θ_s are the dynamic and static contact angles. The value of the coefficient A is $A \sim 94$ if θ is expressed in radians. Figure 1.46 shows a comparison between experimental results for the dynamic contact angle and (1.68).

For microflows, using the approximate values $\eta \sim 10^{-3}$ kg/m/sec, $U \sim 10 \,\mu$m/sec to 10 cm/sec, and $\gamma \sim 50 \cdot 10^{-3}$ N/m, we find that typical values of the capillary number are in the range $2 \cdot 10^{-7}$ to $2 \cdot 10^{-3}$. The capillary number is then small and corresponds to the linear part of the Tanner law. Linearization of (1.68) yields [30]

$$\theta_d = \left(\theta_s^3 + ACa\right)^{\frac{1}{3}} \cong \theta_s \left(1 + \frac{1}{3}\frac{ACa}{\theta_s^3}\right) \tag{1.69}$$

or

$$\theta_d - \theta_s = \frac{1}{3}\frac{ACa}{\theta_s^2} \tag{1.70}$$

Note that the capillary number is signed. The relation (1.70) shows that $\theta_d - \theta_s$ is of the sign of Ca, confirming that the advancing contact angle is larger than the static contact angle and the receding contact angle is smaller than the static contact angle (Figure 1.47).

1.3.3   Plugs Moving Inside a Capillary

In this section, we analyze the motion of one or more liquid plugs inside a cylindrical capillary tube. We use a lumped model, and we show that Bernoulli's equation combined with Tanner's law explains the main features of the behavior of liquid plugs moving inside capillary tubes [31]. Flow regions may be decomposed in two types of

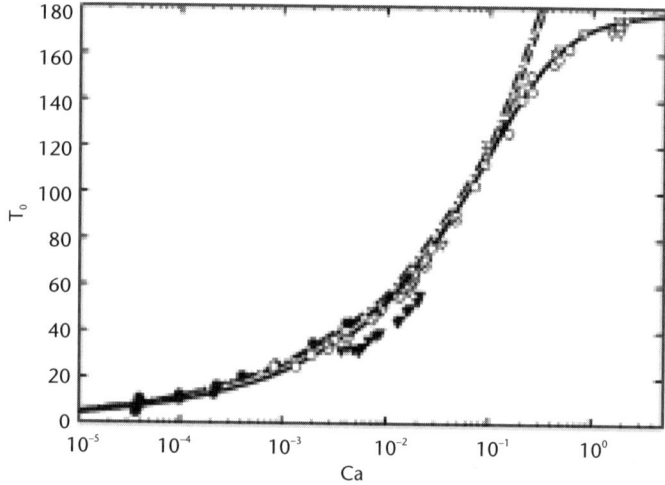

Figure 1.46 Experimental results for the dynamic contact angle versus the capillary number (dots) and Tanner relation (continuous line).

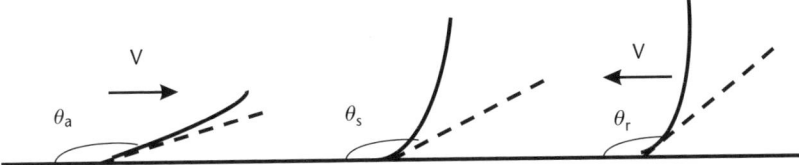

Figure 1.47   Schematic view of advancing, static, and receding contact angles. The advancing contact angle is larger than the static contact angle, which is, in turn, larger than the receding contact angle.

elements (Figure 1.48). The first type of element is a fluid moving inside a capillary and inducing a friction pressure drop; the second type of element corresponds to moving interfaces and induces a capillary pressure drop.

The total pressure drop in the capillary is

$$\Delta P_{channel} = \Delta P_{capillary} + \Delta P_{friction} \qquad (1.71)$$

The pressure drop due to friction on the solid walls is given by the Washburn law [10]

$$\Delta P_{friction} = \frac{8V}{R^2}\left(\eta_1 L_1 + \eta_2 L_2\right) \qquad (1.72)$$

where indices 1 and 2 address liquid 1 (liquid plug) and liquid 2 (surrounding fluid), R is the radius of the capillary, V is the average liquid velocity, and L_1, L_2 is the total length of contact of liquid 1, 2 with the solid wall.

Each interface—advancing and receding—contributes (positively or negatively in function of the contact angles) to the capillary pressure drop. It can be shown that the advancing front contribution is

$$\Delta P_a = -\frac{2\gamma}{R}\cos\theta_a \qquad (1.73)$$

This expression derives directly from the Laplace law, which relates the pressure difference at a spherical interface of curvature radius a (Figure 1.49) by

$$\Delta P_a = \frac{2\gamma}{a} \qquad (1.74)$$

Note that a is signed and the sign of the pressure difference depends on the orientation of the curvature.

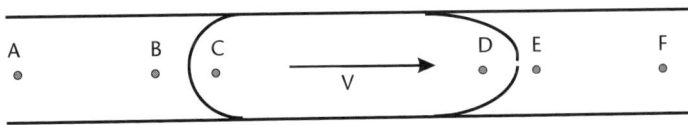

Figure 1.48   Decomposition of a two-phase flow in lumped elements. Between points A and B, C and D, and E and F, the pressure drop is due to friction; between B and C, and D and E, the pressure drop is due to the interface.

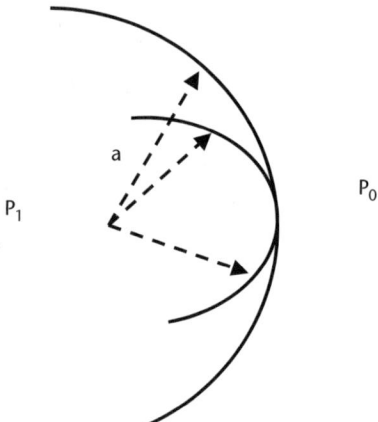

Figure 1.49   Sketch of a spherical interface separating two fluids. The pressure difference between the convex region and the outside region is $P_1 - P_0$ and is given by the Laplace law.

More details on the Laplace law will be given in the next chapter dedicated to microdrops. Assuming that the meniscus have a spherical shape (Figure 1.50), the contact angle is related to the tube radius R and the curvature radius a by

$$\cos\theta = -\frac{R}{a} \qquad (1.75)$$

By substituting (1.75) into (1.74), one obtains (1.73). The receding front contribution is then given by the relation

$$\Delta P_r = \frac{2\gamma}{R}\cos\theta_r \qquad (1.76)$$

Note that in our convention the pressure drop is always taken following the fluid flow. Suppose the two configurations of Figure 1.51. If θ_a is larger than $\pi/2$, there is a positive pressure drop associated with the advancing interface. If θ_r is smaller than $\pi/2$, the receding front contributes positively to the pressure drop [Figure 1.51(a)], and negatively in the opposite case [Figure 1.51(b)].

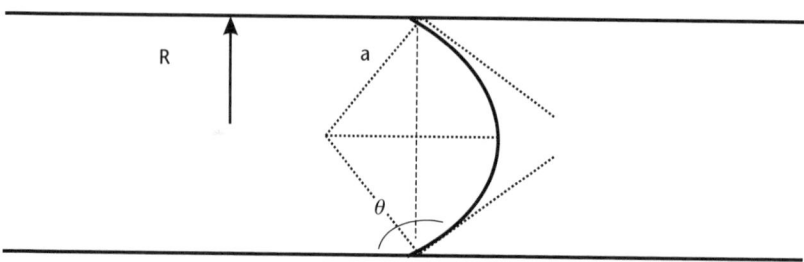

Figure 1.50   Schematic view of the meniscus in a cylindrical capillary tube.

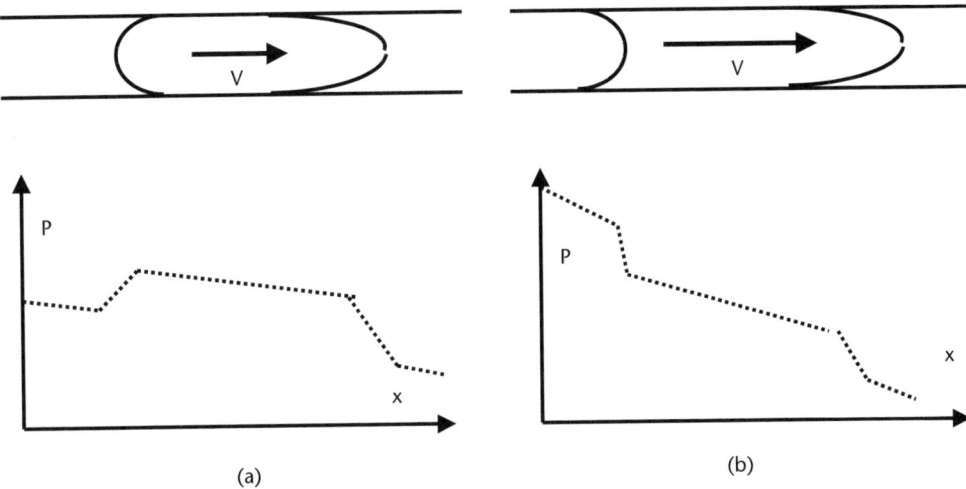

Figure 1.51 Two possible configurations for a plug moving inside a capillary tube: (a) at low velocity, the receding angle is larger than $\pi/2$ and the contribution to the pressure drop is negative; and (b) at high velocity, θ_r is smaller than $\pi/2$ and the contribution to the pressure drop is positive. The slope of the pressure drop inside the different liquid is due to the friction pressure drop.

The capillary pressure drop is due to the difference of the capillary forces between advancing and receding fronts because of the two different contact angles (advancing and receding) θ_a and θ_r

$$\Delta P_{capillary} = \frac{2\gamma}{R}(-\cos\theta_a + \cos\theta_r) \tag{1.77}$$

At low capillary numbers, the pressure drop due to friction is much smaller than that due to the interfaces, and (1.77) shows that too many plugs in the capillary may rapidly block the flow. For an N plugs flow, we will have for some N

$$\Delta P_{capillary} = \frac{2\gamma}{R}N(-\cos\theta_a + \cos\theta_r) > P_i - P_o \tag{1.78}$$

and the flow will come to a stop (P_i and P_o are the inlet and outlet pressures). Now, introduce the linearized Hoffman-Tanner law to find the expression of the capillary pressure drop.

$$\theta_a = \theta_{s,a}\left(1 + \frac{1}{3}\frac{ACa}{\theta_{s,a}^3}\right) \tag{1.79}$$

and

$$\theta_r = \theta_{s,r}\left(1 - \frac{1}{3}\frac{ACa}{\theta_{s,r}^3}\right) \tag{1.80}$$

with

$$Ca = \frac{V\eta_1}{\gamma}$$

where the index s stands for the static contact angle, and $\theta_{s,r}$ and $\theta_{s,a}$ are the two static contact angles. They are equal if there is no static hysteresis (i.e., if the surface is perfectly smooth). The minus sign in (1.80) derives from the fact that the capillary number Ca is positive, even for the receding interface. After substitution of (1.79) and (1.80) in (1.77), and using some algebra, the capillary pressure drop is

$$\Delta P_{capillary} \cong \frac{2\gamma}{R}\left(-\cos\theta_{s,a} + \cos\theta_{s,r}\right)$$
$$+\frac{2AV\eta_1}{3R}\left(\frac{\sin\theta_{s,a}}{\theta_{a,s}^2} + \frac{\sin\theta_{s,r}}{\theta_{r,s}^2}\right) \tag{1.81}$$

If there is no hysteresis of static contact angle ($\theta_{s,a} = \theta_{s,r} = 0$), the first term to the right-hand side of (1.81) vanishes and the capillary pressure drop is proportional to the average velocity V of the flow. As the friction pressure drop is also proportional to V, so is the total pressure drop

$$\Delta P = V\left[\frac{8}{R^2}\left(\eta_1 L_1 + \eta_2 L_2\right) + \frac{4A\eta_1}{3R}\frac{\sin\theta_s}{\theta_{a,s}^2}\right] \tag{1.82}$$

In case of hysteresis of static contact angle, at very low velocities, the pressure drop has an asymptotic value, which is the first term of the right-hand side of (1.81).

1.3.4 Hysteresis of Static Contact Angle

In the following chapter, Young's equation for contact angle will be detailed. This equation predicts the value of the static contact angle as a function of the surface energy of the different materials (liquid plug, surrounding liquid, and solid substrate). Theoretically, it should result in a unique value of the static contact angle. However, it happens frequently that the static contact angle is not uniquely defined. It can be comprised between two values, a first value obtained by slowing down to a stop an advancing front $\theta_{s,a}$, and another value (smaller) obtained by slowing down to a stop a receding front $\theta_{s,r}$, as shown in Figure 1.52.

In fact, Young's law applies only for a homogeneous solid substrate. If the substrate is disordered, it is necessary to modify Young's law, and expressions like the Wenzel or Cassie laws have been derived to bring the necessary changes to Young's law. Thus, hysteresis of contact angle is linked to the smoothness and homogeneity of the substrate [32–34]. An interesting discussion on the different aspects of the Wenzel or Cassie law for contact angle hysteresis is given by He et al. in [35].

1.3.5 Two-Phase Flow Pressure Drop Due to a Sudden Enlargement

The idea of using jagged walls for microchannels (piecewise linear boundaries) has been investigated in the literature. The pressure drop depends very much on the

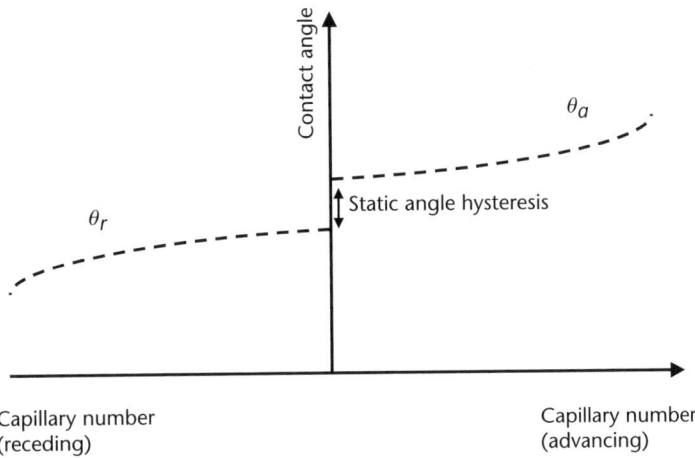

Figure 1.52   Hysteresis of static contact angle and Hoffman-Tanner law for advancing and receding contact angles versus capillary number: the advancing contact angle is larger than the receding contact angle, and there is a static hysteresis at zero velocity.

shape of the walls [36]. In this section, we analyze the simple case of a sudden enlargement. We show how an interface between two immiscible liquids generates a supplementary pressure drop when it reaches a sudden enlargement in a capillary tube. The motion of the interface is sketched in the Figure 1.53.

In the small capillary tube, the curvature radius a_1 is given by (1.75)

$$a_1 = \frac{d_1/2}{(-\cos\theta)}$$

and the pressure difference at the interface ΔP_1 is given by the Laplace equation (1.74)

$$\Delta P_1 = \frac{2\gamma}{a_1} = \frac{-4\gamma\cos\theta}{d_1} \tag{1.83}$$

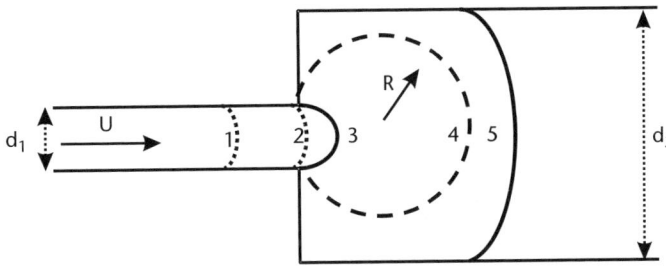

Figure 1.53   Schematic view of an interface passing over a sudden enlargement of a capillary tube (case of hydrophobic walls). At first, the interface is located in the smaller capillary (diameter d_1) and has a curvature radius a_1. When the front arrives at the enlargement, the curvature radius is $a_2 = a_1$. Then a spherical bubble forms, and the curvature radius decreases to a minimum when the interface forms a half-sphere of radius $a_3 = d_1/2 < a_1$; it increases again when the bubble touches the walls of the larger capillary tube (diameter d_2) and the curvature radius reaches the value $a_5 > a_1$.

When the front reaches the edge of the enlargement (location number 2 in Figure 1.53), the situation is at first identical, and

$$\Delta P_2 = \Delta P_1 = \frac{2\gamma}{a_2} = \frac{2\gamma}{a_1}$$

Then the interface is progressively bent and increases its spherical shape. The curvature radius decreases progressively to a value $a_3 = d_1/2$ when the interface forms a half-sphere (Figure 1.54). The corresponding pressure difference has a maximum of

$$\Delta P_3 = \frac{2\gamma}{a_3} = \frac{4\gamma}{d_1} > \frac{2\gamma}{a_1} = \frac{4\gamma(-\cos\theta)}{d_1} = \Delta P_1 \qquad (1.84)$$

Then the "bubble" continues to expand, and the curvature radius starts increasing. When the interface expands to the dimension of the larger capillary tube, the pressure difference is reduced to the value

$$\Delta P_4 = \frac{2\gamma}{a_4} = \frac{-4\gamma\cos\theta}{d_2} \qquad (1.85)$$

The evolution of the pressure difference is plotted in Figure 1.55.

Thus, when an interface passes over a sudden enlargement in a hydrophobic capillary tube, the pressure drop increases first to reach the ratio $\dfrac{\Delta P_{max}}{\Delta P_{ini}} = \dfrac{1}{-\cos\theta}$, and then decreases to a value $\dfrac{\Delta P_{exit}}{\Delta P_{ini}} = \dfrac{d_1}{d_2}$.

An increase in driving pressure is required for an interface to pass through a sudden enlargement. If the driving pressure is not sufficient, the flow can stop at the very location of the enlargement. For small driving pressures, a sudden enlargement can be used as a passive valve. They can also be used as devices to synchronize parallel microflows.

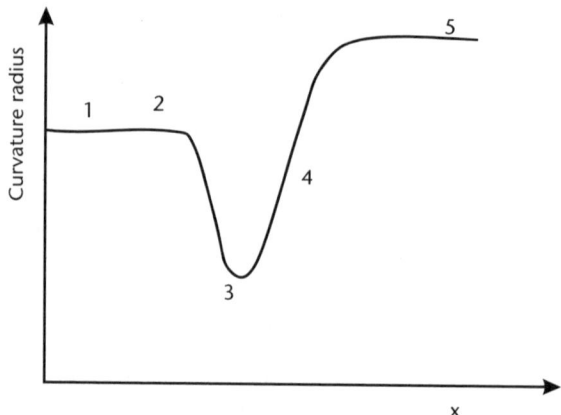

Figure 1.54 Value of the curvature radius of the interface versus the interface location.

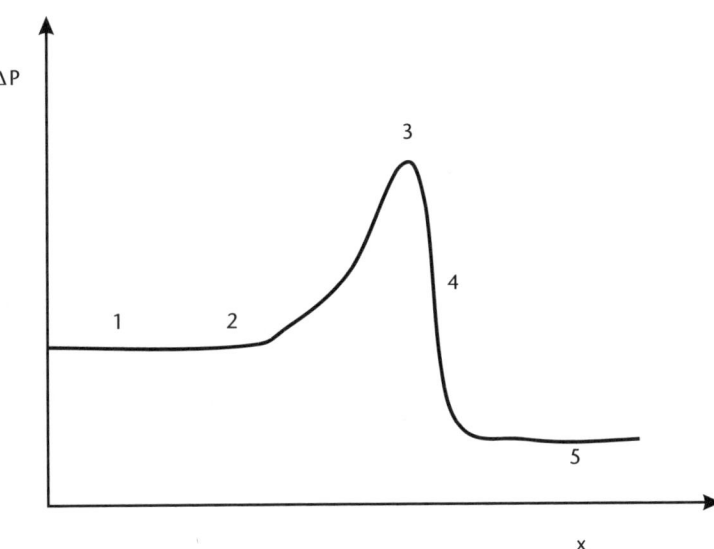

Figure 1.55 Pressure difference at the interface versus the interface location.

1.4 Conclusion

Microflows are an immense subject of study and research. So far, we have only presented their most basic aspects. There is much more to be said, for example, on liquid films and on velocity slip at a solid wall. But the aim of this chapter is to give the reader the bases to tackle the usual microflow problems and to have the prerequisites to deal with more complex situations.

Biotechnology heavily relies on microfluidics and especially on microflows. From continuous single flows for bioanalysis and biorecognition, to plug flows for high-throughput screening, and to microsprays for mass spectrometer analysis, microflows are constantly present.

In the next chapter, we complete the approach to microfluidics with a more recent and fast developing branch of microfluidics called *digital microfluidics*; this chapter details the behavior of microdrops.

References

[1] Nguyen, N. -T., and S. T. Wereley, *Fundamentals and Applications of Microfluidics*, Norwood, MA: Artech House, 2002.

[2] Bejan, A., *Convection Heat Transfer*, Englewood Cliffs, NJ: Prentice-Hall, 1984.

[3] Hemphill, T., W. Campos, and A. Pilehvari, "Yield-Power Law Model More Accurately Predicts Mud Rheology," *Oil & Gas Journal,* Vol. 91, No. 34, 1993, pp. 45–50.

[4] Kolyva, C., http://www.biomedicalphysics.org/PhysCircCourse/2004/Rheology.ppt.

[5] Buckingham, E. "On Physically Similar Systems: Illustrations of the Use of Dimensional Equations," *Phys. Rev.,* Vol. 4, 1914, pp. 345–376.

[6] Rosenow, W. M., and H. Y. Choi, *Heat, Mass, and Momentum Transfer*, Englewood Cliffs, NJ: Prentice-Hall, 1961, p. 48.

[7] Landau, L., and E. Lifchitz, "Mécanique des Fluides," *Editions Mir,* 1971.

[8] Tabeling, P., *Introduction à la microfluidique*, Paris: Belin, 2003.

[9]   Schlichting, H., *Boundary Layer Theory*, New York: McGraw-Hill, 1960, p. 169.

[10]   Washburn, E. W., "The Dynamics of Capillary Flows," *Phys. Rev.*, 1921, pp. 273–283.

[11]   Hartnett, J. P., and M. Kostic, "Heat Transfer to Newtonian and Non-Newtonian Fluids in Rectangular Ducts," *Adv. Heat Transfer,* Vol. 19, 1989, pp. 247–356.

[12]   Shah, R. K., and A. L. London, *Laminar Flow Forced Convection in Ducts*, New York: Academic Press, 1978, p. 197.

[13]   Idel'cik, I. E., *Memento des Pertes de Charge*, Paris: Eyrolles, 1960.

[14]   Bendib, S., and O. Français, "Analytical Study of Microchannel and Passive Microvalve; Application to Micropump Simulation," *Proc. Design, Characterisation, and Packaging for MEMS and Microelectronics 2001*, Adelaide, Australia, pp. 283, 291.

[15]   Feldt, C., and L. Chew, "Geometry-Based Macro-Tool Evaluation of Non-Moving-Part Valvular Microchannels," *J. Micromech. Microeng.*, Vol. 12, 2002, pp. 662–669.

[16]   Zimmerman, W. B., *Process Modelling and Simulation with Finite Element Methods*, Singapore: World Scientific Publishing Co., 2004.

[17]   Adjari, A., "Steady Flows in Networks of Microfluidic Channels: Building on the Analogy with Electrical Circuits," *C. R. Physique*, Vol. 5, 2004, pp. 539–546.

[18]   Pietrabisa, R., et al., "A Lumped Parameter Model to Evaluate the Fluid Dynamics of Different Coronary Bypasses," *Med. Eng. Phys.*, Vol. 18, No. 6, 1996, pp. 477–484.

[19]   Gardeniers, H. J. G. E., et al., "Silicon Micromachined Hollow Microneedles for Transdermal Liquid Transport," *J. Microelectromechanical Systems*, Vol. 12, No. 6, 2003.

[20]   Luttge, R., et al., "Microneedle Array Interface to CE on Chip," *7th International Conference on Miniaturized Chemical and Biochemical Analysts Systems*, Squaw Valley, CA, October 5–9, 2003.

[21]   Rivera, F., et al., "Microdispositif de Diagnostic et de Thérapie in Vivo," *Patent EN*, No. 0350919, November 27, 2003.

[22]   Rivera, F., et al., "In Vivo Transfection Microsystems," *Proc. 26th IEEE Engineering in Medicine and Biology Conference*, San Francisco, CA, September 1–5, 2004.

[23]   Berthier, J., F. Rivera, and P. Caillat, "Numerical Modeling of Diffusion in Extracellular Space of Biological Cell Clusters and Tumors," *Proc. Nanotech 2004 Conference,* Boston, MA, March 7–11, 2004.

[24]   Widera, G., and D. Rabussay, "Electroporation Mediated and DNA Delivery in Oncology and Gene Therapy," *Drug Delivery Technology*, Vol. 2, No. 3, May 2004.

[25]   Berthier, J., et al., "Dimensioning of a New Micro-Needle for the Dispense of Drugs in Tumors and All Clusters," *Proc. Nanotech 2005 Conference*, Anaheim, CA, May 8–12, 2005.

[26]   COVENTOR, http://www.coventor.com/microfluidics.

[27]   *FEMLAB Reference Manual*, Stockholm: COMSOL AB, http://www.comsol.com.

[28]   Sarrut, N., et al., "Enzymatic Digestion and Liquid Chromatography in Micro-Pillar Reactors—Hydrodynamic Versus Electro-Osmotic Flow," *SPIE*, San Jose, CA, Photonics West—MOEMS-MEMS, 2005.

[29]   Hoffman, R. L., "A Study of Advancing Interface," *J. Colloid Interface Science*, Vol. 50, 1975, pp. 228–241.

[30]   Fermigier, M., and P. Jenffer, "An Experimental Investigation of the Dynamic Contact Angle in Liquid-Liquid Systems," *J. Colloid and Interface Science*, Vol. 146, 1990, pp. 226–241.

[31]   Berthier, J., and F. Ricoul, "Numerical Modeling of Ferrofluid Flow Instabilities in a Capillary Tube at the Vicinity of a Magnet," *Proc. 2002 MSM Conference*, San Juan, Puerto Rico, April 22–25, 2002.

[32]   Collet, P., J. de Coninck, and F. Dunlop, "Dynamics of Wetting with a Disordered Substrate: The Contact Angle Hysteresis," *Europhys. Lett.*, Vol. 22, No. 9, 1993, pp. 645–650.

[33]   Chibowski, E., A. Ontiveros-Ortega, and R. Perea-Carpio, "On the Interpretation of Contact Angle Hysteresis," *J. Adherence Sci. Technol.*, Vol. 16, No. 10, 2002, pp. 1367–1404.

[34] Ramos, S. M. M., E. Charlaix, and A. Benyagoub, "Contact Angle Hysteresis on Nano-Structured Surfaces," *Surface Science*, Vol. 540, 2003, pp. 355–362.

[35] He, B., J. Lee, and N. A. Patankar, "Contact Angle Hysteresis on Rough Hydrophobic Surfaces," *Colloids and Surfaces A: Physicchem. Eng. Aspects*, No. 48, 2004, pp. 101–104.

[36] Buguin, A., L. Talini, and P. Silberzan, "Ratchet-Like Topological Structures for the Control of Microdrops," *Appl. Phys. A.*, Vol. 75, 2002, pp. 207–212.

CHAPTER 2

Microdrops

2.1 Introduction

In Chapter 1, we dealt with microflows. Another aspect of microfluidics is the physics of microdrops. Microdrops are a common feature in biotechnology. For example, DNA microarrays are comprised of hundreds to thousands of microwells, each containing a drop of biologic liquid, and a drop dispenser is used to deposit the liquid in each well (Figure 2.1). In such a case, it is essential to understand how the drop relocates in a well.

Another relatively new concept in microfluidics is the use of microdrops to perform the same bioanalysis and biorecognition operations as classical microflows devices do. There are some arguments about whether it is better to use microflows or microdrops in microsystems for biotechnology. The choice is complex because it depends on many factors.

First, microflows are more commonly used, better known, and more familiar to the developers. Micropumps and microvalves are now available, and microchannel etching in silicon, plastic, or glass is now a standard technique. Second, microdrop behavior is often complex and puzzling—as we shall see in this chapter. However, microdrops have the big advantage of minimizing the surface between the liquid and the solid walls. In biotechnology, it is a real advantage since "nonspecific" adsorptions—contact and adherence of the target particles on solid walls at unwanted places in the microsystem—are a constant drawback. If these particles are marked with a fluorescent marker, they constitute a perturbing light source that hampers the detection of the real signal. Another advantage of microdrops in biology is that it is possible to work with very small amounts of liquid, much smaller than it is possible to use in classical microflows. In electrowetting devices, for example, the volume of the drops may be as low as $0.05\ \mu$l. Finally, assuming that the technique to move and control microdrops on a plane surface is mastered, fluidic tracks that resemble the one used to define conducting paths on an electronic semiconductor device may be built (Figure 2.2). One sometimes refers this approach as *digital* or *flatland microfluidics*.

On the other hand, microdrop technology is not adapted to all situations, such as the continuous analysis of large volumes of liquids. Microdrops are not going to replace continuous flow processes but rather complete the panel of existing devices to tackle the many facets of biotechnology.

In this chapter, we deal first with the basic notions of the physics of wetting in order to familiarize the reader with the mechanical behavior of interfaces. Then we will focus on the physics of drops with the important notions of minimal surface

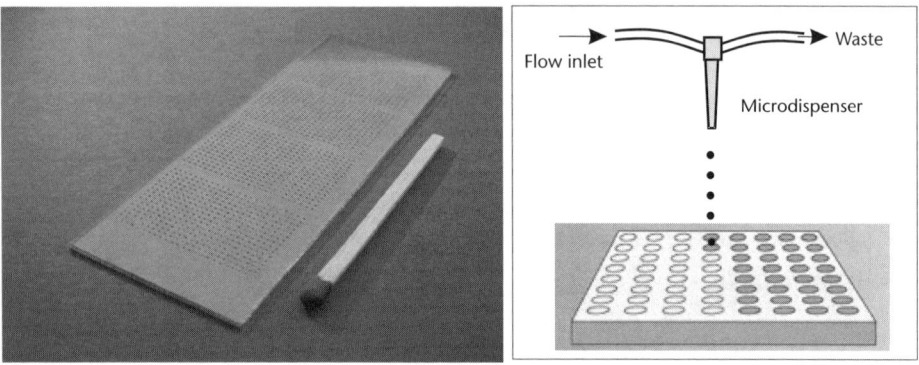

Figure 2.1 (Left) DNA microarray made by SCIENION; and (right) the principle of microdispensing of liquid into wells of a microplate.

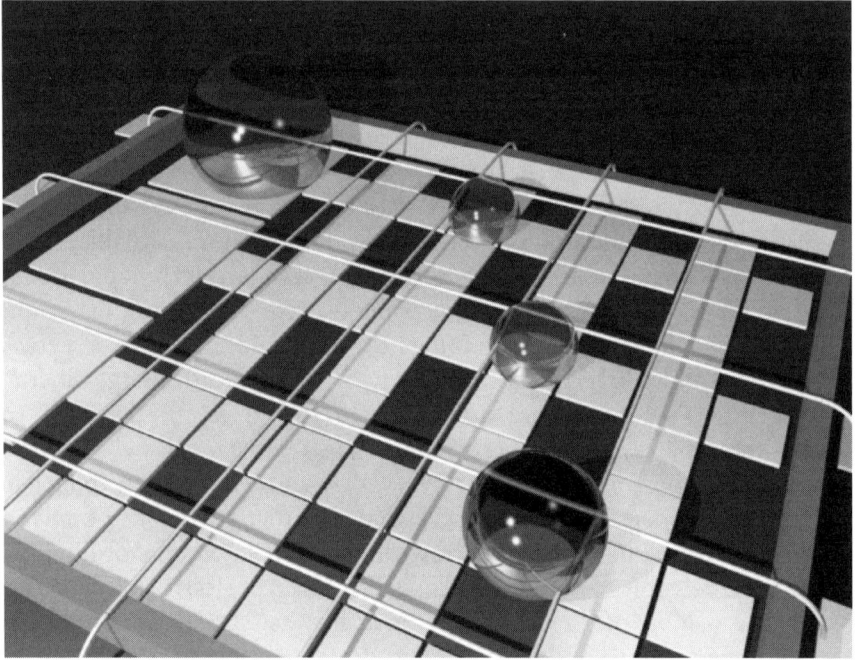

Figure 2.2 Schematic view of a concept of a microdevice using microdrops. (Courtesy of CEA/LETI.)

and contact forces. Finally, we give an example of application of flatland/digital microfluidics using the principle of electrowetting.

2.2 The Physics of Wetting

2.2.1 Capillarity: Surface Tension and Contact Angles

2.2.1.1 Interfaces and Surface Tension

Mathematically speaking, an interface is the geometrical surface that delimitates two fluid domains. This definition implies that an interface has no thickness and is smooth (i.e., it has no roughness). As practical as it is, this definition is in reality a schematic concept. The reality is more complex, and the separation of two immiscible fluids (water/air, water/oil, and such) depends on molecular interactions between the molecules of each fluid [1]. A microscopic view of the interface between two fluids looks more like the scheme of Figure 2.3.

However, as it has been mentioned earlier, even for microdrops, we are more interested in the macroscopic behavior of the interface, and the mathematical concept regains its utility. The former picture can be viewed at a macroscopic size, as shown in Figure 2.4.

In a condensed state, molecules attract each other. Molecules located in the bulk of a liquid have interactions with all neighboring molecules; these interactions are mostly Van der Waals attractive interactions for organic liquids and hydrogen bonds for polar liquids like water. On the other side, molecules at an interface have interactions in one half space with molecules of the same liquid and in the other half space interactions with the molecules of the other fluid or gas (Figure 2.5).

Suppose an interface between a liquid and a gas. In the bulk of the liquid, a molecule is in contact with 4 to 12 other molecules, depending on the liquid (4 for water and 12 for simple molecules); at the interface this number is divided by two. Of

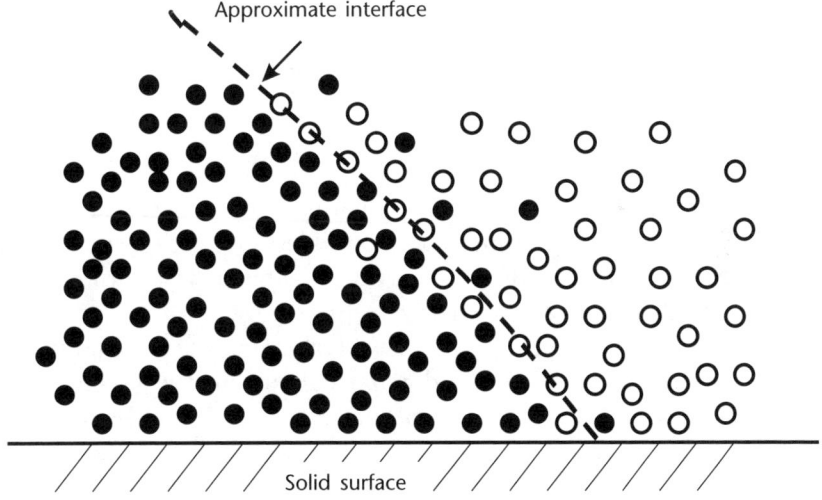

Figure 2.3 Schematic view of an interface at the molecular size.

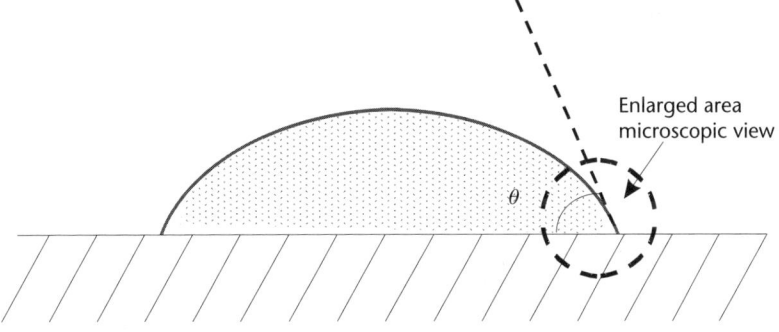

Figure 2.4 Macroscopic view of the interface of a drop.

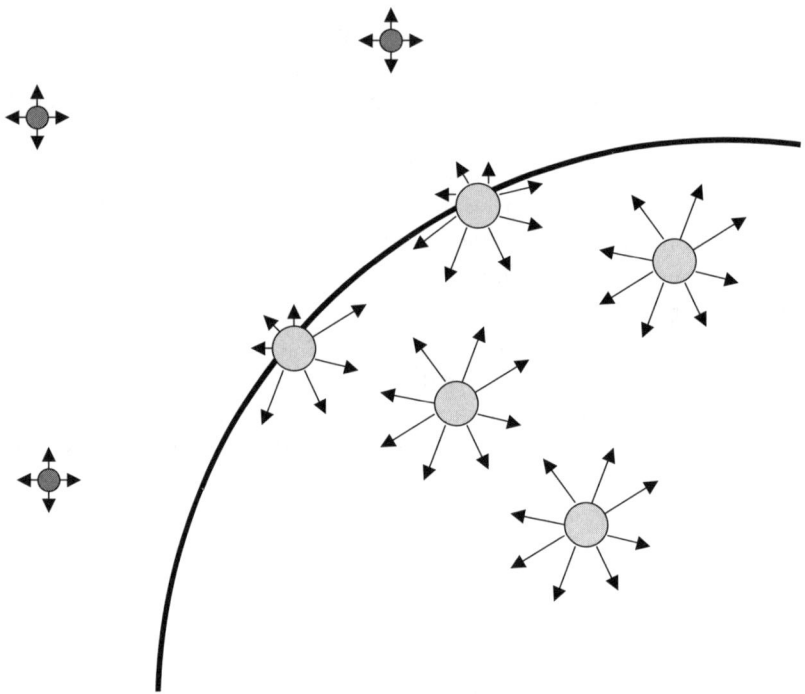

Figure 2.5 Simplified scheme of molecules near an air/water interface. In the bulk, molecules have interaction forces with all of the neighboring molecules. At the interface, half of the interactions have disappeared.

course, the molecule is also in contact with gas molecules, but, due to the low densities of gases, there are less interactions and less interaction energy than in the liquid side. The result is that there is locally a dissymmetry in the interactions, which results in a defect of surface energy.

The same reasoning applies to the interface between two liquids, except that the interactions with the other liquid will usually be more energetic than a gas and the resulting dissymmetry will be less. For example, we will see in Table 2.1 that the contact energy (surface tension) between water and air is 72 mN/m, whereas it is only 50 mN/m between water and oil.

It is also the same for a solid and a liquid. The interface is just the physical contact surface. Molecules in the liquid are attracted toward the interface by Van der Waals forces. But usually these molecules do not "stick" at the wall because of Brownian motion. However, impurities contained in the fluid, like particles of dust or biological polymers like proteins, may well adhere permanently to the solid surface because, at the contact with the solid interface, they experience more attractive interactions. This is because the size of polymers is much larger than that of water molecules, and Van der Waals forces are proportional to the number of contacts.

Table 2.1 Typical Values of Surface Tensions at Room Temperature

Type of Components	Water/Air	Water/Oil	Glycerol/Air	Ethanol/Air	Cyclohexan/Air	Mercury/Air
Surface tension [mN/m]	72	50	63	23	25	485

Source: [2].

At the macroscopic scale, a physical quantity called *surface tension* has been introduced in order to take into account this molecular effect. The surface tension has the dimension of energy per unit surface, and in the International System it is expressed in J/m^2 or N/m (sometimes it is more practical to use mN/m as a unit for surface tension): it is the local defect of energy at the interface divided by the area occupied by a molecule.

In the literature or on the Internet there exist tables for surface tension values [3, 4]. Typical values of surface tensions are given in Table 2.1.

Usually surface tension is denoted by the Greek letter γ, with subscripts referring to the two components on each side of the interface (e.g., γ_{LG} at a liquid/gas interface). Most of the time, if the contact is with air, the subscripts are omitted.

Just because of the definition of the surface tension, for a homogeneous interface (same molecules at the interface all along the interface), the total energy of the surface is

$$E = \gamma S \qquad (2.1)$$

2.2.1.2   Capillary Forces

Surface tension can be looked at as a force per unit length. This can be directly seen from its unit since surface tension is expressed in N/m, which is indeed a force per unit length. But it may be interesting to give a more physical feeling by doing a very simple experiment (Figure 2.6) [2]. Take a solid frame and a solid tube that can roll on this frame. If we form a liquid film of soap between the frame and the tube by plunging one side of the structure in a water-soap solution, the tube starts to move toward the region where there is the liquid film. The surface tension of the liquid film exerts a force on its free boundary.

On the other hand, we can increase the film surface by exerting a force on the tube. The work of this force is given by the relation

$$\delta W = Fdx = 2\gamma L dx$$

The coefficient 2 stems from the fact that there are two interfaces between the liquid and the air. This relation shows that the surface tension γ is a force per unit length, perpendicular to the tube, in the plane of the liquid and directed toward the liquid.

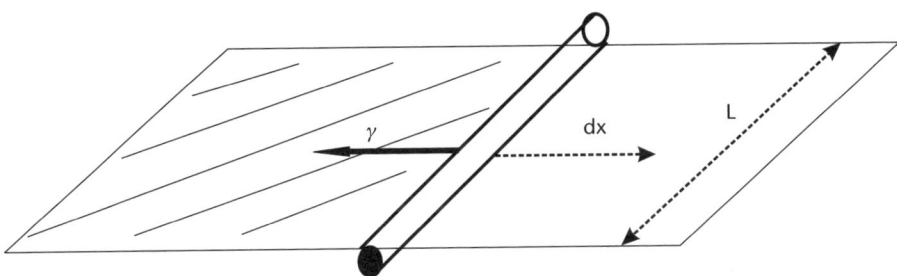

Figure 2.6   A tube placed on a rigid frame where the left part is occupied by a film of glycerin displaced toward the left.

Capillary forces are extremely important at a microscale. For example, it is very difficult to separate two parallel plates separated by a liquid film.

2.2.2 Wetting: Young's Law

The behavior of microdrops on a solid surface is of utmost importance for the conception of microsystems in biology. This solid surface may be the tip of a microneedle or a micropipette; it may be the surface of a microplate with thousands of cusps or that of a Teflon-coated electrowetting device. In any case, we need know the behavior of the drop. The microdrop must detach easily from the tip of the pipette or the needle, stay in the microcusps without flowing by capillarity inside the other neighboring cusps to avoid biological contamination, or follow the electrodes line of the electrowetting device.

Wetting characterizes the contact of a liquid with a solid surface. Generally speaking, there are two types of wetting: total and partial. Total wetting corresponds to the case where a liquid film spreads out on the solid surface, and partial wetting occurs when the liquid stays in drops, as shown in Figure 2.7.

A criterion for total wetting is given by the spreading coefficient S given by

$$S = \gamma_{SG} - (\gamma_{SL} + \gamma_{LG})$$

If S is positive, the drop spreads on the solid surface as a liquid film. If S is negative, it is a situation of partial wetting. In the case of partial wetting, there is a line where all three phases come together. This line is called the *contact line* or sometimes the *triple line*. The contact of a water droplet on a solid is said to be hydrophilic or hydrophobic, depending on the contact angle (Figure 2.8).

We have seen that surface tensions are not exactly forces; their unit is N/m; however, they represent a force that is exerted tangentially to the interface. We can then draw the different forces that are exerted by the presence of a fluid on the triple line (Figure 2.9).

Note that we have noted the wall characteristics by the terms *hydrophilic* or *hydrophobic* with reference to water. In general, we should say that a liquid is "wetting" or "not wetting" a specific surface based on the contact angle of a drop with the surface.

A criterion of wetting may be derived from surface tension considerations: At equilibrium, the resultant of the forces must be zero. We use a coordinate system

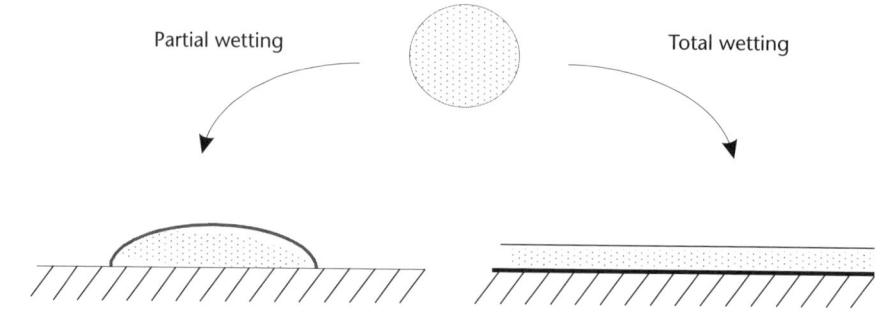

Figure 2.7 Partial wetting and total wetting.

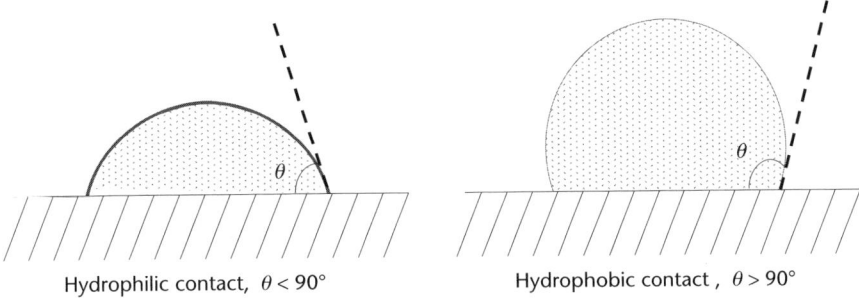

Figure 2.8 Hydrophilic and hydrophobic contact.

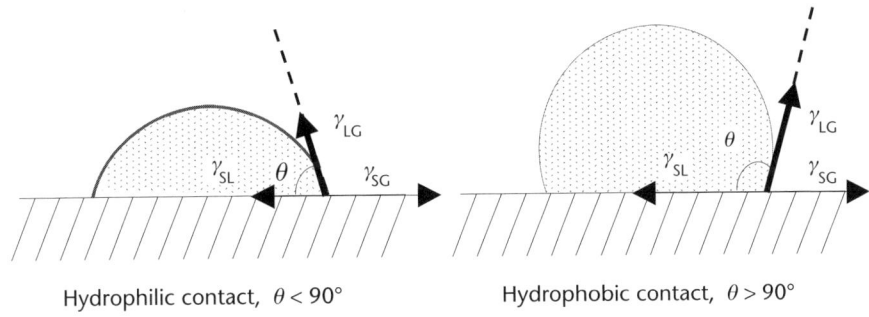

Figure 2.9 Schematic view of the forces at the triple line.

where the x-axis is the tangent to the solid surface at the contact line (in the Figure 2.9 it is horizontal) and the y-axis is the direction perpendicular (vertical). Then the projection of the resultant on the x-axis is zero, and we obtain the relation

$$\gamma_{LG} \cos \theta = \gamma_{SG} - \gamma_{SL} \tag{2.2}$$

This relation is called Young's law and is very useful to understand the behavior of a drop. In particular, it shows that the contact angle is determined by the surface tensions of the three constituents. For a microdrop on a solid, the contact angle is given by the relation

$$\theta = \arccos \frac{\gamma_{SG} - \gamma_{SL}}{\gamma_{LG}}$$

Sometimes in real experimental situations, when we deal with real biological liquids, one observes an unexpected change in the contact angle with time. This is just because biological liquids are inhomogeneous and can deposit a layer of chemical molecules on the solid wall, thus progressively changing the value of the tension γ_{SL}, and consequently the value of θ, as it is stated by Young's law.

One may note that Young's law was obtained by a projection of the forces on the x-axis, but what about the projection on the y-axis? As a matter of fact, there is a vertical force acting on the solid surface and directed away from the solid. It is balanced by the reaction of the solid. If the solid is replaced by a liquid $L2$ nonmiscible with the liquid $L1$ (e.g., oil and water), as shown in Figure 2.10, then the surface is

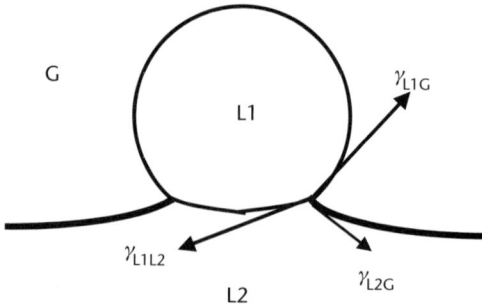

Figure 2.10 Schematic view of a microdrop of oil on top of a water layer. Because of its small dimension, the shape of the drop in the air is spherical. At the triple line, Young's law applies and the resultant of the forces is zero.

distorted so that the resultant of the forces at the triple line is zero. The resultants of the projections of the forces are zero as well on the x-axis and y-axis.

If the dimension of the drop is larger, the contact would be similar, but the shape of the interfaces would be somewhat nonspherical due to the gravitational forces, as shown in Figure 2.11.

Note that there is a singular point at the triple line: the vectors γ_{L1L2} and γ_{L2G} are not collinear, because if they were, the resultant of the forces would not be zero.

Another very interesting derivation of Young's law was done by Shapiro and coworkers [5]. Instead of considering forces at the triple line, Shapiro uses the principle of minimal energy for a drop at equilibrium. Take the example of a sessile drop placed on a horizontal plane (Figure 2.12) and suppose that the dimensions of the drop are small enough so that gravitational forces can be neglected (we will later give a justification for neglecting gravity effect on microdrops), the shape of the drop is then spherical. Due to this spherical shape, the drop volume is a function of R and θ. According to the notations of Figure 2.12, the drop characteristic dimensions h (drop height) and a (drop contact radius) are

$$h = R(1 - \cos\theta)$$

and

$$a = R\sin\theta$$

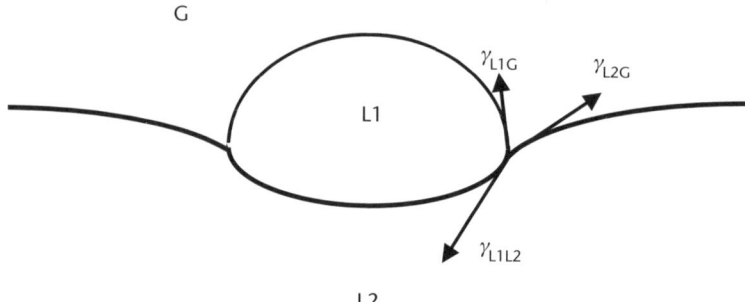

Figure 2.11 Schematic view of a drop of liquid L1 (oil) deposited on liquid L2 (water).

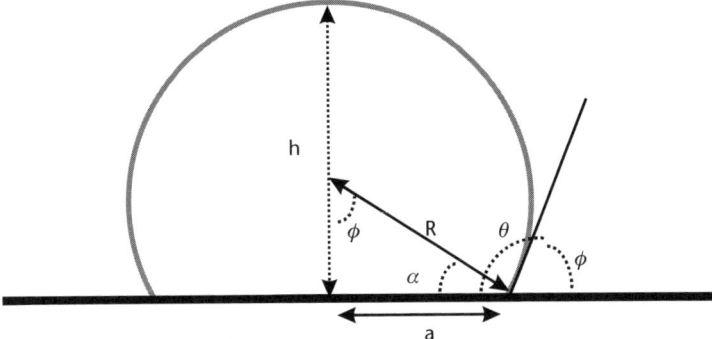

Figure 2.12   Schematic view of a microdrop sitting on a horizontal plane.

The drop volume is then [6]

$$V = \frac{\pi}{3} R^3 \left(2 - 3\cos\theta + \cos^3\theta\right) = \frac{\pi}{3} a^3 \frac{\left(2 - 3\cos\theta + \cos^3\theta\right)}{\sin^3\theta} \tag{2.3}$$

The volume V can be cast into the form

$$V = \pi R^3 \left(\frac{2}{3} - \frac{3\cos\theta}{4} + \frac{\cos 3\theta}{12}\right)$$

Using the principle of minimal energy at equilibrium, and noting that the energy is a function of the two parameters R and θ,

$$E = E(R,\theta)$$

one obtains the following relation

$$dE = \frac{\partial E}{\partial R}(R,\theta)dR + \frac{\partial E}{\partial \theta}(R,\theta)d\theta = 0 \tag{2.4}$$

Since the volume of the drop is constant, its variation must be zero, hence

$$dV = 3\pi R^2 \left(\frac{2}{3} - \frac{3\cos\theta}{4} + \frac{\cos 3\theta}{12}\right)dR + \pi R^3 \left(\frac{3\sin\theta}{4} - \frac{\sin 3\theta}{4}\right)d\theta = 0$$

This later relation can be cast under the form

$$dR = R\left(-\frac{2\cos^2\left(\frac{\theta}{2}\right)\cot\left(\frac{\theta}{2}\right)}{2 + \cos\theta}\right)d\theta = R\ q(\theta)d\theta \tag{2.5}$$

Equation (2.5) can be substituted in (2.4) and we find

$$\frac{\partial E}{\partial R}(R,\theta)Rq(\theta) + \frac{\partial E}{\partial \theta}(R,\theta) = 0 \qquad (2.6)$$

We show now that (2.6) is the Young equation (2.2) written in another form. In this case, where the energy of the drop is only due to solid/liquid, solid/gas, and liquid/gas interfaces, the total energy is

$$E = (\gamma_{LS} - \gamma_{GS})S_{LS} + \gamma_{LG}S_{LG} \qquad (2.7)$$

In (2.7), S denotes the surface. Note that the solid/gas coefficient is negative because if S_{LS} is increased by some amount, then S_{GS} must be decreased by the same amount. Because the drop has a spherical shape, the surfaces appearing in (2.7) are functions of R and θ. Using the expression of the volume of the spherical cap (2.3) and the expression for the contact radius a, we find the following expressions for the contact and free surfaces

$$\begin{aligned} S_{LS} &= \pi R^2 \sin^2 \theta \\ S_{LG} &= 2\pi R^2 (1 - \cos \theta) \end{aligned} \qquad (2.8)$$

In consequence, the interfacial energy is

$$E = \pi R^2 \left[(\gamma_{LS} - \gamma_{GS})\sin^2 \theta + 2\gamma_{LG}(1 - \cos \theta) \right] \qquad (2.9)$$

If we differentiate E relative to R and to θ, and plug the result into (2.6), we obtain exactly, after some trigonometric algebra, Young's law.

This energy approach may seem more complicated than the direct approach using the force balance at the triple line, but it is very powerful because it is more generic, as we shall see in Section 2.2.8.

2.2.3 Wenzel's Law

Let's come back to a contact between a liquid drop and a solid. Roughness of the solid walls modifies the contact between the liquid and the solid. But the effect of roughness on the contact angle is not intuitive. It is a surprising but very useful observation that roughness amplifies the character hydrophilic or hydrophobic of the contact.

Suppose that θ is the angle with the surface with roughness and θ^* is the angle with the smooth surface (in both cases, the solid, liquid, and gas are the same). One very important point here is that we have made the implicit assumption that the size of the roughness is very small, so that the molecules of the liquid are interacting macroscopically with a plane surface but microscopically with a rough surface. This explains why we can use the unique angle of contact θ. As a general rule, the size of the roughness should be smaller than the mean interaction distance between liquid molecules and the solid wall.

Suppose a very small displacement of the contact line (Figure 2.13). Then the work of the different forces acting on the contact line is given by

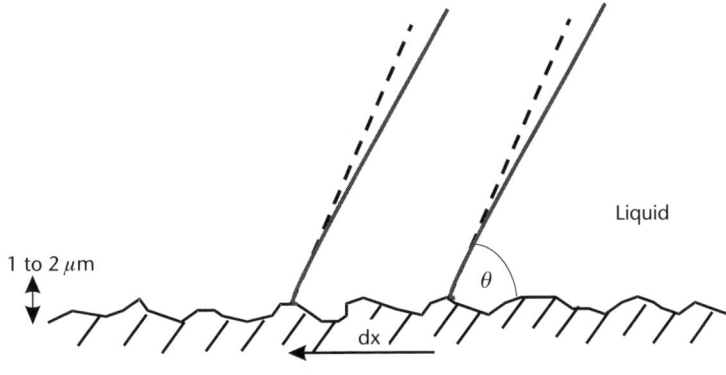

Figure 2.13   Contact of a liquid drop on a rough surface.

$$dW = \sum \vec{F} \cdot \vec{dl} = \sum F_x dx = (\gamma_{SL} - \gamma_{SG})r dx + \gamma_{LG} \cos \theta dx \tag{2.10}$$

where r is the roughness ($r\,dx$ is the real distance on the solid surface when the contact line is displaced of dx). Note that by definition, $r > 1$. Thus, the change in energy is

$$dE = dW = (\gamma_{SL} - \gamma_{SG})r dx + \gamma_{LG} \cos \theta dx \tag{2.11}$$

In fact, if we imagine that the drop finds its equilibrium state after the small perturbation dx, it finally stops at a position where its energy is minimum, so that

$$\frac{dE}{dx} = 0$$

and we obtain the relation

$$\gamma_{LG} \cos \theta = (\gamma_{SG} - \gamma_{SL})r \tag{2.12}$$

If we recall that Young's law is for a smooth surface

$$\gamma_{LG} \cos \theta^* = \gamma_{SG} - \gamma_{SL}$$

then we obtain Wenzel's law

$$\cos \theta = r \cos \theta^* \tag{2.13}$$

Taking into account that $r > 1$, this relation implies that

$$|\cos \theta| > |\cos \theta^*| \tag{2.14}$$

We can deduce that if θ^* is larger than 90° (hydrophobic contact), then $\theta > \theta^*$, and the contact is still more hydrophobic due to the roughness. If θ^* is smaller than 90° (hydrophilic contact), then $\theta < \theta^*$, and the contact is still more hydrophilic due to the roughness.

An important remark at this stage is that the scale of the roughness on the solid surface is *very small compared to that of the drop* [7]. Indeed, if not, it would not be possible to define a unique contact angle anymore; the drop would no longer be axisymmetrical, and the contact could be sketched as in Figure 2.14 (the position of the drop might not be stable).

2.2.4 Superhydrophobicity and Superhydrophilicity

Wenzel's law is in part only the explanation to the so-called superhydrophilic or superhydrophobic contact. According to Wenzel's law, hydrophobicity or hydrophilicity is enhanced by an increase of the surface roughness. However, to obtain superhydrophobicity, the roughness must be increased in such a way that air is trapped in the porosities or rugosities underneath the droplet.

In nature, some plants living in a very wet environment have leaves with such roughness that raindrops just roll along their surface without any adhesion, preventing rotting of the leaves. A microscopic view shows that the leaves have high roughness, trapping air bubbles in the rugosities (Figure 2.15) [8].

In biotechnology, there are cases where it is important to obtain super-hydrophobic contact. It reduces the hydrodynamic drag at the wall, and it prevents cross contamination of one drop by another one moving on the same surface. It is possible to obtain artificially superhydrophobic contact by increasing the surface roughness and mimicking the natural surface of Figure 2.15. It has been shown that a hydrophobic substrate may be rendered super-hydrophobic by etching microgrooves or micropillars [9], as shown in Figure 2.16.

Superhydrophobic surfaces are commonly microfabricated by etching micropillars in a silicon substrate. The shape of a droplet deposited on such pillars has been calculated [10] using the Evolver code [11] (see Section 2.2.7).

There are also cases where a very hydrophilic surface is suitable (e.g., to increase the wetted surface). Uelzen and Müller have shown that most solid smooth surfaces can be roughened by crystallization of pyramidal crystal of tin (Sn) of 1 to 2 μm in size [12]. Then a hydrophilic or hydrophobic layer may be deposited on top, resulting in a substantial increase in hydrophilicity or hydrophobicity. For example, the contact angle of water on gold is 80° but it drops to 60° with this technique. For

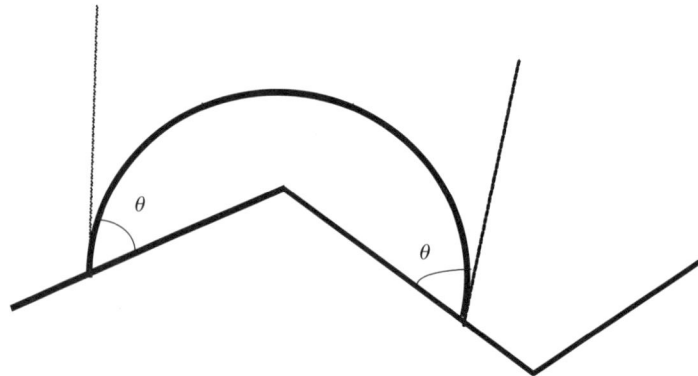

Figure 2.14 Large-scale roughness: schematic view of a drop located on an angle of the solid surface. The position of the drop might not be stable.

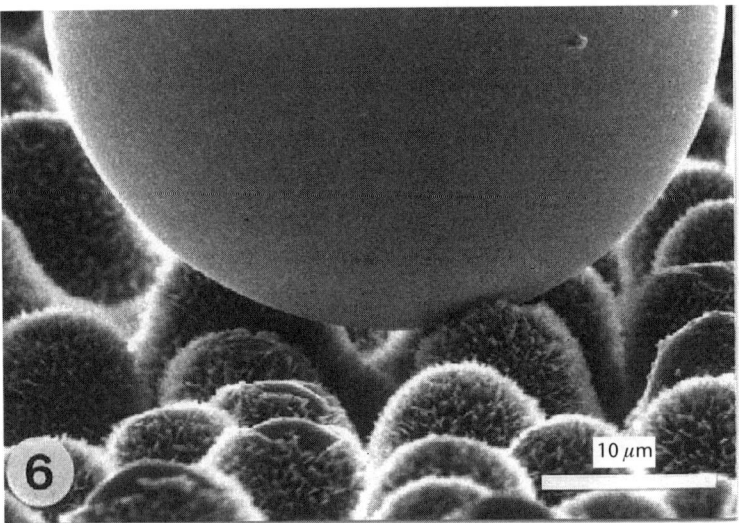

Figure 2.15 Mercury droplet on the papillose adaxial epidermal surface of Colocasia esculenta (*From:* [8]. © 1997 Planta. Reprinted with permission.)

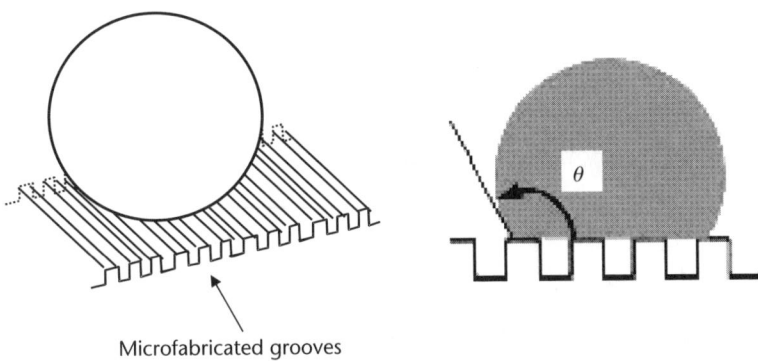

Microfabricated grooves

Figure 2.16 (Left) Superhydrophobic surface fabricated by etching microgrooves in a hydrophobic substrate, and (right) superhydrophobic surface constituted of micropillars.

water on hexamethyldisiloxane (HMDSO), it drops from 30° to 5°, which is a very hydrophilic contact.

2.2.5 Cassie-Baxter Law

The same analysis of Wenzel was done by Cassie for chemically inhomogeneous solid surfaces. As for Wenzel's law, the same requirement of small-size heterogeneities compared to interactions distance between liquid molecules and solid wall applies. For simplicity, we analyze the case of a solid wall constituted by microscopic inclusions of two different materials; if θ_1 and θ_2 are the contact angles for each material at a macroscopic size, and f_1 and f_2 are the surface fractions of the two materials (Figure 2.17), then the energy to move the interface of dx is

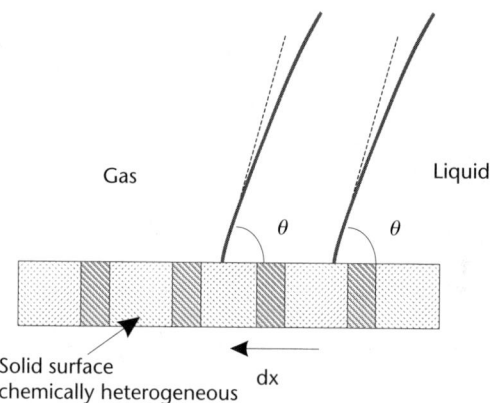

Figure 2.17 Displacement of the contact line of a drop on an inhomogeneous solid surface.

$$dE = dW = \left(\gamma_{SL} - \gamma_{SG}\right)_1 f_1 dx + \left(\gamma_{SL} - \gamma_{SG}\right)_2 f_2 dx + \gamma_{LG} \cos\theta dx \qquad (2.15)$$

The equilibrium is obtained by taking the minimum of E

$$\gamma_{LG} \cos\theta = \left(\gamma_{SG} - \gamma_{SL}\right)_1 f_1 + \left(\gamma_{SG} - \gamma_{SL}\right)_2 f_2$$

and by comparing it with Young's law, we obtain the Cassie-Baxter relation

$$\cos\theta = f_1 \cos\theta_1 + f_2 \cos\theta_2 \qquad (2.16)$$

This relation may be generalized to a more inhomogeneous material

$$\cos\theta = \sum_i f_i \cos\theta_i$$

Note that

$$f_1 + f_2 = 1 \quad \text{or} \quad \sum_i f_1 = 1$$

The Cassie-Baxter relation shows that the cosine of the contact angle on a microscopically inhomogeneous solid surface is the barycenter of the cosine of the contact angles on the different chemical components of the surface.

The Cassie-Baxter law explains some unexpected experimental results: Sometimes, if not enough care was taken during microfabrication, a micro-fabricated surface may present chemical inhomogeneity and the wetting properties are not what they were initially expected. For example, if a uniform layer of Teflon is deposited on a rough substrate, the surface should become hydrophobic. However, if the layer is too thin, the Teflon layer may be porous and the coating inhomogeneous; the wetting properties are then modified according to the Cassie-Baxter law and the gain in hydrophobicity may not be as important as expected.

As for Wenzel's law, an important remark at this stage is that the scale of change of the different chemical materials of the solid surface is *very small compared to that of the drop* [7]. Indeed, if not, it would not be possible to define a unique contact angle anymore, and the contact could be sketched as in Figure 2.18 (in such a case,

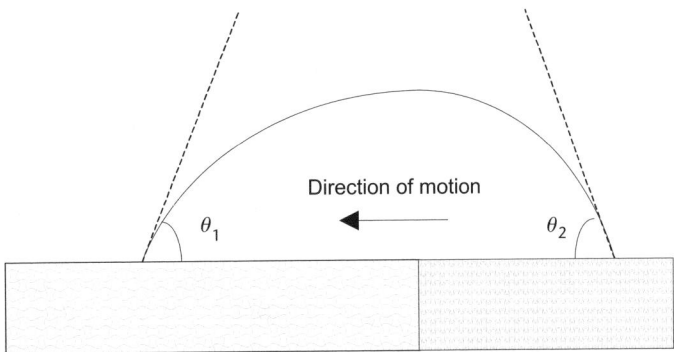

Figure 2.18 Schematic view of the contact on two different solids: the drop is not stable and migrates to the more hydrophilic surface.

we shall show later that the drop is not at equilibrium and migrates to the more hydrophilic region).

2.2.6 Simultaneous Water and Oil Superrepellent Surfaces

It is known that solid surfaces covered by fluoromethyl groups (CF_3) have very low surface energy. The contact angle with oil is 110° and with water is 95°. It is tempting to use the principles of Wenzel's and Cassie-Baxter laws to increase these contact angles and to obtain a water and oil superrepellent surface [13]. By using fractal shape surfaces, Hsieh and coworkers have obtained a rough inhomogeneous surface having contact angles larger than 150° with oil and larger than 160° with water. Note that both the Wenzel and the Cassie-Baxter laws predict the increase of contact angle values. When the roughness parameter r is increased, very small air bubbles are trapped, and the liquid droplet is partly in contact with the fluoromethyl (f_1) and partly with air (f_2). According to Wenzel's law, the contact angle increase with r, and according to the Cassie-Baxter law, it also increases because the contact angle with air is 180°. We can rewrite (2.16) under the form

$$\cos \theta = f_1 \cos \theta_1 - f_2$$

This latter relation shows an increase in contact angle as soon as air bubbles are trapped in rugosities. This shows that the problem of air trapped by very rough surfaces is still a subject of investigation.

2.2.7 The Effect of Surfactants

Surfactant is the short form for *surface active agent*. They are long molecules characterized by a hydrophilic head and a hydrophobic tail, and for this reason they are called amphiphilic molecules. More details about the chemical structure of surfactants are given in Chapter 6. Very often surfactants are added to biological samples in order to prevent the formation of aggregates and to prevent target molecules to stick to the solid walls of the microsystem (remember that microsystems have extremely large ratios between the wall surface and the liquid volume). Due to their amphiphilic nature, surfactants gather on the interface between the liquid and the surrounding gas, as sketched in Figure 2.19(a), lowering the surface tension of the liquid [14]. Above a critical concentration, the interface is saturated with surfacatants and surfactants molecules in the bulk of the fluid group; together they form micelles.

Surface tension is reduced by the presence of surfactants at the interface, as shown in Figure 2.19(b). For example, pure water has a surface tension 72 mN/m, and water at the critical micelle concentration (CMC) has a surface tension 30 mN/m.

How is this effect of surfactants on surface tension measured? Usually one uses the pendant-drop method on small but not microscopic drops, so that the shape of the drop is a balance between surface tension and gravity forces [15]. The pendant-drop method consists of forming a drop pending from the tip of a vertical capillary tube, taking a picture of the drop, and analyzing the shape of the contour of the drop (Figure 2.20).

It can be shown that the contour is linked to the value of the surface tension. The calculation of the interfacial tension is achieved using the two following equations.

First, the Laplace equation (see Section 2.3.2):

$$\Delta P = \gamma \left(\frac{1}{R} + \frac{1}{R'} \right)$$

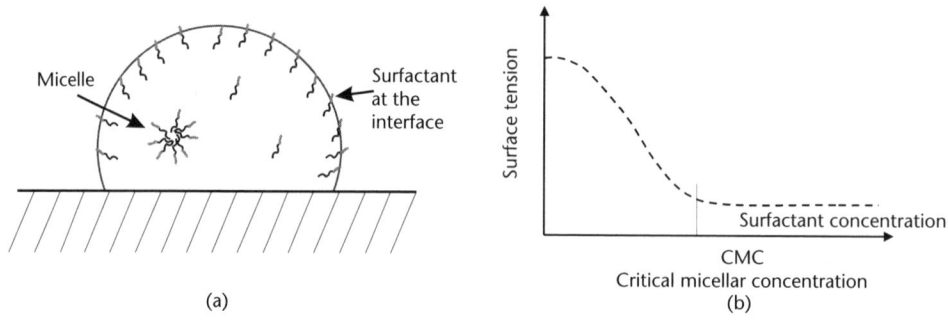

Figure 2.19   (a) Schematic view of surfactants in a liquid drop; and (b) relation between surface tension and surfactant concentration.

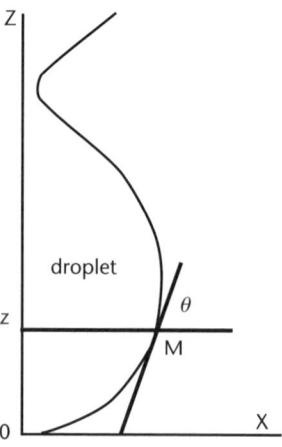

Figure 2.20   Typical drop contour in the pendant-drop method.

Second, the equilibrium equation of the drop between the interfacial tension, gravity, and pressure:

$$2\pi R\gamma \sin\theta = V(\rho_b - \rho_l)g + \pi R^2 P$$

where ΔP is the pressure difference through the interface, resulting from the curvature of the interface, R and R' are the main curvature radii of the interface, x and z are the coordinates of M on the contour of the image of the drop, θ is the angle of the tangent at M to the contour of the image of the drop, V is the volume of the fluid under the plane of altitude z, ρ_b and ρ_l are the volumic masses of the two fluids, and g is the gravitational acceleration. Substitution of the force equilibrium equation into the Laplace equation yields a relation between γ, R, R', V, and θ. The analysis of the image provides the values of R, R', V, and θ. A more precise value of the surface tension is obtained by averaging the results for many different values of z.

In the case of a liquid containing surfactants, successive pictures of the drop show an evolution of the contour related to the adsorption of surfactants on the interface [16], as shown in Figure 2.21. When CMC is reached, the shape of the drop does not change anymore.

2.2.8   Shape of a Drop on Solid Surface

We have seen in the preceding section that, in absence of other forces, the shape of a liquid drop is determined by a balance between gravitational force and surface tension. It is the same when the drop is deposited on a solid surface. It is a common observation that large drops are not spherical, but small drops are. Take a drop of water of 0.05 cm^3 volume placed on a horizontal hydrophilic plate. The real shape of the drop is shown in Figure 2.22(a); if gravitational forces are not taken into account, the shape of the drop would be spherical, as shown in Figure 2.22(b). As can be expected, the effect of gravity is to flatten the drop.

It is worth remarking that the contact angle of the liquid with the solid surface is macroscopically not the same if gravitational forces have an effect on the drop

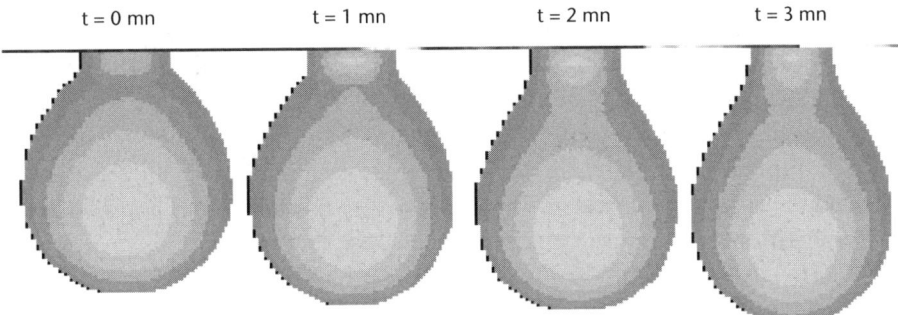

t = 0 mn   t = 1 mn   t = 2 mn   t = 3 mn

Figure 2.21   Different shapes of pendant-drop during surfactant concentration adsorption on the interface. The drop is in equilibrium between the interfacial tension and the gravity. These forces are opposed. The interfacial tension gives the drop a spherical shape, whereas the gravity elongates it. Then the drop seems to be pear shaped. The influence of the interfacial tension on the shape of the drop can be observed during the adsorption of surfactant. The drop becomes more and more elongated and the area of the interface increases when the interfacial tension decreases with time.

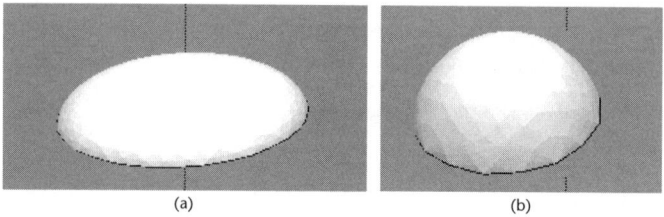

Figure 2.22   (a) The real shape, and (b) with $g = 0$ m/s^2.

shape. To this point we come back to Shapiro's energy approach [5]. Suppose a drop sufficiently small to be close to spherical, but showing a small deformation due to gravitational forces. It can be shown that the total energy of the drop (the surface energy plus the potential energy) is given by

$$E = (\gamma_{LS} - \gamma_{GS})R^2\pi \sin^2\theta + \gamma_{LG}2\pi(1 - \cos\theta) + R^4\rho g\frac{2\pi}{3}(3 + \cos\theta)\sin^6\left(\frac{\theta}{2}\right) \quad (2.17)$$

The two first terms in (2.17) are the surface energy (2.9), and Shapiro has shown that the third term is the potential energy. The interfacial term is at minimum when θ is equal to the nongravity contact angle, and the gravity term is at minimum when $\theta = 0$, so it tends to flatten the drop. Using the same derivation principle as before, a modified Young's equation is obtained:

$$\cos\theta - \left(\frac{\gamma_{SG} - \gamma_{SL}}{\gamma_{LG}}\right) + \left(\frac{\rho g R^2}{\gamma_{LG}}\right)\left[\frac{\cos\theta}{3} - \frac{\cos 2\theta}{12} - \frac{1}{4}\right] = 0 \quad (2.18)$$

The two first terms of this equation are just the Young's equation; the third term is a correction due to gravity. In this third term, a nondimensional number appears; that is usually called the Bond number

$$Bo = \frac{\rho g R^2}{\gamma_{LG}} \quad (2.19)$$

The Bond number represents the ratio of the gravitational forces to the surface tension. For a low Bond number, the gravity has no effect on the drop, the shape is spherical, and the contact angle is given by Young's equation. For increasing Bond numbers, the shape is less and less spherical.

Typically, for water drops used in microsystems, $\rho = 1,000$ kg/m^3, $\gamma = 72$ mN/m, and $R = 1$ mm at the most. This leads to a maximum Bond number of 0.15, so that gravitational force may usually be neglected.

2.3   The Physics of Drops

2.3.1   Minimization of Surface Energy

As a general rule, the equilibrium state of a physical system corresponds to a minimum of energy of the system. This rule may be applied to liquid drops. The equilibrium shape of a drop minimizes its surface energy, taking into account all the other forces that act on the drop. This is why a drop tends to the spherical shape—when it is possible.

To show the effect of the minimization of surface energy, we give examples obtained with the public domain Surface Evolver software by Ken Brakke [11], well adapted for drops at equilibrium or near equilibrium.

Take the example of a thin liquid film on a horizontal plane, and assume that the Young's contact angle of the liquid with the solid is 80°. The results of the Evolver code are shown in Figure 2.23. The drop wants to minimize the surface energy, and it takes a spherical cap shape.

Another example is the case of a drop forming on a spherical solid sphere (Figure 2.24). We will show a very striking application of this very simple principle in Chapter 7 (dedicated to the study of magnetic beads). Actually, the liquid behavior is exactly the same as in the case of the flat plane (for sizes less than 1 μm, the gravity force does not modify the shape of the drop); just the geometry is now spherical.

2.3.2   Laplace Law

Minimization of energy also has theoretical consequences, as we are going to see with the Laplace law. The Laplace law is a relation between the pressure difference across an interface and the radius of curvature of the interface. First, for simplicity, consider a spherical drop (Figure 2.25). Assume an infinitesimal change in the radius R. The energy variation is then [3]

$$dE = -P_1 dV_1 - P_0 dV_0 + \gamma dA \tag{2.20}$$

taking into account the values of the volumes and surface

$$dV_1 = -dV_0 = 4\pi R^2 dR$$
$$dA = 8\pi R dR \tag{2.21}$$

and applying the minimization theory

$$\frac{dE}{dR} = -4\pi (P_1 - P_0) R^2 + 8\pi \gamma R = 0$$

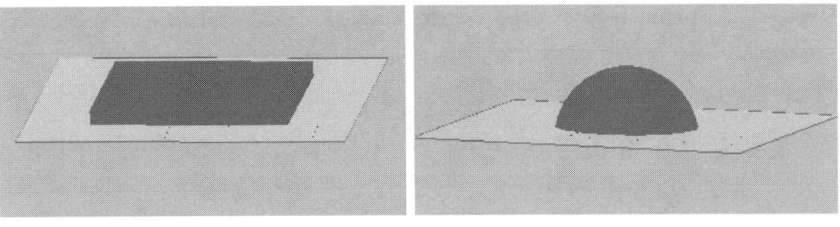

Figure 2.23 A liquid layer initially deposited on a flat solid surface is not at equilibrium; it evolves until it reaches a minimum for the surface energy. The shape of the drop is then a spherical cap. The first (left) view is not physical; the final (evolved) shape is the equilibrium shape.

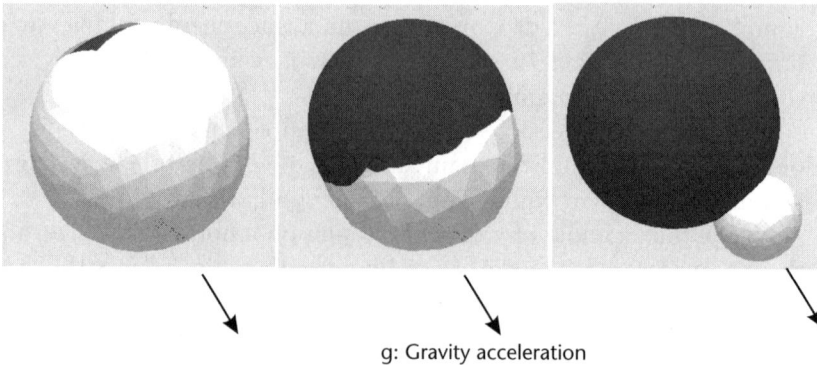

g: Gravity acceleration

Figure 2.24 A liquid layer initially covers nearly entirely a solid sphere (far left), but due to the constraint of a 120° contact angle, the layer "peels off" and forms a spherical cap attached to the solid sphere.

and we find the Laplace equation for spherical drops

$$P_1 - P_0 = 2\frac{\gamma}{R}$$ (2.22)

Note that in the general case, there are two radiuses of curvature at the drop surface R and R'. In this case, the derivation of the Laplace law requires more algebra. We just indicate the result

$$P_1 - P_0 = \gamma\left(\frac{1}{R} + \frac{1}{R'}\right)$$ (2.23)

Laplace law signifies that there is a discontinuity in pressure when crossing an interface. The result may seem simple, but it hides some complications: first note that the pressure inside a microdrop equilibrates very quickly because of the size of the drop. So what to think of drops for which curvature radius is not constant? Take the example of Figure 2.25 (right part). There is a change of curvature radius when the bubbles are facing each other. Because the inner pressure is uniform, Laplace law indicates that the pressure in the liquid film between the two bubbles P_i is larger than the pressure P_0.

The use of Laplace law is often a very elegant way to solve drop deformation [17, 18], and it has been thought to apply this law to in-vivo situations, like for the lung alveoli or vesicles. But in these two latter cases, it has been shown that incorrect use of Laplace law can be made, mostly because these objects are deformed by external unknown pressure [19, 20].

2.3.3 Motion of Drops Under the Action of Hydrophilic/Hydrophobic Forces

It is well established that microdrops can be put into motion by hydrophilic and hydrophobic forces. Let's suppose for an instant that a water droplet is placed on a perfectly smooth horizontal plane (Figure 2.26) at the boundary between two different chemical coatings: hydrophilic on one side and hydrophobic on the other

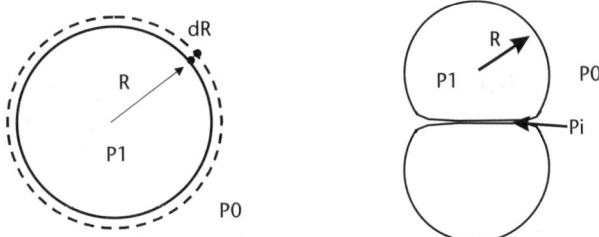

Figure 2.25 Scheme of a spherical drop and an elementary change in radius (left), and schematic view of two bubbles in a foam (right).

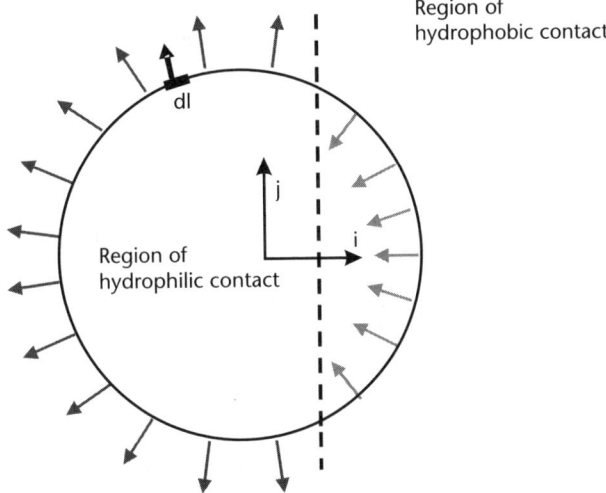

Figure 2.26 Schematic view of a water droplet standing above a hydrophilic/hydrophobic contact. There is a resulting force directed toward the hydrophilic region.

one. In such a case, this scheme can be used for the contact forces (as we do not know the exact shape of the drop at the very instant it is placed on the surface, we have drawn an approximate shape, close to—but not exactly—a circle; however, the reasoning will stand, whatever the shape).

From this scheme, it results that the droplet is displaced toward the hydrophilic surface. If L_1 and L_2 are the contact lines, respectively, in the hydrophilic and hydrophobic plane, and θ_1 and θ_2 are the contact angles, the force acting on the drop is

$$F_x \int_{L_1} = \left(\gamma_{SG} - \gamma_{SL}\right)_1 \left(\vec{i} \cdot d\vec{l}\right) + \int_{L_2} \left(\gamma_{SG} - \gamma_{SL}\right)_2 \left(\vec{i} \cdot d\vec{l}\right) = \int_{L_1} \gamma_{LG} \cos\theta_1 \left(\vec{i} \cdot d\vec{l}\right)$$

$$+ \int_{L_2} \gamma_{LG} \cos\theta_2 \left(\vec{i} \cdot d\vec{l}\right) < 0$$

(2.24)

So the resulting force is directed toward the left in the scheme of Figure 2.26, and the drop moves to the left (assuming a perfectly smooth surface). The motion stops when the resultant of the contact forces is zero (i.e., when the drop is entirely on the hydrophilic region, as shown in Figure 2.27). It would be the same if there were wettability gradient [21].

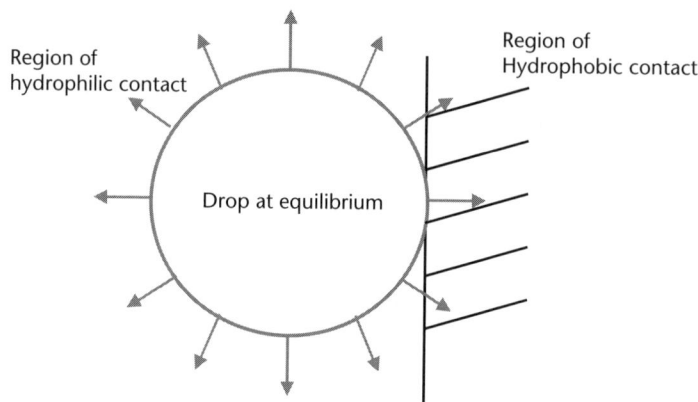

Figure 2.27 The drop is at equilibrium when it is entirely located on the hydrophilic region.

Experimental evidence confirms the preceding analysis: In Figure 2.28, a drop is deposited with a micropipette on a flat horizontal surface at the boundary of two regions with different contact angles.

The preceding analysis is also confirmed by a calculation with the Evolver code (Figure 2.29). We can start with any unphysical volume of liquid spread over a hydrophilic/hydrophobic boundary. After a few iterations, the drop is formed, but it is not at equilibrium because of the global force directed toward the hydrophilic region. The drop evolves to find its equilibrium location, which is located just at the boundary of the transition line but on the hydrophilic side.

Note that the direction of the motion of a water drop is toward the hydrophilic region, whereas the motion of an oil droplet would be toward the hydrophobic region in the same geometrical conditions.

Another striking example of capillary forces is that they can be sufficient to make drops move upward. Chaudhury and coworkers [22] have shown that a drop can go up a slightly inclined plate, presenting a wettability gradient as shown in Figure 2.30.

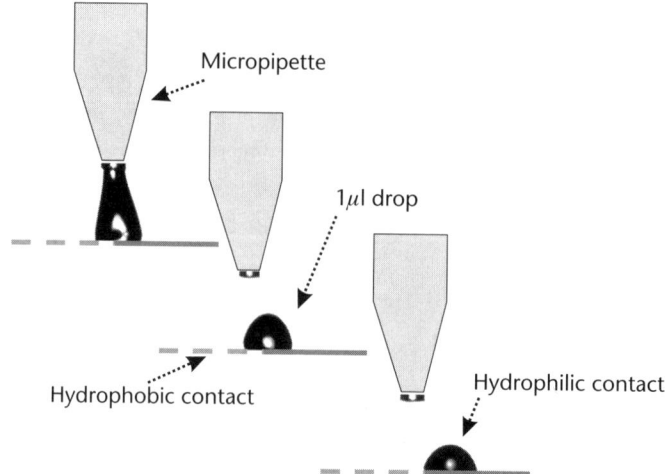

Figure 2.28 Experimental view of the relocation of a microdrop (1 μl) deposited on a hydrophilic/hydrophobic boundary.

Figure 2.29   Motion of a drop toward the hydrophilic plane (simulation with Surface Evolver). The microdrop is initially deposited over the hydrophilic/hydrophobic transition line, and it is not an equilibrium state. Drop moves to find an equilibrium state on the hydrophilic plate, just at the boundary between the two regions.

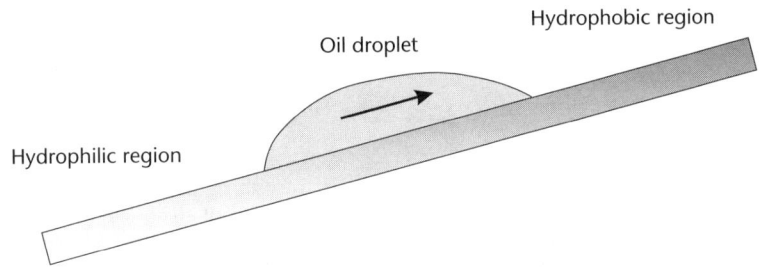

Figure 2.30   An oil drop may run uphill toward the more hydrophobic region.

Another interesting demonstration of the power of capillary forces—occurring in microsystems—may be shown by making a drop move up a step: in such a case, a microdrop is initially located on a step at the boundary of a hydrophilic region (on top of the step) and a hydrophobic region (at the base of the step). The calculation with the Evolver code shows that the drop progressively moves toward the hydrophilic region, even if this region is located at a higher level (Figure 2.31). Capillary forces dominating gravity is this example.

Note that in this section, we have considered the surface perfectly smooth, and we have neglected the effect of hysteresis. In reality, a drop does not move as soon as there is a gradient of wettability. The drop moves as soon as the gradient of wettability is sufficient for the capillary forces to dominate the hysteresis reaction force.

2.3.4   Marangoni Effect

As a general rule, surface tension is not constant; it depends on temperature and concentration of chemical species at the surface. The classical relation between surface tension and temperature is [23]:

$$\gamma = \gamma_0 \left(1 - \beta \left(T - T_0 \right) \right) \tag{2.25}$$

This is a first-order relation, but it is sufficient to describe the variation of surface tension with temperature. For an interface water/air, $\gamma_0 = 72$ mN/m and (β γ_0) ~ 0.1 mN/m. This change of surface tension with temperature has important consequences in microphysics of drops. The most common example is that of

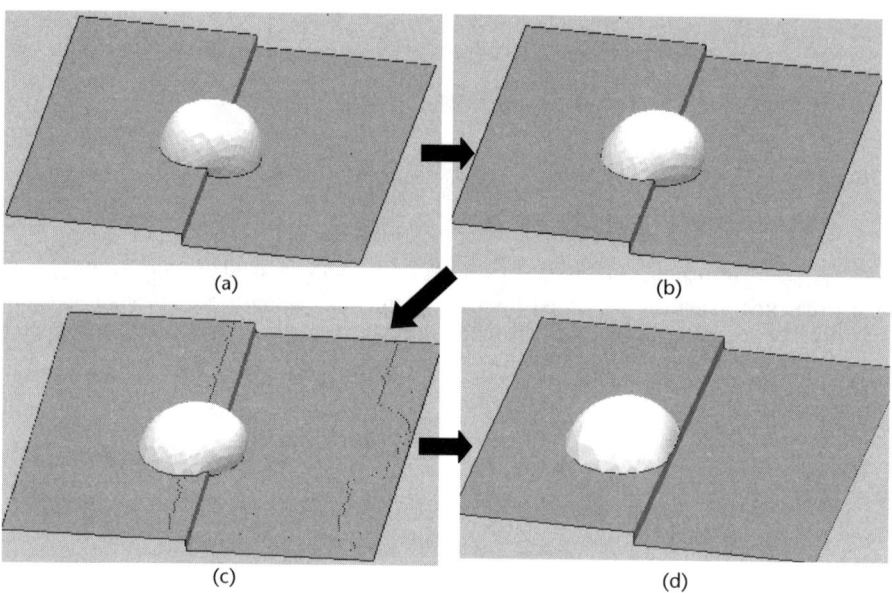

Figure 2.31 Motion of a drop up a step toward the hydrophilic plane (simulation made with Surface Evolver): (a) the drop is deposed on the step; (b) as (a) is not an equilibrium state, the drop moves up the step pulled by the hydrophobic forces of the upper plate and pushed by the hydrophobic forces of the lower plate; (c) motion continues; and (d) equilibrium state is reached when the drop is entirely on the upper plate.

Marangoni convection—also called thermocapillarity or surface tension–induced convection [23]. Each time a drop is not isothermal, there is a gradient of temperature at the interface and subsequently, due to (2.25), a gradient of surface tension. Reminding that surface tension can be looked at as a force, the surface tension distribution at the liquid/gas interface induces a tangential force distribution on the interface. These tangential forces act as a motor at the fluid interface and may lead to a convective motion inside the drop (Figure 2.32). This phenomenon is very common in microfluidics and is called Marangoni convection.

A classical example is that of a drop maintained between a solid surface and the tip of a needle. If the solid surface is maintained at a temperature T_1, and the tip of the needle is at a temperature T_2, then a convective motion appears in the drop, as schematized in Figure 2.33.

The intensity of the Marangoni convection is linked to a nondimensional number, the Marangoni number Ma defined by

$$Ma = \frac{\Delta \gamma R}{\rho v \alpha} \tag{2.26}$$

where $\Delta \gamma$ is the variation of the surface tension, R is the radius of the drop, ρ is the density of the liquid, v is the kinematic viscosity, and α is the thermal diffusivity.

Another example of the Marangoni effect is found in microwells of DNA arrays. DNA arrays are designed for the recognition of DNA segments. The principle is quite simple: it is based on the matching between a target DNA and its complementary sequence. When the target DNA finds its complementary sequence,

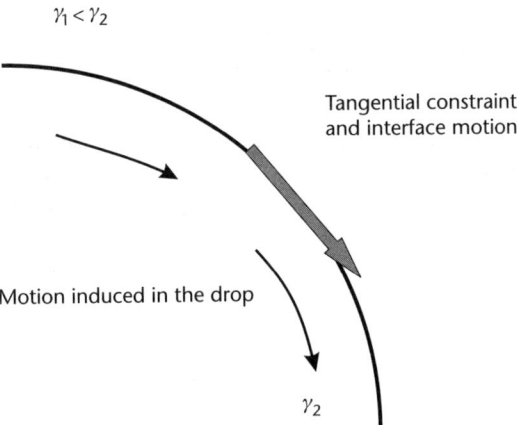

Figure 2.32   Surface tension gradient exerts a tangential constraint on the interface, resulting in a tangential motion. Viscosity diffuses the motion inside the droplet. This phenomenon is called Marangoni convection.

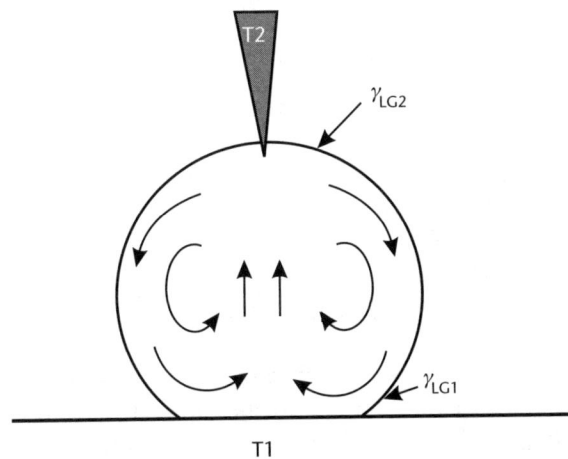

Figure 2.33   Schematic view of a Marangoni-driven convection inside a droplet. Here $T_1 < T_2$ and the surface tension γ_{LG1} is larger than γ_{LG2}, so that the convective motion on the surface is directed from the needle tip to the solid plate.

there is a binding between the two segments due to hydrogen bonds. Because we don't know the target—suppose it is the DNA of some virus that we want to identify—we use a microplate with many microwells or cusps. Each well is grafted with a predetermined DNA sequence so that each well aims at a specific target. The target can be identified by fluorescence when it binds in a well. Grafting of complementary sequences in the bottom of a cusp requires successive operations of deposition of liquid drops and heating for evaporation. As the heating of the drop is not uniform, surface tension is lower at the walls, and the liquid rises along the walls due to increased capillarity (Figure 2.34). Cusps should then be designed in a way that the liquid cannot exit by capillarity and overflow in the neighboring cusps.

Another type of Marangoni convection may occur in biological microsystems due to surfactants concentration. Figure 2.35 shows the well-known picture of a

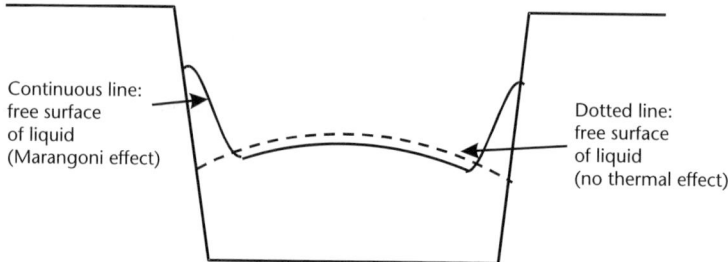

Figure 2.34 The Marangoni effect in a microwell of a DNA array. Note that in the case of a uniform temperature, the surface tension is constant, and the interface is a spherical cap with the usual contact angles, whereas in the case where the drop is heated, the interface is deformed by the Marangoni effect.

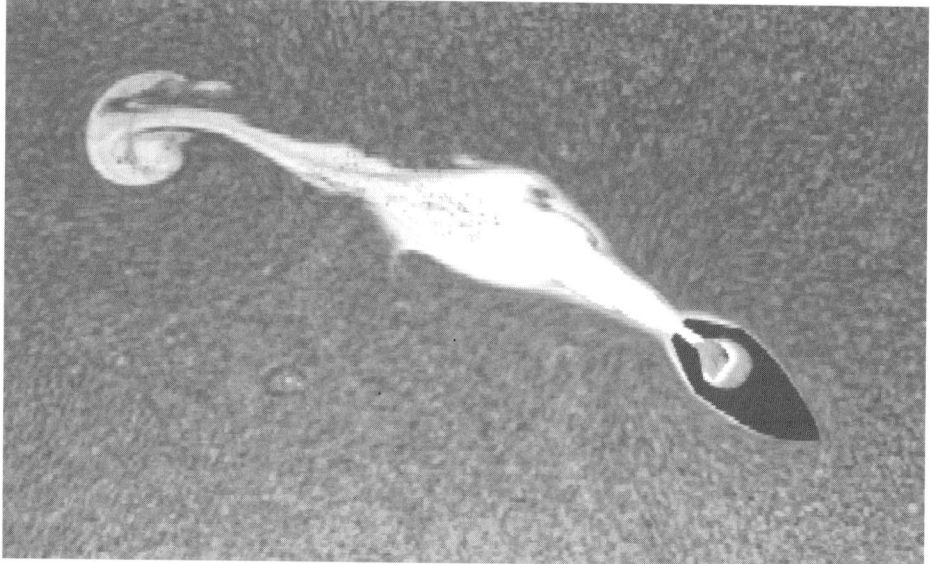

Figure 2.35 The soap boat: A floating body contains a small volume of soap; the soap exits the rear of the boat, decreasing locally the surface tension. The resulting gradient of surface tension at the surface of the liquid makes the water move toward the high surface tension region, and this motion of the water propels the boat [24]. (Courtesy of John Bush, David Hu, and Brian Chan, Department of Mathematics, MIT.)

soap boat: a floating body releases soap at the surface of water, creating a gradient of surface tension; water moves from low surface tension regions to high surface tension regions to equilibrate the surface tension; consequently the boat moves forward, toward the high surface tension region.

We have already seen that the surface tension changes with surface concentration of surfactants [25]. When surfactants are added to the biofluid, they migrate to the interface, and a gradient of interface concentration may occur, leading to a convective motion of the interface (Figure 2.36).

In the case of concentration-induced Marangoni convection, it is possible to define a nondimensional Marangoni number by

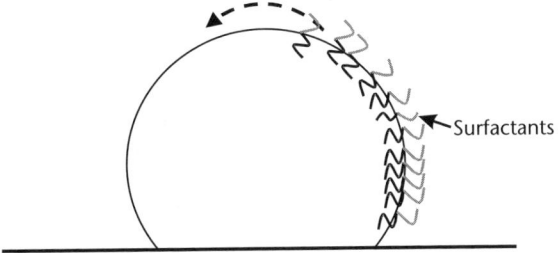

Figure 2.36   Surfactants migrate on the surface in order to homogenize their surface concentration.

$$Ma = \frac{\Delta \gamma R}{\rho \nu D} \qquad (2.27)$$

where D is the diffusion coefficient.

In a more general approach, it can be shown that Marangoni convection can have three different causes: thermal gradient, concentration gradient, and electrical gradient [26]. If we write the surface tension under the form

$$\gamma = \gamma(T, c, V)$$

where T is the temperature, c is the concentration, and V is the electric potential, then

$$\nabla \gamma = \frac{\partial \gamma}{\partial T} \nabla T + \frac{\partial \gamma}{\partial c} \nabla c + \frac{\partial \gamma}{\partial V} \nabla V$$

2.3.5   Microdrops Evaporation

Microdrops have surface-to-volume ratios much larger than usual macroscopic drops. For a spherical drop, the surface-to-volume ratio is

$$\frac{S}{V} = \frac{3}{R} \qquad (2.28)$$

and, for a spherical drop placed on a solid horizontal surface, using relation (2.3) and (2.8)

$$\frac{S}{V} = \frac{3}{R} \frac{2}{\left(2 - \cos\theta - \cos^2\theta\right)} \qquad (2.29)$$

It can be easily shown that for any contact angle θ, the ratio S/V is larger than

$$\frac{S}{V} \geq \frac{8}{3R} \qquad (2.30)$$

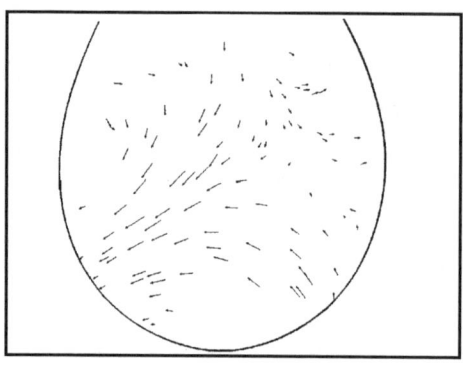 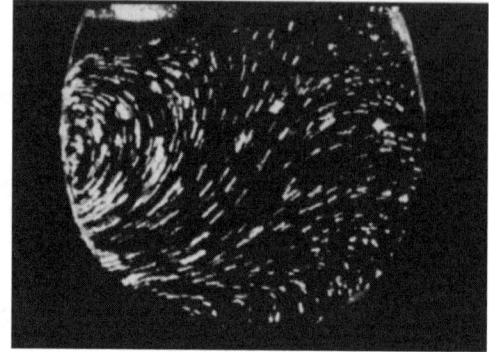

Figure 2.37 Marangoni convective motion due to evaporation in a water droplet. (*From:* [27]. © 1996 American Physical Society. Reprinted with permission.)

This relation shows that microdrops have a large surface-to-volume ratio, and they rapidly tend to evaporate in an open environment. It is important when working with droplets to estimate the lifetime of a droplet. The classical Maxwell model for a drop lifetime assumes that the evaporation process is controlled by a mass diffusion process exterior to the droplet and leads to a total lifetime

$$\tau = \frac{R_0^2 \rho}{2D\Delta c} \tag{2.31}$$

where R_0 is the initial radius, ρ is the density of the evaporating media, and Δc is the concentration change from droplet surface to ambient concentration (far from droplet).

However, it was shown [27] that this lifetime is largely overestimated, because it does not take into account convective motion inside the droplet. It is established that this convective motion is due to a Marangoni effect. The evaporation process induces a heat flux at the drop surface and consequently a temperature gradient inside the drop. This conductive state usually becomes unstable: Imagine a very small area at the surface where the evaporation rate is stronger than the average; the local temperature is then smaller than average surface temperature, resulting in a surface tension gradient. This surface tension gradient leads to a Marangoni-type convection (Figure 2.37). This convective motion may be very strong (velocities of 1 mm/sec in a 2-mm diameter drop for $\Delta T = 1°C$) and has a random behavior because the location of the "cold spot" on the surface changes randomly.

Finally the convection inside the drop causes a much stronger evaporation rate than that predicted by the conductive Maxwell model. A water drop with an initial radius of 200 μm and a $\Delta T = 1°C$ will evaporate in 5 minutes (instead of a value of more than 10 hours predicted by the conductive theory). An approximated lifetime was derived by Hegseth and coworkers based on the Marangoni number and the Jacob number

$$\tau \approx \frac{R_0^2}{\alpha Ja} \frac{1}{(1 + Ma)} \tag{2.32}$$

where α is the thermal diffusivity of the liquid, and Ja is the Jacob number given by

$$Ja = \frac{C_p \Delta T}{h_{LG}} \qquad (2.33)$$

Ja is nondimensional; it is the ratio between the conducted heat and the latent heat of vaporization (per unit mass). The lifetime of a droplet is largely overestimated by the formula in (2.31), based on a diffusion-only process. In reality, the Marangoni convection, due to spatial instabilities of evaporation rates, leads to a much shorter lifetime of the droplet, as expressed by (2.32).

The preceding paragraph dealt with spherical or nearly spherical drops (wetting angle sufficiently important). For drops having a small contact angle on a solid surface, evaporation takes another aspect. Very often, one observes a ring when the drop has completely dried, as shown in Figure 2.38 [28]. The reason is that evaporation is stronger at the boundary of the drop, and a convective motion carries particles from the center to the periphery; these particles are then deposited as a ring on the initial periphery of the drop.

In conclusion, evaporation is always a concern when working with microdrops. Most of the time in biotechnology, biologic liquids have a water base, and there are some means to limit evaporation. One of them is to maintain a controlled environment with a partial pressure of vapor; another is to include the microdrops in an organic nonmiscible layer, like oil, or to add glycerol to the solution when it is chemically acceptable. Finally, as we will see in the next section, drops may be maintained between two parallel plates, and their contact surface with air is largely reduced (Figure 2.39).

2.4   The Example of Electrowetting

We have so far analyzed the behavior of drops; in this section, we give an example of microsystems using drop microfluidics.

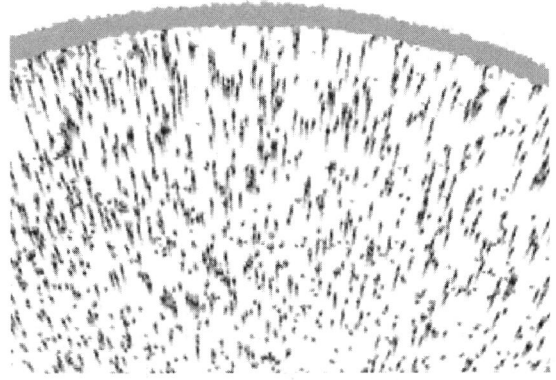

Figure 2.38   Circular ring left by an evaporating droplet. (*From:* [28]. © 2000 American Physical Society. Reprinted with permission.)

Figure 2.39 Drop confined between two parallel plates (calculation by Surface Evolver); the upper plate has been dematerialized for visualization.

In the preceding paragraphs we saw that the shape of a liquid drop on a surface is determined by the nature of the liquid, by the nature and morphology of the underlying solid, and by the surrounding fluid. It has been observed that when an electrical potential is applied, ions and dipoles redistribute in the liquid and to a lesser extent in the solid. This redistribution can cause a change in the wetting properties of the drop, according to Lippmann's law [29]. A hydrophobic surface (like Teflon) can behave like a hydrophilic surface when an electric potential is actuated: this phenomenon is called electrowetting. It is then possible to use time-regulated electric potential to displace, merge, and divide drops to perform biologic operations [30–34].

2.4.1 Principle and Theory

Lippmann's law states that the surface tension γ_{SL} of an electrically conductive liquid (surface tension between the liquid and the substrate) changes when the drop is placed in an electric field. Electric charges migrate to the liquid/substrate interface and consequently toward the contact angle. Lippmann's law can be expressed by the relation

$$\gamma_{SL} = \gamma_{SL,0} + \frac{1}{2}CV^2 \tag{2.34}$$

where C is the capacitance of the material layers in the substrate, V is the electric potential, and the index 0 refers to the nonactuated state. Combining Lippmann's law with Young's law, we obtain the well-known Lippmann-Young relation established first by Berge [35]

$$\cos\theta = \cos\theta_0 + \frac{1}{2}\frac{C}{\gamma_{LG}}V^2 \tag{2.35}$$

showing that the contact angle of a drop on a substrate can be changed by applying an electric field. Relation (2.35) has been experimentally verified on a device of the type sketched in Figure 2.40. Experimental observations of electrowetting effect are shown in the Figure 2.41.

With this in mind, drop displacement can be achieved by a difference in the electric field between two opposite sides of a drop, produced by actuated and nonactuated electrodes embedded in a solid, hydrophobic, and electrically

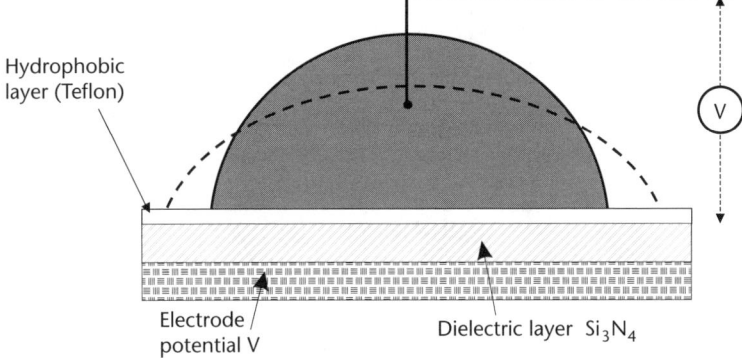

Hydrophobic
layer (Teflon)

V

Electrode
potential V

Dielectric layer Si_3N_4

Figure 2.40   Principle of electrowetting on dielectric (EWOD): The wetting angle decreases when the drop is placed in an electrical field (dc or ac). Electric charges appear in the liquid at the contact of the substrate, and the contact angle changes.

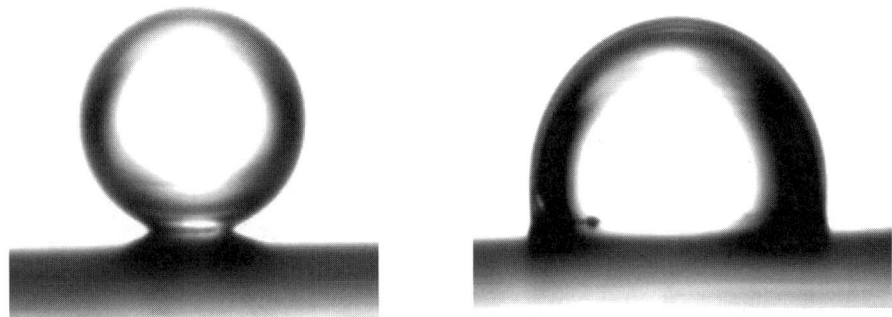

Figure 2.41   Microdrop on a hydrophobic substrate (left), and change of contact angle after actuation of an electric field (right).

insulating substrate (Figure 2.42). The contact angle is then different on the two opposite sides of the drop. A gradient of wettability is then created between two opposite sides of the droplet. Assuming there is no hysteresis (i.e., the surface is perfectly smooth) the drop moves in the direction of the smaller contact angle.

The Lippmann-Young equation can be derived by an energy minimization approach [5]. The total energy of the drop submitted to an electrical field is the sum of the surface energy and the electric potential energy

$$E = R^2 \left[\left(\gamma_{LS} - \gamma_{GS} - \frac{CV^2}{2} \right) \pi \sin^2 \theta + \gamma_{LG} 2\pi (1 - \cos \theta) \right] \qquad (2.36)$$

In (2.36), the right-hand side is the sum of the interfacial energy from (2.9) and a term corresponding to the electric energy. It can be shown [5] that minimization of the total energy E leads to

$$\cos \theta - \left(\frac{\gamma_{LS} - \gamma_{GS}}{\gamma_{LG}} + \frac{CV^2}{2\gamma_{LG}} \right) = 0 \qquad (2.37)$$

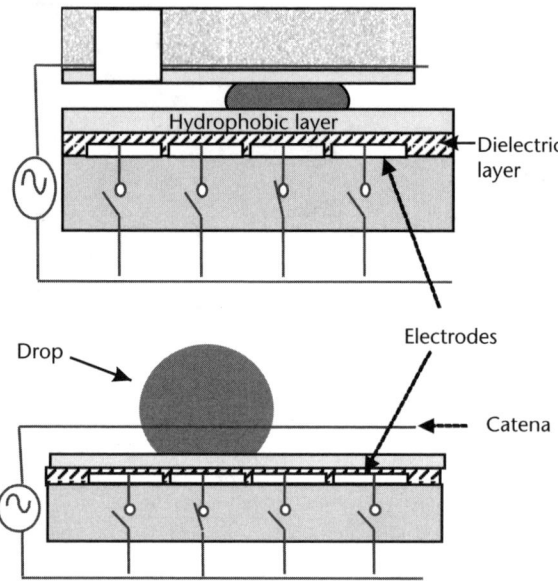

Figure 2.42 Top: "closed" or "covered" EWOD system; the drop is squashed between two horizontal parallel plates. The upper plate is at zero potential, whereas the potential of the lower plate may be adjusted. Bottom: "open" EWOD system. A catena is used to fix the zero potential. (*From:* [39]. © 2005 NSTI. Reprinted with permission.)

This equation is identical to Lippmann-Young equation (2.35), showing that the drop shape is again obtained by energy minimization.

Figure 2.43 shows a comparison between the theoretical Lippmann-Young equation and measurements. The results agree at first, for small values of the potential, but show a saturation limit not predicted by the theory (Figure 2.44). This saturation limit is currently the object of many investigations, and different explanations have been proposed [36–38]. For the rest of this section, it is worth

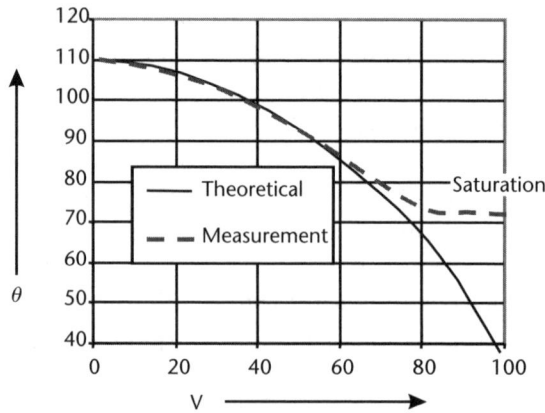

Figure 2.43 Contact angle versus electric potential: continuous line corresponding to the Lippmann-Young equation, and dotted line corresponding to the experimental results. Saturation occurs above a value of the electric potential, and the contact angle cannot be reduced further by increasing the difference of potential.

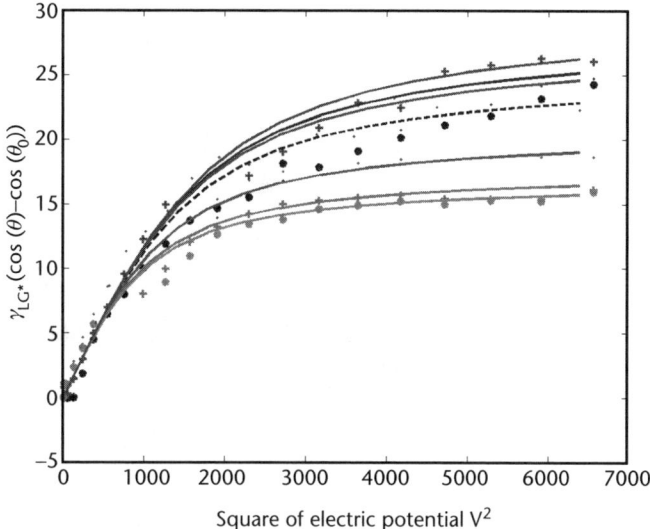

Figure 2.44   Saturation of the electrowetting effect: according to Lippmann's law (2.34) a plot of γ $\cos\theta - \gamma \cos\theta_0$ versus V^2, the plot should be a linear curve. This is the case for low potentials but not for large potentials.

remembering that above a certain value of the potential, the contact angle does not decrease any more.

EWOD microdevices are aimed at performing manipulations on microdrops. Figure 2.45 shows an electrode cross where a microdrop can be moved anywhere on the cross.

Among the different EWOD systems, the "covered" system appears to be more workable. In such systems, the drop is confined between two parallel horizontal plates, as shown in Figures 2.42 (top) and 2.46.

It can be shown that this is the solution requiring the less energy and for which the different operations of dispensing, dividing, and merging drops are the easiest [39].

2.4.2   Modeling Electrowetting

Different approaches to the modeling of electrowetting are found in the literature, most of which involve the dynamic approach using the Navier-Stokes equations, incorporating Young's constraint, and a volume of fluid (VOF) numerical formulation [40, 41]. However, inertia and viscous forces can often be neglected in front of surface tension.

Suppose a sessile drop of water placed on an electrode row. The surface energy of the drop is given by

$$E_s = \gamma S = \gamma 2\pi R^2 \left(1 - \cos\theta\right)$$

where R is the radius of the sphere (curvature), and θ is the contact angle. On the other hand, the kinetic energy is given by

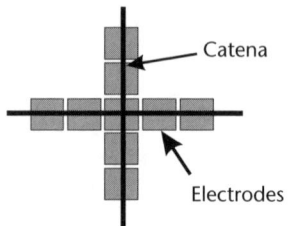

Figure 2.45 The electrode cross is constituted of nine electrodes. By switching on one electrode at a time, the drop moves and can be addressed anywhere in the cross. (Courtesy of CEA/LETI.)

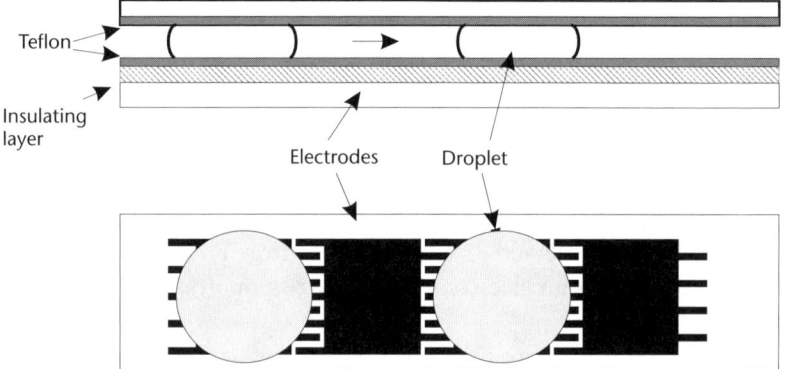

Figure 2.46 (Top) Side view of a covered EWOD device; and (bottom) top view of an electrodes alignment.

$$E_K = \frac{1}{2}mv^2 = \frac{1}{2}\rho V v^2 = \frac{1}{2}\rho \frac{\pi R^3 \left(2 - 3\cos\theta + \cos^3\theta\right)}{3}v^2 \qquad (2.38)$$

where v is the velocity of displacement of the drop between two electrodes. The ratio between these two energies is

$$W_{EM} = \frac{E_K}{E_S} = \frac{\rho R\left(2 - 3\cos\theta + \cos^3\theta\right)v^2}{12\gamma\left(1 - \cos\theta\right)} = \frac{\rho R v^2}{\gamma}g(\theta) \qquad (2.39)$$

where

$$g(\theta) = \frac{\left(2 - 3\cos\theta + \cos^3\theta\right)}{12\left(1 - \cos\theta\right)} = \frac{2 - \cos\theta - \cos^2\theta}{12}$$

We note immediately that the nondimensional W_{EM} number is very similar to the nondimensional Weber number for a spherical drop of diameter D, given by

$$We = \frac{\rho D v^2}{\gamma} \tag{2.40}$$

We can look at W_{EM} as a "modified" Weber number. Note that the Weber number is used to describe the behavior of a jet of drops (like ink jet printers); it characterizes the ratio between inertia and surface tension and—together with the Reynolds number—governs the "splash" of the drop on the solid surface.

Suppose now that $\theta = 90°$, then $W_{EM} = \frac{\rho R v^2}{6\gamma}$. Typical displacement velocities in EWOD electrowetting are approximately of the order of 1,000 μm in 0.1 second (i.e., 0.01 m/s). For a water drop in contact with air, a typical value is $W_{EM} \approx \frac{R}{4.3}$. Thus, a 1-mm radius drop has Weber number of about 1/4,000. This shows that in most cases, the surface tension is what governs the physics of the displacement of drops in EWOD processes.

It can be shown that viscosity effects can also be negated by calculating the ratio between viscous forces and capillary forces given by the nondimensional Ohnesorge number

$$On = \frac{\mu}{\sqrt{\rho l \gamma}} \tag{2.41}$$

where l is a typical length for the drop; usually $l = R$.

Typically the Ohnesorge number is less than 0.025, which is lower than the critical value of 0.1 [42].

These results have an important consequence. It leads us to understand that the behavior of drops in electrowetting devices is mostly a morphological problem. The behavior of drops in EWOD microsystems is dominated by surface tension forces and Young's constraints at the apparent contact of the solid walls. The problem is mostly topological with the drop shape and behavior adapting to the morphology of the electrodes. Minimization of drop surface energy is a well-adapted method to solve such problems. With this approach, a quasi-steady state model using the Evolver numerical software [11] is adequate to predict drop behavior [39].

Figures 2.47 and 2.48 show the results of the modeling of drop division (splitting in two) and drop dispense (extraction from a reservoir) compared to experimental results.

2.5   Conclusion

Microdrops are an unavoidable feature in biotechnology, and it seems that microarrays using microdrops fluidics are gaining importance, especially for DNA analysis and cells manipulation. In this chapter, we analyzed the physics of liquid/solid contact and exposed some of the behaviors of microdrops. Understanding the mechanical behavior of drops is not an easy task, and we dealt with only some elements of the problem—our concern was mostly static or quasi-static aspects occurring when using microplates and electrowetting.

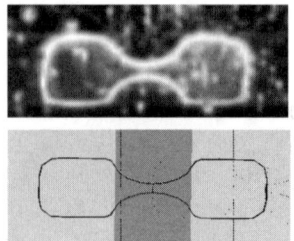

Figure 2.47 Drop division: a comparison between experimental results (top) and numerical results (bottom). Drop division is obtained by a combined effect of stretching by hydrophilic forces acting on two opposite sides of the drop (electrode actuated) and pinching in the center by hydrophobic forces (electrode not actuated). (Courtesy of D. Jary, CEA/LETI.)

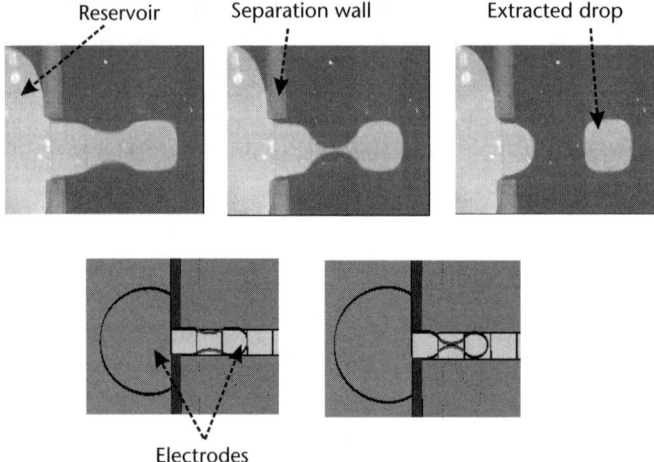

Figure 2.48 Formation of microdrops by electrowetting (drop dispense from a reservoir): experimental view (top), and numerical results (bottom). Note that the microdrop tends to mimic the form of the square electrode. Extraction is a difficult operation [17] and requires at the same time stretching by hydrophilic forces applied on the right side of the drop, pinching by hydrophilic forces at the "pinching" electrode, and back pumping from the reservoir (reservoir electrode actuated). (*From:* [39]. © 2005 NSTI. Reprinted and revised with permission.)

The first two chapters dealt with microfluidics in its two forms—microflows and microdrops—in order to predict the behavior of the carrier fluid; in the next two chapters, we present the microparticles that are convected by—or diffusing in—the carrier biofluid.

References

[1]   Israelachvili, J., *Intermolecular and Surface Forces*, New York: Academic Press, 1992.

[2]   de Gennes, P. G., F. Brochard-Wuart, and D. Quéré, *Gouttes, Bulles, Perles et Ondes*, Paris: Belin, 2002, pp. 13–14.

[3]   Table of surface tension: http://www.gewater.com/library/tp/772_Hydrophilicity_and.jsp.

[4]   Table of surface tension for chemical fluids: http://www.surface-tension.de.

[5]   Shapiro, B., et al., "Equilibrium Behavior of Sessile Drops Under Surface Tension, Applied External Fields, and Material Variations," *J. Applied Physics*, Vol. 93, No. 9, 2003, pp. 5794–5811.

[6]   Spherical cap volume and surface: http://mathworld.wolfram.com/SphericalCap.html.

[7]   Wolansky, G., and A. Marmur, "Apparent Contact Angle on Rough Surfaces: The Wenzel Equation Revisited," *Colloids and Surfaces A: Physicochemical and Engineering Aspects*, Vol. 156, 1999, pp. 381–388.

[8]   Barthlott, W., and C. Neinhuis, "Purity of the Sacred Lotus, or Escape from Contamination in Biological Surfaces," *Planta,* Vol. 202, 1997, pp. 1–8.

[9]   Callies, M., et al., "Microfabricated Textured Surfaces for Super-Hydrophobicity Investigations," *Microelectronic Engineering*, Vol. 78–79, 2005, pp. 100–105.

[10] Patankar, N. A., and Y. Chen, "Numerical Simulation of Droplet Shapes on Rough Surfaces," *Proc. 2002 Nanotech Conference*, Puerto-Rico, April 21–25, 2002, pp. 116–119.

[11] Brakke, K., "The Surface Evolver," *Exp. Math.*, Vol. 1, 1992, p. 141.

[12] Uelzen, T., and J. Müller, "Wettability Enhancement by Rough Surfaces Generated by Thin Film Technology," *Thin Solid Films*, Vol. 434, 2003, pp. 311–315.

[13] Hsieh, C. -T., et al., "Influence of Surface Roughness on Water and Oil Repellent Surfaces Coated with Nanoparticles," *Applied Surface Science*, Vol. 240, 2005, pp. 318–326.

[14] Hiemenz, P., and R. Rajagopalan, *Principles of Colloid and Surface Chemistry*, New York: Marcel Dekker, 1997.

[15] Drop tensiometer and interface rheology: http://www.itconceptfr.com/Pageshtml/Tracker_Desc_VA.html.

[16] Wege, H. A., et al., "Development of a New Langmuir-Type Pendant Drop Film Balance," *Colloids and Surfaces B: Biointerfaces*, Vol. 12, 1999, pp. 339–349.

[17] Berthier, J., et al., "An Analytical Model for the Prediction of Microdrop Extraction and Splitting in Digital Microfluidics Systems," *Proc. Nanotech 2005 Conference,* Anaheim, CA, May 8–12, 2005.

[18] Raccurt, O., J. Berthier, and P. Clementz, "Time Dependent Surface Tension Modification due to Surfactant During Electrowetting Basic Droplet Operation," *Proc. Euromech Colloquium on Microfluidics and Transfer*, Grenoble, France, September 6–8, 2005.

[19] White, S. H., "Small Phospholipid Vesicles: Internal Pressure, Surface Tension, and Surface Free Energy," *Proc. Natl. Acad. Sci. USA*, Vol. 77, No. 7, 1980, pp 4048–4050.

[20] Prange, H. D., "Laplace's Law and the Alveolus: A Misconception of Anatomy and a Misapplication of Physics," *Advan. Physiol. Edu.*, Vol. 27, 2003, pp. 34–40.

[21] Moumen, N., Subramanian R. S., and J. McLaughlin, "The Motion of a Drop on a Solid Surface Due to a Wettability Gradient," *Proc. AIChE 2003 Annual Meeting*, San Francisco, CA, 2003.

[22] Chaudhury, M. K., and G. M. Whitesides, "How to Make Water Run Uphill," *Science*, No. 256, 1992, pp. 1539–1541.

[23] Goldstein, R. J., et al., "Heat Transfer—A Review of 2001 Literature," *Int. J. Heat and Mass Transf.*, Vol. 46, 2003, pp. 1887–1992.

[24] MIT, Lecture 4, Marangoni flows: http://web.mit.edu/1.63/www/Lec-notes/Surfacetension/Lecture4.pdf.

[25] Velankar, S., et al., "CFD Evaluation of Drop Retraction Methods for the Measurement of Interfacial Tension of Surfactants-Laden Drops," *J. Colloid and Interface Science,* Vol. 272, 2004, pp. 172–185.

[26] Colin, S., *Microfluidique*, Paris, France: Hermes Science, 2004.

[27] Hegseth, J. J., N. Rashidnia, and A. Chai, "Natural Convection in Droplet Evaporation," *Physical Review E.,* Vol. 54, No. 2, August 1996, pp. 1640–1644.

[28] Deegan, R. D., "Pattern Formation in Drying Drops," *Physical Review E.*, Vol. 61, No. 1, 2000, pp. 475–485.

[29]   Lippmann, G., "Relations Entre les Phénomènes Électriques et Capillaries," *Annales de Chimie et de Physique*, Vol. 5, pp 494–549, 1875.

[30]   Pollack, M. G., R. B. Fair, and D. Shenderov, "Electrowetting-Based Actuation of Liquid Droplets for Microfluidic Applications," *Applied Physics Letters*, Vol. 77, No. 11, 2000, pp. 1725–1726.

[31]   Cho, S. K., et al., "Splitting a Liquid Droplet for Electrowetting-Based Microfluidics," *Proc. ASME International Mechanical Engineering Congress and Exposition*, New York, November 11–16, 2001.

[32]   Moon, H., et al., "Low Voltage Electrowetting-on-Dielectric," *J. Applied Physics*, Vol. 92, No. 7, October 2002, pp. 4080–4087.

[33]   Cho, S. K., H. Moon, and C.-J. Kim, "Creating, Transporting, Cutting, and Merging Liquid Droplets by Electrowetting-Based Actuation for Digital Microfluidics Circuits," *J. Microelectromechanical Systems*, Vol. 12, No. 1, 2003, pp. 70–80.

[34]   Paik, P., et al., "Electrowetting-Based Droplet Mixers for Microfluidics Systems," *Lab-on-a-Chip*, Vol. 3, 2003, pp. 28–33.

[35]   Quilliet, C., and B. Berge, "Electrowetting: A Recent Outbreak," *Current Opinion in Colloid & Interface Science*, Vol. 6, No. 1, 2001, pp. 34–39.

[36]   Peykov, V., A. Quinn, and J. Ralston, "Electrowetting: A Model for Contact Angle Saturation," *Colloid Polym. Science*, Vol. 278, 2000, pp. 789–793.

[37]   Verheijen, H. J. J., and M. W. J. Prins, "Reversible Electrowetting and Trapping of Charge: Model and Experiments," *Langmuir*, Vol. 15, 1999, p. 6616.

[38]   Vallet, M., and B. Berge, "Limiting Phenomena for the Spreading of Water on Polymer Films by Electrowetting," *Eur. Phys. J. B*, Vol. 11, 1999, p. 583.

[39]   Berthier, J., et al., "Mechanical Behavior of Micro-Drops in EWOD Systems: Drop Extraction, Division, Motion and Constraining," *Proc. Nanotech 2005 Conference*, Anaheim, CA, May 8–12, 2005.

[40]   Shapiro, B., et al., "Equilibrium Behavior of Sessile Drops Under Surface Tension, Applied External Fields, and Material Variations," *J. Applied Physics*, Vol. 93, No. 9, May 2003, pp. 5794–5811.

[41]   Bedekar, A. S., J. W. Jenkins, and S. Sundaram, "A Computational Model for the Design of Electrowetting on Dielectric (EWOD) Systems," *Proc. Nanotech 2005 Conference*, Anaheim, CA, May 8–12, 2005.

[42]   Duan, R.-Q., S. Koshizuka, and Y. Oka, "Two-Dimensional Simulation of Drop Deformation and Break-Up Around the Critical Weber Number," *Nuclear Engineering and Design*, Vol. 225, 2003, pp. 37–48.

CHAPTER 3

Diffusion of Nanoparticles and Biochemical Species

3.1 Introduction

In Chapters 1 and 2, we have analyzed the motion of a liquid under the form of microflow as well as under the form of microdrop (digital microfluidics). In reality the fluid we have studied in these two chapters is a buffer fluid containing the micro- and nanoparticles or macromolecules in which we are interested.

In the rest of this book, we focus on the behavior of the microparticles themselves in the buffer fluid. Different forces may act on the particles. At a microscopic scale, diffusion is always present, assuming that the particles are sufficiently small. Often, other physical phenomena superpose with diffusion. In Chapter 4, we deal with advection/diffusion problems; in Chapter 5, with biochemical reactions; in Chapter 7, with magnetic forces on magnetic beads; and in Chapter 8, with electric forces on charged and neutral microparticles. In this chapter we present the basis of the diffusion theory and the different approaches to solve diffusion problems.

First we analyze the basis of diffusion that is the random walk of particles due to the Brownian agitation of the fluid molecules. Next we introduce the diffusion equation of concentration and present some examples of applications in the biotechnology domain, and then we introduce the discrete Monte Carlo approach and some applications to the diffusion of macromolecules in the human body.

3.2 Brownian Motion

Because Brownian motion is a movement on a microscopic scale, it is no wonder that the Brownian motion was first discovered by biologists J. Ingenhousj and R. Brown [1], the latter after the observation of pollen grains floating at the surface of a drop of water.

In a gas or a liquid, there is an agitation of the molecules linked to temperature. A molecule moves in a straight line until it collides with another molecule, resulting in a change of direction. The average linear displacement between two collisions is called the mean free path.

In biotechnology, we deal with macromolecules and microparticles larger than the fluid molecules. The basic scheme of displacement is the same; the molecules of the carrier liquid collide with the macromolecules to make them perform a random walk (Figure 3.1).

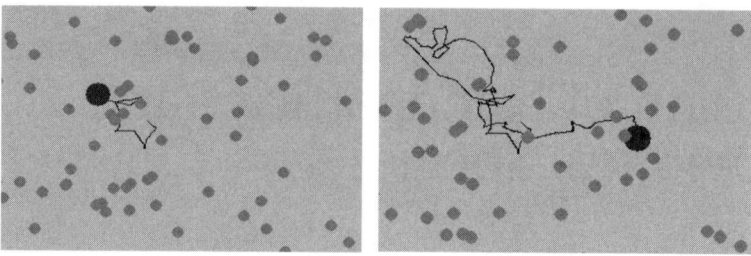

Figure 3.1 Brownian motion of a nanoparticle in a fluid at two different times. The continuous line shows the trajectory of the particle.

Brownian motion and the random walk theory is the basis of diffusion: Imagine a very small volume where the diffusing particles are initially confined. Each particle originated in this volume affects a random walk with the time, and the particles are progressively dispersed in the buffer liquid (Figure 3.2). It can be shown that the concentration follows a Gaussian profile that smears out with time.

There are two main different approaches to calculate the diffusion of these macromolecules/microparticles. The first one is the concentration approach based on the continuum hypothesis, and the second one is discrete methods where particles are followed individually.

3.3   Macroscopic Approach: Concentration

Suppose an elementary volume of liquid δV located at a coordinate (x,y,z), as shown in Figure 3.3.

The concentration c of biological species or microparticles contained in this volume at the time t is defined by the relation

$$c(x,y,z,t) = \frac{\delta m}{\delta V} \tag{3.1}$$

and a mass flux of substance through a surface is defined by [2]

$$J = \frac{\delta m}{\delta t \delta A} \tag{3.2}$$

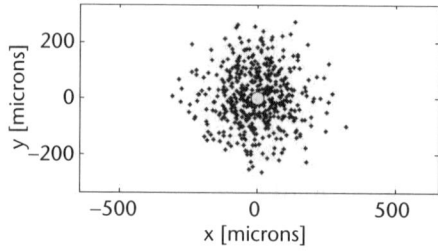

Figure 3.2 Example of diffusion from a point source. The dots are the final location of the diffusing particles at some time t.

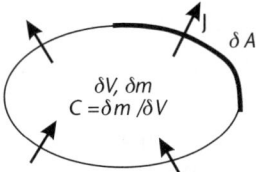

Figure 3.3   Schematization of an elementary volume with a concentration c of diffusing substance.

where the mass flux J traverses the elementary surface δA in a time interval δt. The international unit for concentration is the kilo per cubic meter (kg/m^3). However, biologists and chemists often express a concentration in mole/μL. In this case, the concentration is defined by the number of moles inside an elementary volume. The SI unit for mass flux is kg/m^2/s, and we will use also a more adapted unit (i.e., mole/μm^2/s). A fundamental law that links the mass flux to the concentration gradient is called Fick's law.

3.3.1   Fick's Law

Fick's law can be expressed as

$$J = -D\nabla c \tag{3.3}$$

where D is the diffusion constant or diffusion coefficient. The SI unit of D is m^2/s, the same as for the cinematic viscosity ν or the thermal diffusivity α. All of these quantities are coefficients in a transport equation either of concentration, velocity or enthalpy and characterized by a *flux*.

3.3.2   Concentration Equation

3.3.2.1   Differential Diffusion Equation

Using Fick's law (3.3) and evaluating the mass balance of a substance in an elementary volume of carrier liquid yields the diffusion equation

$$\frac{\partial c}{\partial t} = div\left(D\nabla c\right) + S \tag{3.4}$$

In (3.4), the term S stands for a source or sink term of concentration. For example, if there is a biochemical reaction in some part of the domain, concentration of substance may locally appear or disappear. We will come back to this point in Chapter 5.

Equation (3.4) is sometimes called Fick's second law [3]. In every subdomain where D is constant (does not depend on the spatial coordinates), (3.4) may be rewritten as

$$\frac{\partial c}{\partial t} = D\Delta c + S \tag{3.5}$$

showing that the diffusion equation is of parabolic type. Equation (3.5) is a differential equation with a solution that describes the concentration of a system as a function of time and position. The solution depends on the boundary conditions of the problem as well as on the parameter D. If the concentration c in diffusing particles or molecules is small—which is the most usual case—the diffusion coefficient D does not depend on c, and (3.5) is linear. Note that the magnitude of D is 10^{-9} m²/s for self diffusion (diffusion of the molecules of the buffer fluid) and typically 10^{-11} m²/s for colloidal substances.

3.3.2.2 Diffusion Coefficient

An expression of the diffusion coefficient of a particle in a carrier fluid was first obtained by Einstein. This expression may be derived by two different approaches. The first one is based on thermodynamics: The starting point is Gibbs free energy [3]. The magnitude of the driving force of diffusion is

$$F_{diffusion} = -\frac{1}{N_A}\frac{\partial \mu}{\partial x} \tag{3.6}$$

where μ is the chemical potential of the of the diffusing species and N_A is the Avogadro number. Thermodynamics show that

$$\mu = \mu_0 + RT\ln(\gamma c) \tag{3.7}$$

where γ is the activity coefficient. For dilute systems, $\gamma = 1$, and we obtain, after substitution of (3.7) in (3.6),

$$F_{diffusion} = -\frac{k_B T}{c}\frac{\partial c}{\partial x} \tag{3.8}$$

Under stationary state conditions, the diffusion force is balanced by the viscous resistance

$$F_{diffusion} = -\frac{k_B T}{c}\frac{\partial c}{\partial x} = F_{friction} = C_D v \tag{3.9}$$

where C_D is the friction factor and v is the stationary velocity. Thus,

$$v = -\frac{k_B T}{C_D c}\frac{\partial c}{\partial x} \tag{3.10}$$

If we note that the flux of material through a cross section is

$$J = vc$$

then by comparison with Fick's law, we obtain the important result

$$D = \frac{k_B T}{C_D} \tag{3.11}$$

The second approach [1] is based on Langevin's formula (similar to Newton's law, but with a complementary term for the Brownian motion)

$$m \frac{dv}{dt} = -C_D v + F(t) \tag{3.12}$$

where m is the mass of the particle, and $F(t)$ is a randomly fluctuating force representing Brownian motion. By multiplying (3.12) by x and taking the time average, we can rewrite (3.12) under the form

$$m < \frac{d}{dt}\left(x \frac{dx}{dt}\right) >= m < \left(\frac{dx}{dt}\right)^2 > -C_D < x \frac{dx}{dt} > + < xF(t) >$$

where the brackets correspond to the time averaging operation. However, because the variables x and $F(t)$ are independent,

$$< xF(t) >=< x >< F(t) >= 0$$

and the isotropic distribution of the energy yields

$$\frac{1}{2} m < \left(\frac{dx}{dt}\right)^2 >= \frac{1}{2} k_B T$$

then the Langevin equation is reduced to a differential equation

$$m < \frac{d}{dt}\left(x \frac{dx}{dt}\right) >= k_B T - C_D < x \frac{dx}{dt} > \tag{3.13}$$

The solution of (3.13) for times much larger than C_D/m is [1]

$$< x^2 >= 2 \frac{k_B T}{C_D} t = 2Dt$$

and

$$D = \frac{k_B T}{C_D}$$

3.3.2.3 Anisotropic Media

In free space, diffusion is isotropic. However, in a confined space, diffusion may be anisotropic if the media containing the fluid is anisotropic [4]. Anisotropic media have different diffusion properties in different directions. Some common examples are textile fibers, polymer films, and laminated microlayers in which the molecules have a preferential direction of orientation (Figure 3.4). It is also shown [5] that at the very vicinity of a surface, diffusion become anisotropic.

The diffusion constant D must be replace by a coefficient matrix $[D]$ where

$$[D] = \begin{bmatrix} D_{11} & D_{12} & D_{13} \\ D_{21} & D_{22} & D_{23} \\ D_{31} & D_{32} & D_{33} \end{bmatrix} \tag{3.14}$$

The mass flux is then anisotropic and Fick's law can be written under the form [4]

$$\begin{Bmatrix} J_x \\ J_y \\ J_z \end{Bmatrix} = -[D]\nabla c \tag{3.15}$$

Finally the diffusion equation is

$$\frac{\partial c}{\partial t} = D_{11}\frac{\partial^2 c}{\partial x^2} + D_{22}\frac{\partial^2 c}{\partial y^2} + D_{33}\frac{\partial^2 C}{\partial z^2} + (D_{23} + D_{32_e})\frac{\partial c}{\partial y \partial z}$$
$$+ (D_{31} + D_{13})\frac{\partial c}{\partial z \partial x} + (D_{12} + D_{21})\frac{\partial c}{\partial x \partial y}$$

if the Ds are taken constant. We may rotate the axis in order to transform the rectangular coordinates (x,y,z) into the rectangular coordinates (ξ, η, ζ) characterizing the principal axes of diffusion, and the preceding equation becomes

$$\frac{\partial c}{\partial t} = D_1\frac{\partial^2 c}{\partial \xi^2} + D_2\frac{\partial^2 c}{\partial \eta^2} + D_3\frac{\partial^2 c}{\partial \zeta^2} \tag{3.16}$$

It is possible to make the further transformation

$$\xi_1 = \xi\sqrt{\frac{D}{D_1}}, \eta_1 = \eta\sqrt{\frac{D}{D_2}}, \zeta_1 = \zeta\sqrt{\frac{D}{D_3}}$$

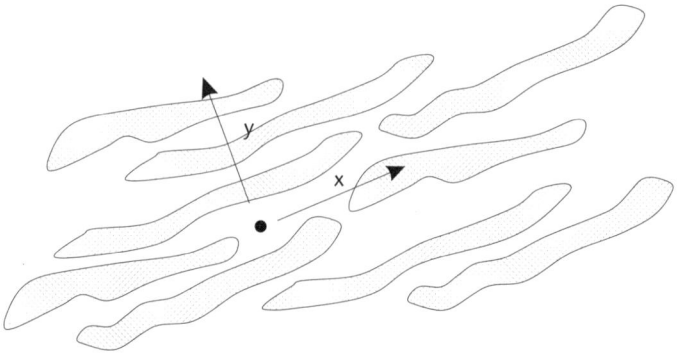

Figure 3.4   Sketch of an anisotropic media. The preferred direction of diffusion is the x-direction.

where D is arbitrary, to obtain

$$\frac{\partial c}{\partial t} = D\left[\frac{\partial^2 c}{\partial \xi_1^2} + \frac{\partial^2 c}{\partial \eta_1^2} + \frac{\partial^2 c}{\partial \zeta_1^2}\right] \tag{3.17}$$

Equation (3.17) has an isotropic value for the diffusion constant D to the price of a rotation plus a homothetic transformation of the axes. In Section 3.3.8, for some reasons that we explain, we proceed in an opposite way: We transform a very elongated (anisotropic) computational domain into a more regular domain. This transformation requires that the initially isotropic diffusion coefficient be changed into an anisotropic diffusion matrix.

3.3.3   Spreading from a Point Source—1D Case

We analyze here the diffusion of a substance (tracers or nanoparticles) in a one-dimensional geometry. Suppose that a very small spot of concentration of tracer particles has been initially placed in a rectangular capillary of very small cross section (Figure 3.5).

In such a case, the diffusion may be considered one-dimensional and depends one two variables: the time t and the axial coordinate x. The initial condition may be approximated by:

$$c(x,t_0) = c_0 \delta(x) \tag{3.18}$$

where $\delta(x)$ is the Dirac function. With such an initial condition, the solution to (3.5) is

$$c(x,t) - \frac{c_0}{\sqrt{4\pi Dt}} e^{-\frac{x^2}{4Dt}} \tag{3.19}$$

The solution (3.19) shows that the distribution profile of concentration in tracers is Gaussian in x. In Figure 3.6, we have plotted the solution of (3.19) with scaling c/c_0 for $D = 10^{-10}$ m²/s at three different times (0.2, 1, and 10 seconds).

Note that the characteristic nondimensional group $\dfrac{x^2}{4Dt}$ appears in the solution (3.19) of (3.5). This group represents in fact a characteristic diffusion length. The characteristic diffusion length may be defined by

$$x_c \approx \sqrt{4Dt} \tag{3.20}$$

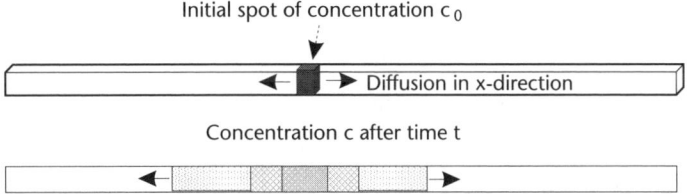

Initial spot of concentration c_0

Diffusion in x-direction

Concentration c after time t

Figure 3.5   Schematic view of the diffusion of tracers in a one-dimensional geometry.

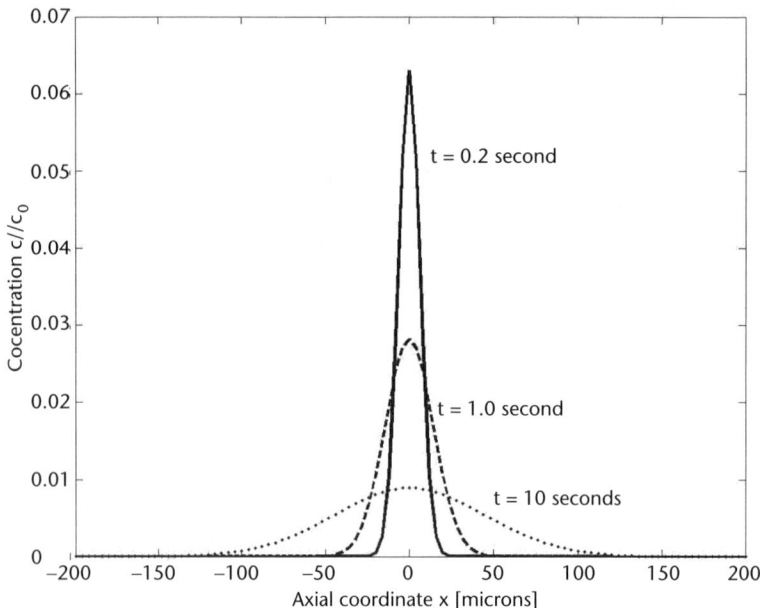

Figure 3.6 Gaussian profiles of diffusion from a point source according to (3.19).

In the preceding example, one finds by using (3.20): $x_c(t = 0.2) \sim 10 \ \mu m$; $x_c(t = 1.0) \sim 20 \ \mu m$; and $x_c(t = 10) \sim 60 \ \mu m$.

3.3.4 The Ilkovic's Solution for a Semi-Infinite Space

It is seldom that the diffusion equation (3.4) can be solved analytically. There are some one-dimensional cases where an analytical solution may be found (we have seen one in the preceding section); but usually, as soon as the geometry of the diffusion problem is two-dimensional, or if the one-dimensional problem presents complex boundary conditions, the use of a numerical approach is required to solve the diffusion equation (3.4). We expose here the analytical solution of the diffusion equation in the simple case of diffusion of species in a half space.

Suppose a half-space with an initial concentration of c_0. Suppose also that any microparticle or macromolecule that contacts the solid wall limiting the half-space domain is immediately immobilized. Then the concentration at the wall is zero at any time.

The solution for the concentration equation is then

$$c = c_0 \, erf\left(\frac{x}{\sqrt{4Dt}}\right) \tag{3.21}$$

where x is the distance from the wall, and the error function erf is defined by

$$erf(x) = \frac{2}{\sqrt{\pi}} \int_0^x e^{-u^2} \, du$$

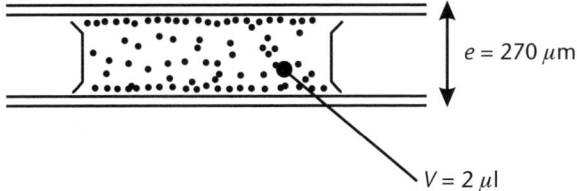

Figure 3.7 Schema of the drop and the two glass plates.

This function has the following characteristic values: $erf(0) = 0$ and $erf(\infty) = 1$ and its derivative is $\dfrac{d(erf)}{dx} = \dfrac{2}{\sqrt{\pi}} e^{-x^2}$. Thus, the space derivative of (3.21) is

$$\frac{\partial c}{\partial x} = c_0 \frac{2}{\sqrt{\pi}} e^{-\frac{x^2}{4Dt}} \frac{1}{\sqrt{4Dt}}$$

With this in mind, the mass flux per surface unit, given by Fick's law, may be written under the form

$$\vec{J} = -D\nabla c\big|_{wall} = -D\frac{\partial c}{\partial x}\Big|_{x=0} = -Dc_0 \frac{2}{\sqrt{\pi}} \frac{1}{\sqrt{4Dt}}\hat{i} = -c_0\sqrt{\frac{D}{\pi t}}\hat{i} \qquad (3.22)$$

where $\hat{i}$ is the unit vector perpendicular to the wall. This latter relation is called the Ilkovic's solution to the diffusion problem. It shows that the mass flux is proportional to the concentration far from the wall and to the square root of the diffusion coefficient. Strictly, from (3.22), the initial mass flux is infinite. In fact, on a practical point of view, such a situation is not possible: There is always a transition time during which the fluid with the concentration c_0 is brought in contact with the wall, and the initial time for diffusion is always approximate.

3.3.5 Example of Diffusion Between Two Plates

In this section, we show the limitation of Ilkovic's solution for the problem of diffusion between two plates. Suppose that we insert a small volume of liquid ($V = 2\,\mu l$) between two parallel horizontal glass plates (Figure 3.7) separated by a distance of $270\,\mu m$. The liquid contains nanoparticles (hydrodynamic diameter $D_H = 100$ nm, diffusion coefficient $D = 0.21 \cdot 10^{-11}$ m^2/s) in concentration c_0.

At the beginning, the particles are uniformly dispersed in the liquid at rest; progressively, the particles closer to the walls are immobilized by contact with the walls under the action of the Brownian motion, and a concentration depletion progresses from the walls toward the drop center.

It is possible to count the number of particles immobilized at any time on the photographs of the upper plate taken under the microscope (Figure 3.8). It may be shown that the particle size is so small that sedimentation can be completely neglected, and we assume that there is statistically the same number of particles immobilized on the upper and lower plate.

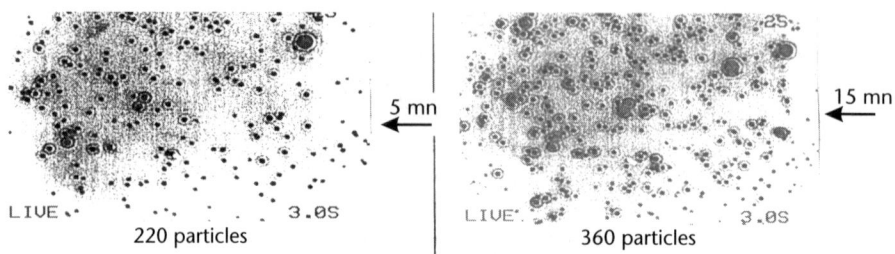

Figure 3.8 Photographs of the upper plate taken under the microscope after 5 mn and 15 mn. (Courtesy of D. Marsi, CEA/LETI.)

Assuming that the drop is cylindrical (which is the equilibrium shape—see Chapter 2), diffusion process is governed by the two-dimensional axisymmetrical equation

$$\frac{\partial c}{\partial t} = div(D \ grad \ c) = D\left(\frac{1}{r}\frac{\partial c}{\partial r} + \frac{\partial^2 c}{\partial r^2} + \frac{\partial^2 c}{\partial z^2}\right) \tag{3.23}$$

In this particular case, we can express the concentration in particles per unit volume, and the mass flux at the wall J is then expressed in particles per seconds per unit surface. The mass flux at each wall (defined by $z = 0$ and $z = e$) is given by Fick's law

$$|J| = |J_{upper}| + |J_{lower}| = -D\left.\frac{\partial c}{\partial z}\right|_{z=e} + D\left.\frac{\partial c}{\partial z}\right|_{z=0} \tag{3.24}$$

If we suppose that the radial dimension R of the drop is large in front of the distance between the plates e, we can approach (3.23) by the one-dimensional equation

$$\frac{\partial c}{\partial t} = div(D \ grad \ c) = D\frac{\partial^2 c}{\partial z^2} \tag{3.25}$$

A first approach to the problem consists of solving (3.25) for times less than $\tau = \dfrac{\left(\dfrac{e}{2}\right)^2}{4D}$. For these times, the depletion of concentration has not reached the center plane of the drop. The problem is then similar to that of Ilkovic for each plate, and the solution of (3.24) is

$$J = 2c_0\sqrt{\frac{D}{\pi t}} \tag{3.26}$$

This solution breaks down when the concentration at the center plane starts decreasing from its initial value c_0. So when the time is larger than $\tau = \dfrac{\left(\dfrac{e}{2}\right)^2}{4D}$,

Ilkovic's solution is no more valid. In Figure 3.9, we compare the experimental results (dots) to the Ilkovic's solution and to the results of a simple one-dimensional numerical scheme (finite differences method).

Vertical concentration profiles at different times—obtained by the numerical method—are plotted in Figure 3.10. At the beginning, the profile is still flat at the center with a concentration c_0. At times t larger than τ, the concentration at the center plane decreases below the value c_0.

3.3.6 Radial Diffusion

In the case of a purely radial diffusion from a source point or a central sphere (Figure 3.11), the diffusion equation is

$$\frac{\partial c}{\partial t} = D\left[\frac{\partial^2 c}{\partial r^2} + \frac{2}{r}\frac{\partial c}{\partial r}\right] \tag{3.27}$$

On putting [4]

$$u = cr \tag{3.28}$$

Equation (3.27) becomes

$$\frac{\partial u}{\partial t} = D\frac{\partial^2 u}{\partial r^2} \tag{3.29}$$

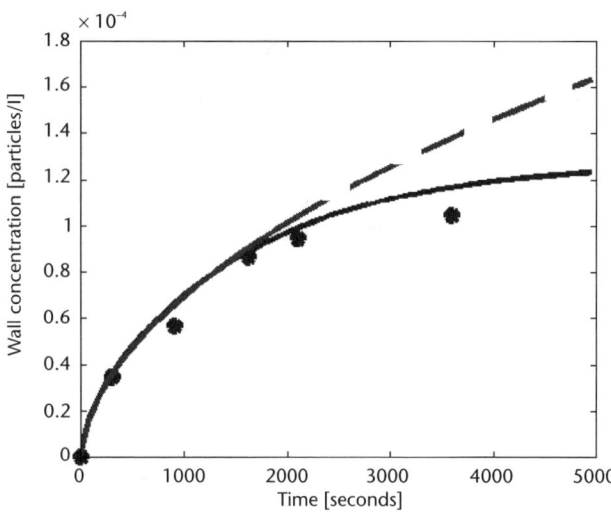

Figure 3.9 Wall concentration of immobilized nanoparticles (particles/μm^2) as a function of time. Comparison between measurements (dots), Ilkovic's solution, and numerical results. Before $t < \tau = \dfrac{\left(\dfrac{e}{2}\right)^2}{4D} = 2166\,s$, all of the results are close together. After the time τ, the Ilkovic solution departs from the experimental results, whereas the numerical results still agree with the experimental results.

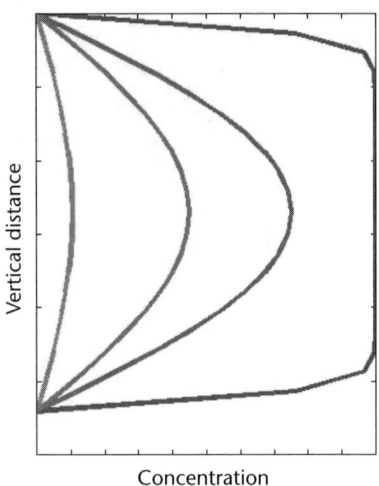

Figure 3.10 Vertical profiles of concentration c/c_0 versus time.

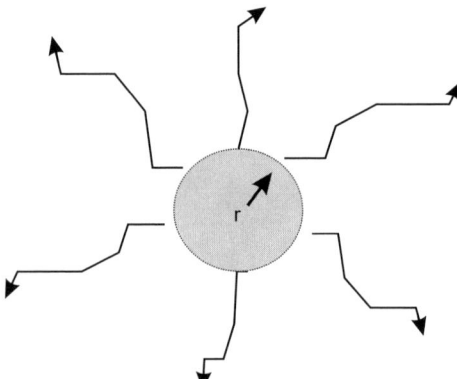

Figure 3.11 Radial diffusion from an initial spherical volume.

This equation is formally the same as the one-dimensional diffusion equation and can be solved by using the same methods.

3.3.6.1 Steady State Solution

In the case of a steady state problem, (3.29) becomes

$$\frac{d}{dr}\left(r^2\,\frac{dc}{dr}\right) = 0 \tag{3.30}$$

of which the general solution is

$$c = \frac{A}{r} + B \tag{3.31}$$

where A and B are constants to be determined from the boundary conditions.

3.3.6.2   Transient Solution

In the case of a spherical surface with a constant concentration $c_1(r = a)$ and an external concentration $c_0(r > a)$, the solution is derived directly from the one-dimensional Ilkovic's solution

$$\frac{c - c_0}{c_1 - c_0} = \frac{a}{r}\left(1 - erf\frac{r - a}{\sqrt{4Dt}}\right)$$

(3.32)

3.3.7   Diffusion Inside a Microchamber

The standard biodiagnostic procedure for DNA recognition is the polymerase chain reaction (PCR). However, there have been recently developments of new microdevices using direct detection of DNA strands by fluorescence (Figure 3.12). The principle is to bring the DNA strands inside the microchamber (e.g., using magnetic particles) and then let them diffuse so that the DNA strands can hybridize on a labeled surface. Because the system must be very sensitive and work with very few DNA strands, it is important to stop any back diffusion toward the inlet channel.

A very simple analysis of the diffusion inside the microchamber may be done by considering the diffusion equation in the two-dimensional geometry defined by Figure 3.13 and using standard numerical techniques. The results presented in Figure 3.14 have been obtained by discretization of the diffusion equation by using a Crank-Nicholson formulation [6]. The two-dimensional diffusion equation may be written under the form

$$\frac{c_{i,j}^{n+1} - c_{i,j}^n}{\Delta t} = \frac{D}{2}\left[\frac{c_{i+1,j}^{n+1} - 2c_{i,j}^{n+1} + c_{i-1,j}^{n+1}}{(\Delta x)^2} + \frac{c_{i,j+1}^{n+1} - 2c_{i,j}^{n+1} + c_{i,j-1}^{n+1}}{(\Delta y)^2}\right]$$

$$+ \frac{D}{2}\left[\frac{c_{i+1,j}^n - 2c_{i,j}^n + c_{i-1,j}^n}{(\Delta x)^2} + \frac{c_{i,j+1}^n - 2c_{i,j}^n + c_{i,j-1}^n}{(\Delta y)^2}\right]$$

(3.33)

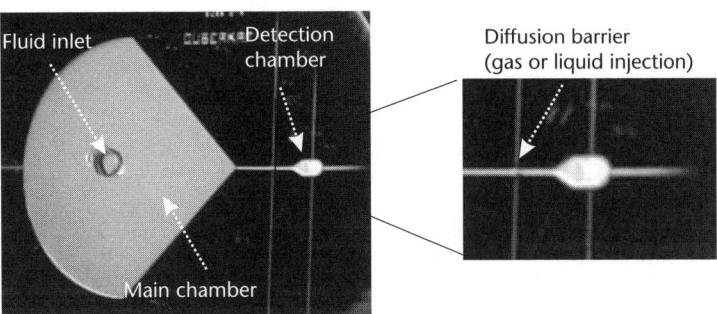

Figure 3.12   Biodiagnostic detection device. Left: view of the main and detection microchambers. Right: close up on the detection chamber. (Courtesy of F. Ginot and R. Campagnolo, LETI/BioMérieux.)

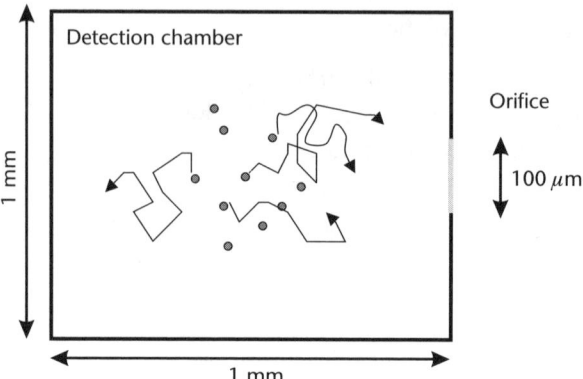

Figure 3.13 Schematic view of the computation domain.

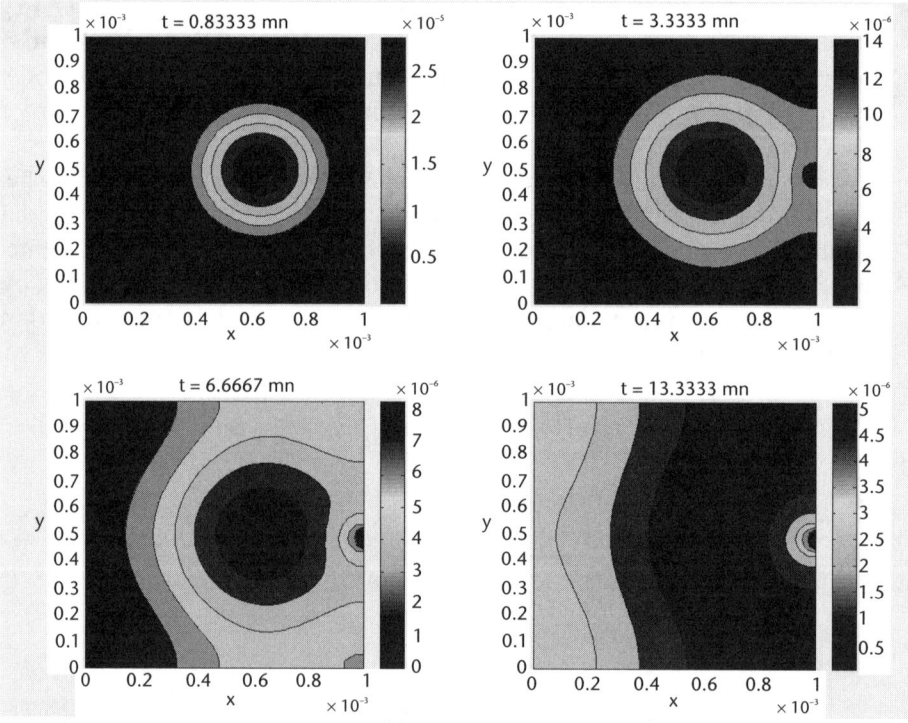

Figure 3.14 Diffusion of macromolecules initially concentrated in the middle of the detection microchamber (the macromolecules are released from the aggregate of magnetic beads). The right side of the chamber is considered an exit toward the inlet channel.

where i and j are the indices corresponding to the space location, and n is the time index. Equation (3.33) can be written in a matrix form and inverted to obtain the solution for the concentration at each time step t_n at every point of the computational grid.

An analysis of the results shows that it is necessary to stop back diffusion inside the inlet channel. This is done by injecting a gas in secondary reservoirs (by thermal expansion for example) (Figure 3.15).

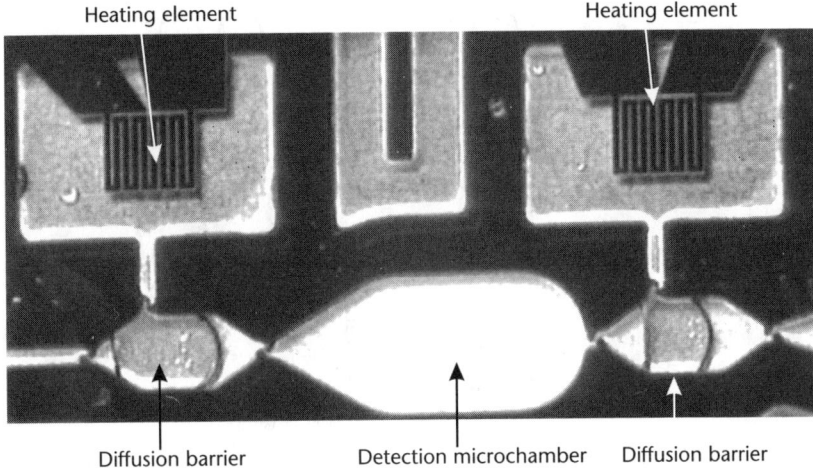

Heating element   Heating element

Diffusion barrier   Detection microchamber   Diffusion barrier

Figure 3.15   Enlarged view of the detection chamber and diffusion blocking reservoirs. (Courtesy of N. Sarrut, CEA/LETI.)

3.3.8   Diffusion Inside a Capillary: The Example of Simultaneous PCRs

3.3.8.1   Introduction

Biorecognition and detection of DNA by PCR is still the most often used procedure. In order to parallelize PCR reactions—and other similar biological analyses—it has been thought to perform these operations at the same time at different locations in a capillary tube [7]. In this example of the simultaneous PCRs, the biological sample to be analyzed is brought into the tube under the action of capillary forces (or by pipetting). At the solid wall, at different locations, different primers (reverse and forward) have been immobilized (Figure 3.16).

When the liquid has filled the tube and is at rest, the primers are released by optical methods (insolation) and then diffuse locally inside the tube (Figure 3.17). The presence of both reverse and forward primers is necessary for PCR amplification. In the regions of the tube where a sequence of the DNA contained in the liquid is corresponding to a type of primer, DNA amplification occurs, and detection is made by fluorescence methods.

Within the time frame of the PCR cycles for amplification, the primers should not diffuse to the next region in a substantial way (Figure 3.18). If they do, detection would be inaccurate. Thus, it has been planned to introduce neutral gaps between

Capillary tube, $\Phi = 100 \ \mu m$

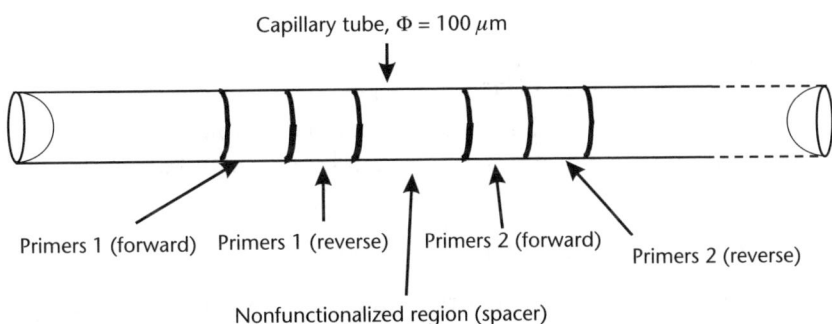

Primers 1 (forward)   Primers 1 (reverse)   Primers 2 (forward)   Primers 2 (reverse)

Nonfunctionalized region (spacer)

Figure 3.16   Schematic view of the capillary with the different labeled regions.

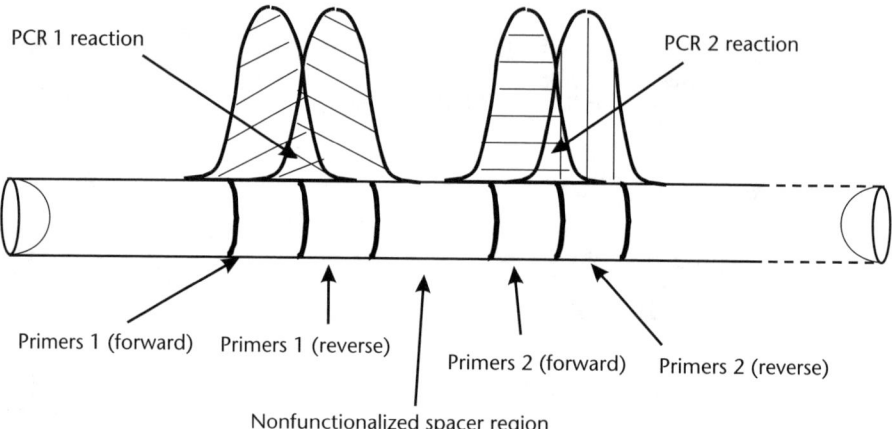

Figure 3.17 Schematic view of the primers concentration and the PCR reaction regions (drawing not to scale).

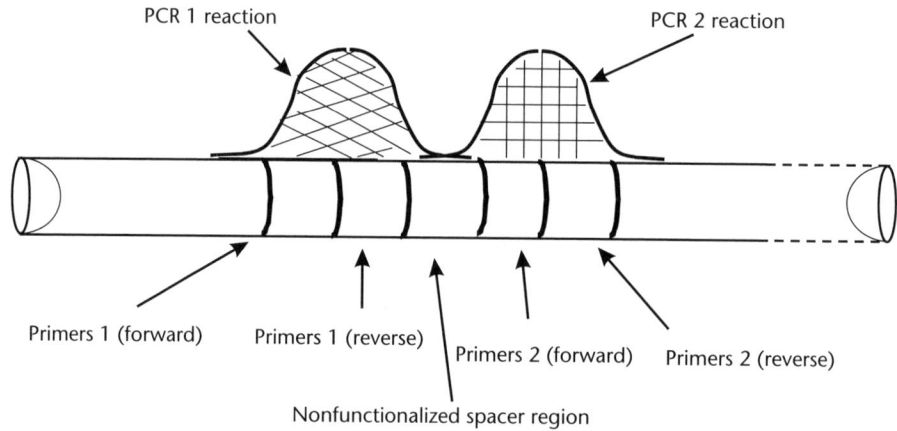

Figure 3.18 Schematic view of the concentration profile of the different PCR reaction products.

the primers regions so that there is no diffusion of primers between two regions. However, the aim is a compact microsystem, and these gaps should be reduced as much as possible. The problem consists of calculating the primers diffusion inside the tube and determining their concentration as a function of time.

In the following sections, we present two approaches to calculate the diffusion inside the capillary. The first one is an analytical approach, based on the simplification that the concentration in primers—after a few seconds—is uniform in a cross section of the tube. The second approach consists of numerically solving the diffusion equation after it has been rendered nondimensional. The advantage of this second solution is that it gives an insight on diffusion barrier.

3.3.8.2 Analytical Model for Diffusion Inside a Tube

First, taking into account that the ratio L/R between the tube length and the tube radius is large, we show that we can assume the concentration in primers uniform in

Free Subscription

Artech Direct email newsletter

New Title News • Special Offers • Author Insights

☐ Yes! Please enter my free subscription to *Artech Direct* and keep me up-to-date with emailed news of product and service information from Artech House/Horizon House Publishers.

email address:
_____

You may also make my email address available to selected industry organizations and companies. ☐ Yes ☐ No

Please indicate your areas of interest

☐ Telecommunications/Wireless/Networking   ☐ Signal Processing

☐ Software Engineering/Computer Security   ☐ Sensors/MEMS/Nanotechnology

☐ Microwave   ☐ Antennas & Propagation

☐ Radar/Remote Sensing/Electronic Defense   ☐ Engineering Management

☐ Biomedical Engineering

Mailing address:

Name:
_____

Company:
_____

Address:
_____
_____

Fax or mail this card to the Artech House office nearest you. Please see other side.

 ARTECH HOUSE BOSTON | LONDON

any cross section in a very short time after their release from the wall. A characteristic diffusion time for a primer to diffuse on the length R is

$$\tau \approx \frac{R^2}{4D} \tag{3.34}$$

Primers have a diffusion coefficient of the order $D = 10^{-10}\,\mathrm{m^2/s}$, and suppose that the radius of the tube is $R = 50\,\mu\mathrm{m}$. The characteristic time is then $\tau = 6$ seconds. During the time τ, axial diffusion is not significant, as we can show by using a very simple Monte Carlo approach (Figure 3.19).

So we may assume a uniform concentration in primers in any annular volume delimited by the initial functionalized regions (volume $V = \pi R^2 a$, where a is the length of a region).

After the time τ, the primers are homogeneously scattered in the different annular volumes. The second step consists of calculating the concentration of the primers as a one-dimensional diffusion phenomenon. For each primer i, we have

$$\frac{\partial c_i}{\partial t} = D_i \frac{\partial^2 c_i}{\partial z^2} \tag{3.35}$$

An analytical solution to (3.35) is given by the following combination of error functions [4]:

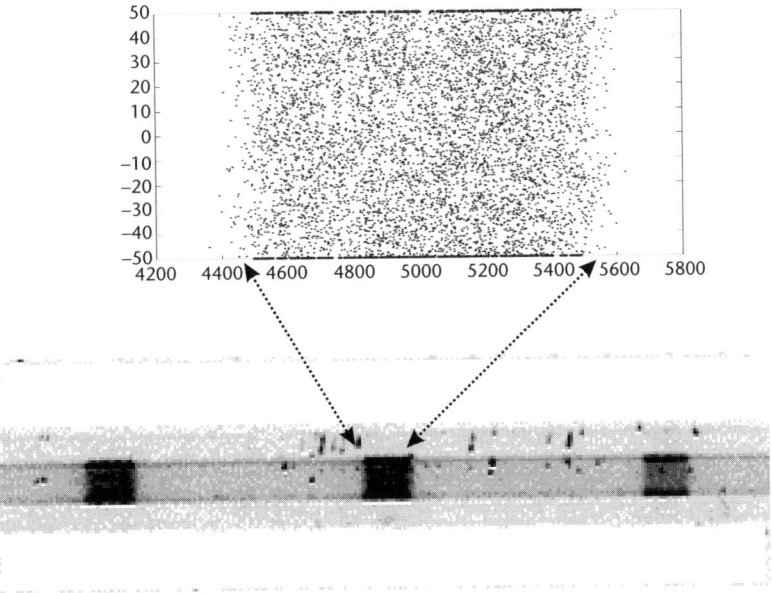

Figure 3.19 Diffusion of primers from the wall after 6 seconds (obtained by a Monte Carlo simulation). The starting location of the primers has been randomly chosen on the walls. The diffusion outside the annular volume is still negligible at that time. Bottom: experimental results. (Courtesy of CEA/LETI.)

$$c_i = \frac{1}{2} c_{0,i} \left[erf\left(\frac{a_i - z}{2\sqrt{D_i t}} \right) + erf\left(\frac{a_i + z}{2\sqrt{D_i t}} \right) \right] \tag{3.36}$$

In Figure 3.20, concentration profiles of two neighboring primers with different coefficients of diffusion have been plotted. It is immediately seen that a spacer gap must be introduced between the two regions to prevent cross mixing.

The advantage of the analytical solution is that we can produce an expression of the required spacing between the functionalized regions. Suppose that the concentration in primers i in region $i + 1$ should not be larger than a threshold concentration c_{max}, at a time t_f defined by the kinetics of amplification; then the minimum distance z_{min} between the two regions i and $i + 1$ is given by the implicit relation

$$\frac{2c_{max}}{c_{0,i}} = \left[erf\left(\frac{a_i - z_{min}}{2\sqrt{D_i t_f}} \right) + erf\left(\frac{a_i + z_{min}}{2\sqrt{D_i t_f}} \right) \right] \tag{3.37}$$

The solution of (3.37) requires us to find the zero of a function, which is a standard procedure to most mathematical software.

3.3.8.3   Dimensional Analysis

This analytical method is a fast and simple way to find an approximate solution to the problem. However, a dimensional analysis reveals more of the physics of axial diffusion and will be the basis for a numerical approach. Start from the axisymmetrical diffusion equation for each primer

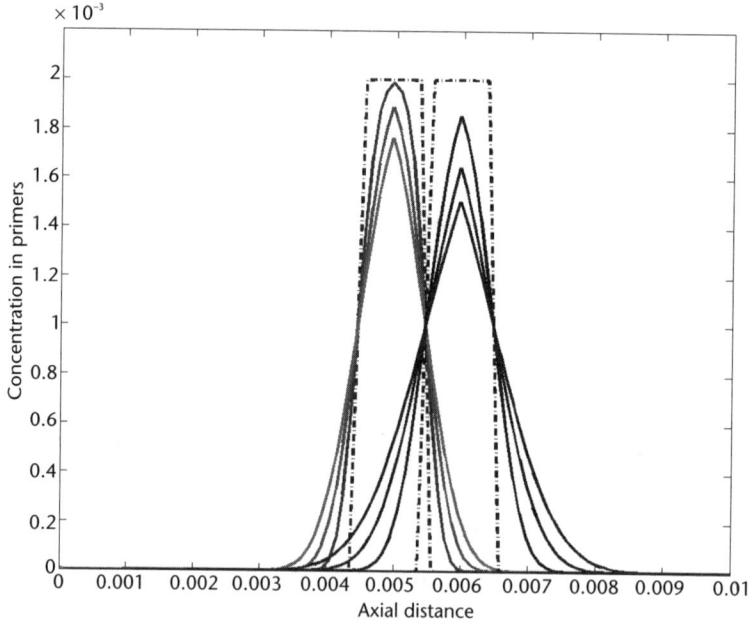

Figure 3.20   Concentration plot of the solution of (3.36) at different times. The two primers have different diffusion coefficients.

$$\frac{\partial c}{\partial t} = div(D \ grad \ c) = D\left(\frac{1}{r}\frac{\partial c}{\partial r} + \frac{\partial^2 c}{\partial r^2} + \frac{\partial^2 c}{\partial z^2}\right) \tag{3.38}$$

and note that the capillary length L is very large before the capillary radius R. If we want to set up a numerical calculation, we have to deal with a computational domain with a very large aspect ratio L/R. We can introduce the new variables

$$z^* = \frac{z}{L}, r^* = \frac{r}{R} \tag{3.39}$$

so that the transformed computational domain is defined by $L^* = 1$, $R^* = 1$. Let's introduce the other nondimensional variables

$$c^* = \frac{c}{c_0}, t^* = \frac{t}{\frac{RL}{D}} \tag{3.40}$$

It is straightforward to see that the nondimensional diffusion equation is

$$\frac{\partial c^*}{\partial t^*} = \frac{L}{R}\left(\frac{1}{r^*}\frac{\partial c^*}{\partial r^*} + \frac{\partial^2 c^*}{\partial r^{*2}}\right) + \frac{R}{L}\frac{\partial^2 c^*}{\partial z^{*2}} \tag{3.41}$$

This equation is an axisymmetrical diffusion equation with the anisotropic diffusion coefficients

$$D^*_z = \frac{R}{L}, D^*_r = \frac{L}{R} \tag{3.42}$$

In order to compensate for the change in geometry, the diffusion coefficients are now strongly anisotropic; the equivalent diffusion coefficient in the direction r is large, whereas that in the direction z is small. The ratio between the r and z diffusion coefficient is

$$\frac{D^*_r}{D^*_z} = \frac{L^2}{R^2} \gg 1$$

Note that (3.41) verifies Buckingham's Pi theorem [8]. There are four independent parameters in (3.38): c_0, L, R, and D. These parameters depend on three different units: kilos or moles (if we count the concentration in kilos or moles), meters, and seconds. According to Buckingham's theorem, there should be $4 - 3 = 1$ dimensionless parameter in the nondimensional equation. This parameter is evidently L/R.

3.3.8.4   Numerical Solution

Equation (3.41) may be solved by using a standard finite element method. The computational domain is defined by $r^* \in \{0,1\}, z^* \in \{0,1\}$. At the wall, the condition of impermeability yields $\left.\dfrac{\partial c^*}{\partial r^*}\right|_{r^*=1} = \left.\dfrac{\partial c}{\partial r}\right|_{r=R} = 0$. An initial condition $c^*_0 = 1$ is imposed

in a very small volume at the periphery of the tube. Concentration contours obtained at two different times are shown in Figure 3.21.

Note that the nondimensional calculations results have to be back transformed into dimensional values. It is a straightforward process if we use (3.39) and (3.40), and if we note that the initial concentration c_0 is obtained by converting the surface concentration in immobilized primers into a volume concentration.

3.3.8.5 Diffusion Barriers

It can be checked that the analytical and numerical results agree (Figure 3.22). Most of the time, analytical results when sufficiently accurate should be preferred. However, in the present case, the numerical approach—although more complex to set up—is more powerful. For example, it is possible to investigate whether axial diffusion inside the tube can be reduced. Although this problem is discussed in Chapter 5 (biochemical reactions), it is a diffusion barrier problem and we will mention it here. We found that if the gap between the annular regions initially labeled with the primers were adequately functionalized, the recapture of the

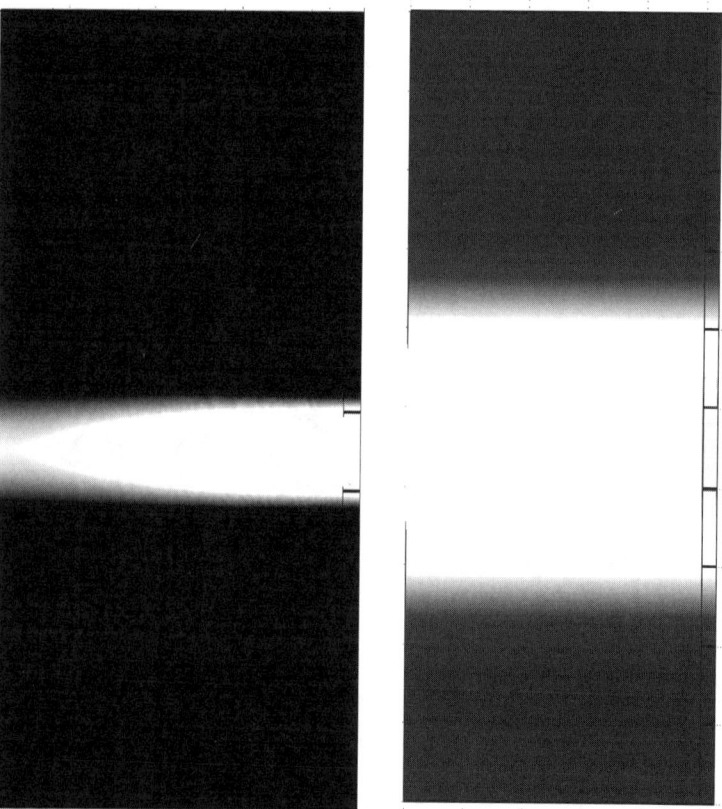

Figure 3.21 Concentration contours at two different times calculated with the numerical software FEMLAB [9], showing the same behavior as the analytical approach.

Concentration profile in two different primers after 5 mn

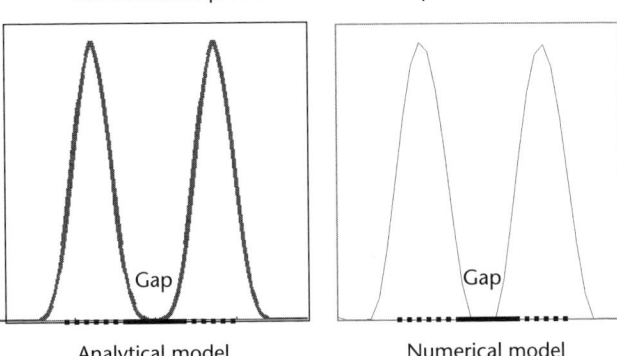

Analytical model   Numerical model

Figure 3.22   Comparison of concentration profiles between analytical and numerical models after the same time interval.

primers during diffusion would limit the axial diffusion. In Figure 3.23, we compare the axial diffusion of primers for a simple gap and a labeled gap.

3.3.9   Particle Size Limit: Diffusion or Sedimentation

The following question stems out of an inspection of (3.5): Does gravity force—which is not present in the equation—affect the diffusion of the microparticles? In other words, is the apparent weight of the particles negligible? One conceives easily that if the particles are small enough they will not sediment, and they will diffuse in the available volume; if they are sufficiently large, they will sediment despite the molecular agitation [3]. We derive here a criterion to estimate the sedimentation of the microparticles and to decide if the diffusion equation is valid.

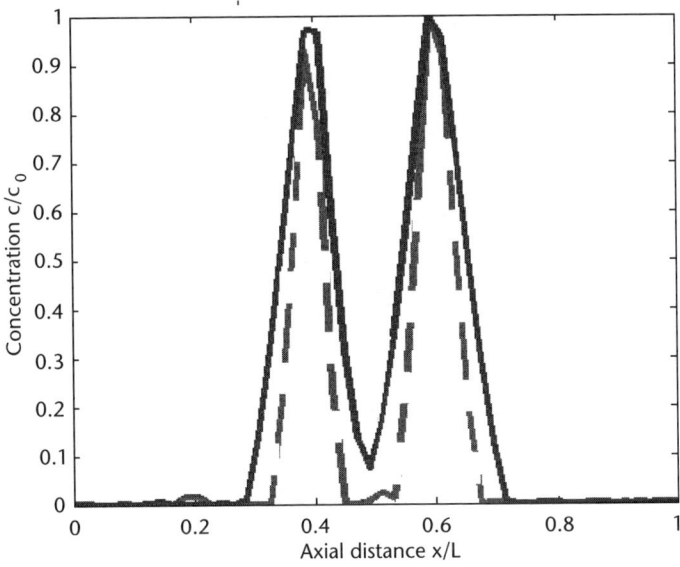

Figure 3.23   Comparison of the concentration profiles between a labeled (dotted line) and a nonlabeled (continuous line) gap. Axial diffusion is remarkably reduced if the gaps are labeled.

The settling velocity is defined as the uniform vertical velocity of a particle in a liquid at rest. The settling velocity can be calculated by writing the balance between gravitational force and hydrodynamic drag. If C_D be the friction factor (hydrodynamic drag coefficient), C_D is defined by

$$C_D = 6\pi\eta R_H \qquad (3.43)$$

Then, the hydrodynamic drag force on the particle is

$$F_{friction} = C_D v = 6\pi\eta R_H v \qquad (3.44)$$

and the settling velocity V_s is obtained by the force balance

$$C_D V_s = \Delta\rho g Vol_p \qquad (3.45)$$

where $\Delta\rho$ is the buoyancy density (difference between the volumic mass of particle and liquid), η is the dynamic viscosity of the fluid, and g is the gravity acceleration (9.8 m/s^2). For a spherical particle, the sedimentation velocity is given by

$$V_s = \frac{2}{9}\frac{\Delta\rho g R^2}{\eta} \qquad (3.46)$$

Suppose now that a typical dimension of the problem is d. For example, the vertical dimension of a biodiagnostic microchamber is of the order of $d = 50\,\mu m$. Let us compute the times τ_1 and τ_2 for the particle to move on the distance d by sedimentation and Brownian motion

$$\tau_1 = \frac{d}{V_s}$$

and

$$\tau_2 \approx \frac{d^2}{4D}$$

then, the ratio $\beta = \tau_1/\tau_2$ is

$$\beta = \frac{\tau_1}{\tau_2} \approx \frac{d}{V_s}\frac{4D}{d^2} = 4\frac{k_B T}{\Delta\rho g d Vol_p} = 4\frac{k_B T}{\Delta m g}\frac{1}{d} \qquad (3.47)$$

If we introduce the characteristic Boltzmann length scale L [5], (3.47) becomes

$$\beta = 4\frac{L}{d}$$

The ratio (3.47) is an energy ratio between the energy of the Brownian motion and the potential energy of the particle. If $\beta \ll 1$, sedimentation dominates, and the particles will descend with the settling velocity. As a general rule, particles larger than 1 μm tend to be affected by sedimentation. For example, cells usually sediment because their size is larger than 10 μm. A more detailed analysis may be found in [3].

The preceding analysis shows that it is very important not to let the particles aggregate. If so, the Brownian motion will cease, and the aggregate will sediment. This is why surfactants or poly-ethylene-glycol (PEG) [10] are usually added to buffer liquids.

3.4 Microscopic (Discrete) Approach

In the preceding sections, we have followed an approach based on the continuum: this approach assumes that in every elementary volume of liquid there are a number of microparticles sufficiently important to define a concentration. This approach is very convenient because it introduces a PDE that can be solved by ordinary discretization techniques, like the finite element method or the finite volume method. However, very complex geometries of the diffusion domain may not be easily treated with such methods.

We present here another approach that is very useful when dealing with microscopic scales. Contrary to the preceding continuum approach, this approach is discrete, meaning that the displacement of every particle is calculated. This approach is well adapted to complex geometrical domains and small number of particles.

3.4.1 Monte Carlo Method

The Monte Carlo method is based on the mimicking of the random walk of particles. Because the mean free path is very short, it is possible without changing the statistical randomness to allow for longer linear displacement steps, at the condition that they remain small compared to the free space defined by the surrounding geometry.

3.4.1.1 Two-Dimensional Case

In the two-dimensional case, a particle moves in a time step Δt from the location (x, y) to the location ($x + \Delta x$, $y + \Delta y$), where the space increments are defined by

$$\Delta x = \sqrt{4D\Delta t}\,\cos(\alpha)$$
$$\Delta y = \sqrt{4D\Delta t}\,\sin(\alpha) \tag{3.48}$$
$$\alpha = random(0,2\pi)$$

In (3.48) the function "random" is a choice of uniformly distributed random numbers in the interval [0, 2π]. The validity of the method depends on the quality of randomness of the angle α. In the following we have used the MATLAB command "rand" [11]. The length of the displacement has been scaled by the real diffusion-length scale.

Example of Random Walk from a Source Point

We show here two examples of random walk calculations. The first case is the two-dimensional diffusion from a source point (Figure 3.24).

Using the algorithm defined by (3.48), we find the pictures of Figure 3.25. Diffusion is isotropic around the initial spot and has a Gaussian shape along a radius.

In Figure 3.26, it can be verified that the average square distance is related to time by the relation

$$< d^2 > = 4Dt$$

Example of Random Walk in a Microchannel

The same algorithm can be applied to the case of a microchannel. The results are shown in Figure 3.27. In this case also, the particle distribution follows a Gaussian profile.

3.4.1.2   Three-Dimensional Case

In the three-dimensional case, a particle moves in a time step Δt from the location (x, y, z) to the location ($x + \Delta x$, $y + \Delta y$, $z + \Delta z$). The random walk algorithm is the following

$$\Delta x = \sqrt{4Ddt} \cos(\alpha)\sin(\beta)$$
$$\Delta y = \sqrt{4Ddt} \sin(\alpha)\sin(\beta)$$
$$\Delta z = \sqrt{4Ddt} \cos(\beta) \tag{3.49}$$
$$\alpha = random(0,2\pi)$$
$$\beta = \arccos\left(1 - 2\ random(0,1)\right)$$

The angles α and β are defined in Figure 3.28. Note the definition of the angle β in (3.49). If we had taken simply $\beta = random\ (0, 2\pi)$, the z-direction would be a preferred direction of displacement. If we want a uniformly distributed direction angle, we have to take a random α angle between 0 and 2π and a random z-coordinate between −1 and +1. This random z-coordinate is obtained by the function 1–2 random(0,1) and the angle β is equal to arcos(1–2 random(0,1)).

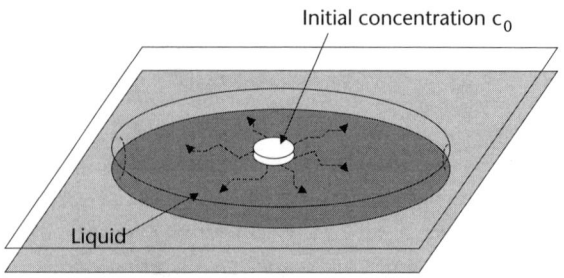

Figure 3.24   Two-dimensional diffusion of tracers originated at a source point.

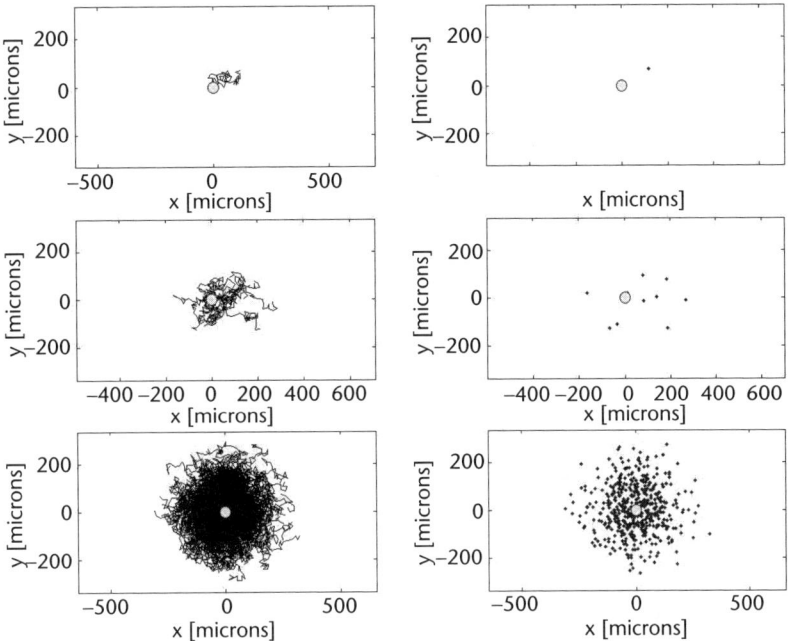

Figure 3.25 Random walk of 10-nm particles originated at the location (0,0). Left, from top to bottom: trajectories for 1, 10, and 500 nanoparticles in a time interval of 50 seconds. Right, from top to bottom: end point of 1, 10, 500 nanoparticles at $t = 50$ seconds.

3.4.2 Diffusion in Confined Volumes: Drug Diffusion in Human Body

3.4.2.1 Introduction

One of the most useful applications of the Monte Carlo method is the diffusion inside a confined domain. It is striking to see how diffusion occurs in very small volumes. In chemistry, there is this example of a solid alloy composed of aluminum with lead inclusions. At a temperature where the lead is molten and the aluminum still solid, one can follow the random walk of the lead molecules inside the aluminum matrix (Figure 3.29) [12].

In biology, there are many examples of diffusion in very confined media: proteins diffuse in cells [13], and macromolecules diffuse in cells' interstitial spaces. Due to its importance, diffusion processes in confined media is the object of many studies, and there is an abundant literature on this topic. We present here an important example in biology: the diffusion in ECSs of cell clusters.

In the biological field, the delivery of drugs in the human body, especially in clusters of cells, is a problem of utmost importance, requiring the knowledge of diffusion in a complex, confined geometry. In this case, there are two different media: the ECS and the cells; these two media are separated by the cell membrane. Diffusion of biological molecules first takes place inside the ECS. After the molecules have penetrated the cells through the cell membranes (which is called uptake), the molecules diffuse inside the cells. In this section, we focus on the diffusion inside the ECS.

Because cellular uptake is not immediate, biological cluster of cells may be seen as porous media, where the cells are the *solid grains* and the ECS is the *pores*

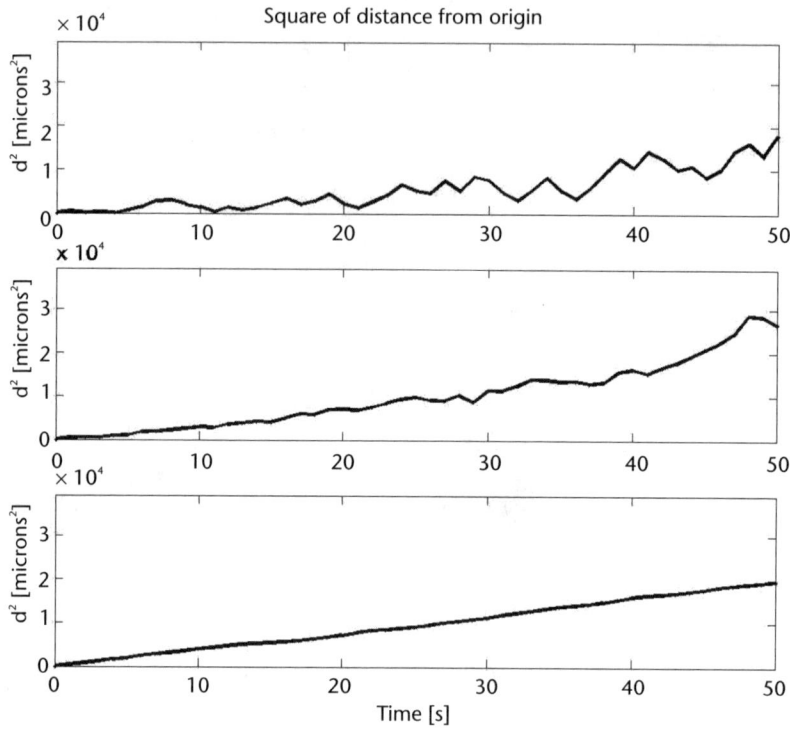

Figure 3.26 Square of average distance of particles versus time. From top to bottom, 1, 10, and 500 nanoparticles. In the third figure at the bottom, the curve is nearly linear, and its slope is approximately $2 \cdot 10^4/50 = 400 \, \mu m^2/s$. If we relate this value to the relation $<d^2> = 4Dt$, the value of the slope must be 4D, and we find $D = 10^{-10} \, m^2/s$, which is the input value in the model.

(Figures 3.30 and 3.31). In the particular case of tumoral cells, the extracellular path is called the tumor interstitial matrix (IM) [14].

We will show in the following that diffusion in the ECS is much slower than free diffusion. It is typical to use an apparent (or effective) diffusion coefficient (ADC) to determine the speed of diffusion of drugs in the tumor ECS [15]. Speed of diffusion based on the apparent diffusion coefficient is equal to that of the real diffusion coefficient in the restricted geometry of the ECS. The apparent diffusion coefficient depends on the morphology of the ECS, especially on the tortuosity (Figure 3.32)—ratio of the real distance to the straight line distance between two points—and also to special features of the ECS, like intercleft spaces and constrictions. Real drug delivery time will be determined by adding to the diffusion characteristic time in the ECS the uptake characteristic time (time for the macromolecule to enter the cell) plus the diffusion time inside the cell [13].

Note that it is of great importance in cancer treatment to be able to estimate the value of the apparent diffusion coefficient [16]. If the delivery time is too long—it may reach 48 hours—or if some cells are not delivered, the balance between destruction and multiplication of cancerous cells will be unfavorable. Note also that in some cases, a change of the ADC reflects a change in the cells' shape and arrangement [17], so that an evolution of a disease may be followed.

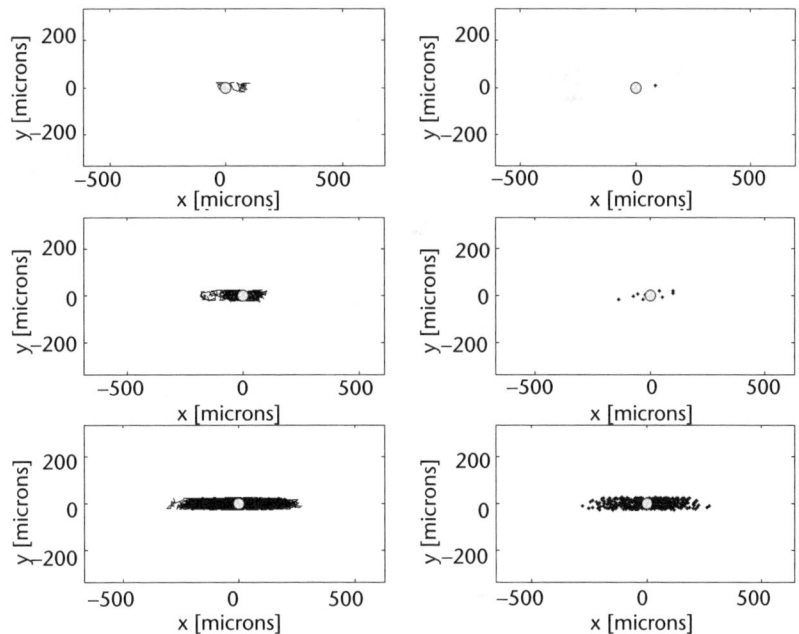

Figure 3.27 Monte Carlo diffusion in a quasi-one-dimensional geometry. Left, from top to bottom: trajectories of 1, 10, and 500 nanoparticles in a time interval of 50 seconds. Right, from top to bottom: end point of the trajectories at $t = 50$ seconds.

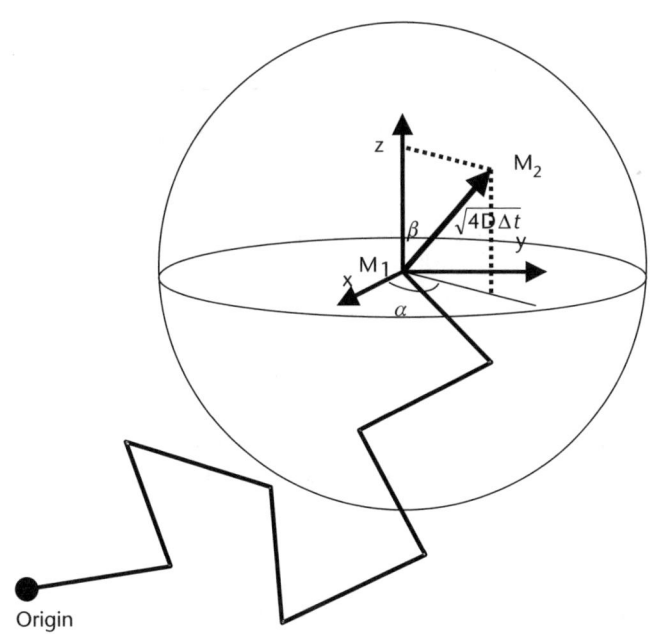

Figure 3.28 Schematic view of the linear motion M_1M_2 during the random walk.

We suppose that the fluid flow in the ECS is negligible in front of the molecular diffusion, so we have to calculate the diffusion of a substance in a very complex

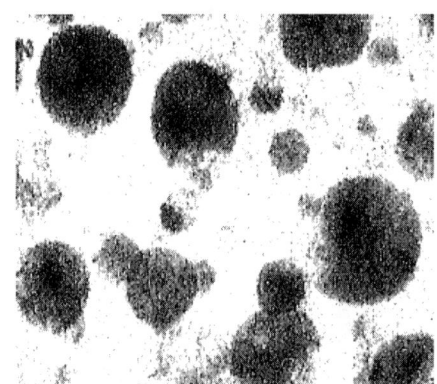
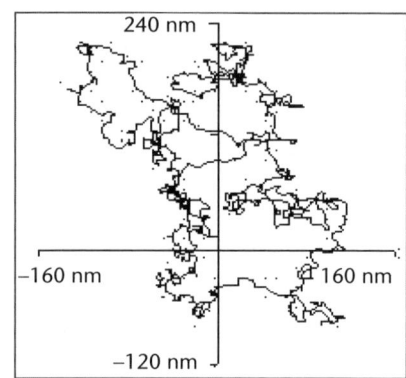

Figure 3.29 Example of very confined diffusion. Lead molecules diffusing inside a aluminum matrix (*From:* [12]. © 2002 Erik Johnson. Reprinted with permission.)

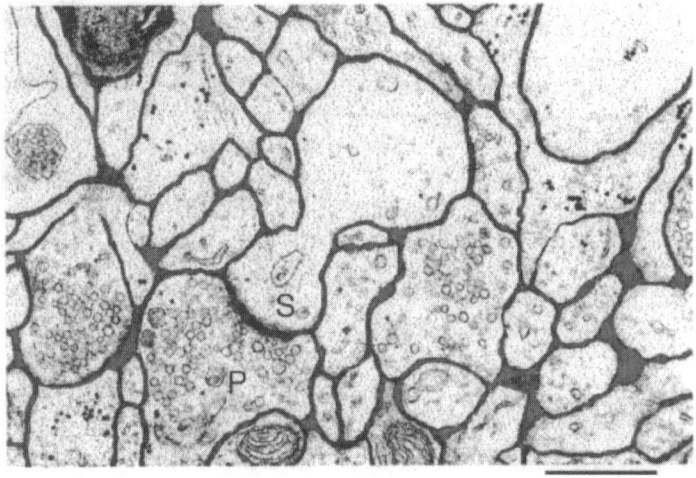

Figure 3.30 Geometry of ECS from [15]. Electromicrograph of small region of rat cortex. The ECS is in dark on the picture. *Lakes,* or intercleft spaces, can be seen at the bottom right where the ECS widens. (*From:* [18]. © 1998 TINS. Reprinted with permission.)

geometry. Different types of numerical approaches have been proposed for regular repetitive patterns like squares and triangles: the homogenization theory [19], which is based on the calculation of the diffusion in a motif and extending the result to the whole domain, and the Monte Carlo method [20] in geometry, where the boundaries are defined by analytical linear functions. At a certain point, it has been thought that regular patterns calculation could be sufficient to approximate an average ADC [21]; however, it is not always the case if the ECS has intercleft spaces or constrictions, particularly if one wants to estimate the local uptake rates [22] or if any change in cell shape and arrangement takes place [19]. So far there have been very few investigations for irregular and disordered clusters, mostly because of the difficulty in describing the geometry [23].

However, recent progress has been made in tackling the problem of diffusion in the ECS of clusters of cells. In the following, we present a two-dimensional numerical approach based on a two-step calculation: first, the calculation of the cells

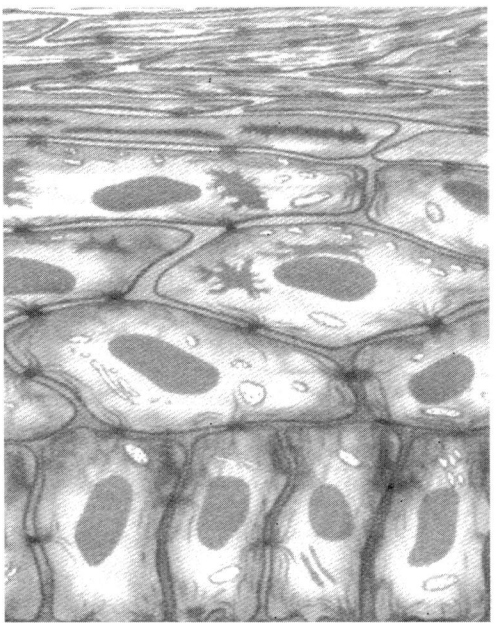

Figure 3.31 Cell arrangement in the human skin from [24]. The shape of the cells is regular, but the anisotropy of the ECS changes from top to bottom. The typical width of the ECS is a few microns. (Courtesy of CEA.)

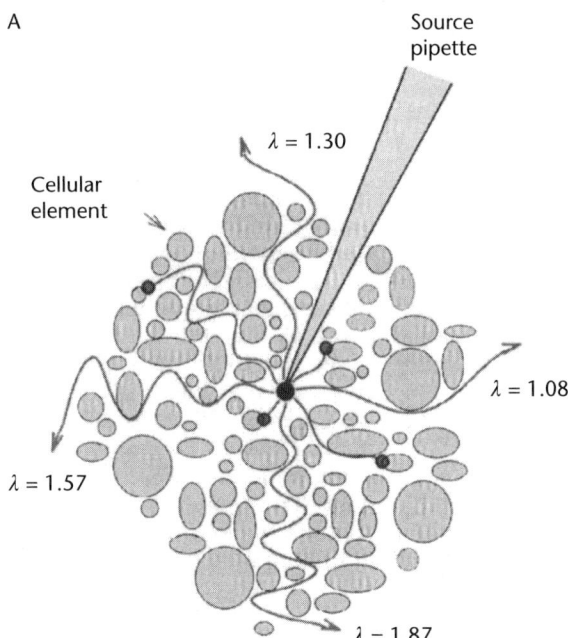

Figure 3.32 Schematic view of the diffusion paths in a porous media, depending on the tortuosity.

boundaries; second, a Monte Carlo numerical scheme for the diffusion in the ECS morphology defined at the preceding step.

3.4.2.2 Cell Boundaries

First, cells arrangement may be mimicked: cells rearrange inside the boundaries of the cell cluster in a function of constraints, like the surface tension of the membranes and their volume (depending on the growth or the shrinking of cells). A numerical software like the Surface Evolver [25] (see Chapter 2) is well adapted to calculate the morphology of the cell cluster. In order to describe a cell cluster morphology, a given set of points (vertices), segments (edges) delimiting the initial cells, is introduced in the Evolver numerical program. Depending on line (surface) tensions and cell volumes, the shape of the cells evolves until convergence to a minimum energy arrangement, mimicking real cell arrangement, is reached. It is assumed here that cell membranes behave similarly to an interface with surface tension. The initial edges are then refined and deformed depending on the specified constraints. A calculated arrangement of cells mimicking a real cluster of cells has been plotted in the Figure 3.33.

In the Evolver approach, the computational nodes are located on the cells' edges and are referenced by their coordinates (x, y) and by the corresponding edge number. Besides, each cell is referenced by its oriented edges. In order to prepare step 2, all of this information is memorized and stored.

3.4.2.3 Monte Carlo Numerical Scheme

Particles—or macromolecules—are initially placed in a central microregion, simulating the injection point at the tip of the microneedle. Diffusion is then simulated by following the particles executing random walks inside the ECS. In a two-dimensional system, the displacement $(\Delta x, \Delta y)$ of any particle in the time step Δt is given by the relations

$$\Delta x = \sqrt{4D\Delta t}\,\cos(\alpha)$$
$$\Delta y = \sqrt{4D\Delta t}\,\sin(\alpha)$$
$$\alpha = random(0,2\pi)$$

where D is the *free* diffusion coefficient, given by Einstein's law

$$D = \frac{k_B T}{6\pi\eta R_H}$$

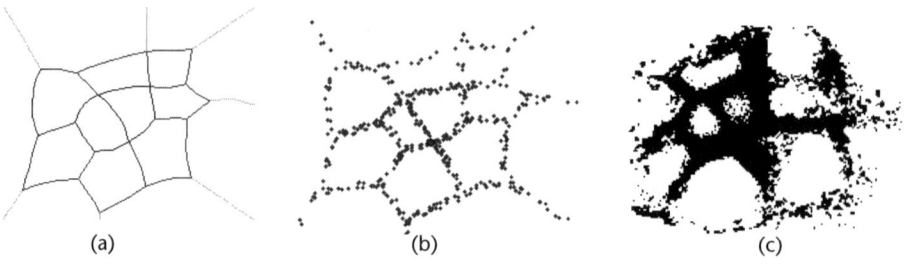

Figure 3.33 (a) Cell arrangement is calculated with the Evolver numerical program; (b) cell cluster morphology is enlightened by the calculated location of pharmaceutical molecules after they have diffused in the ECS; and (c) real cell cluster is observed by fluorescence imaging.

where k_B is the Boltzman constant ($1.38 \cdot 10^{-23}$ J/K), T is the temperature (K), η is the dynamic viscosity of the carrier fluid, and R_H is the hydraulic radius of the particle.

3.4.2.4 Uniform and Narrow ECS

Suppose first that the width of the ECS is approximately constant (as in Figure 3.33). The cell edges defined in the preceding step are widened to the desired width to define a real ECS. Particles location inside the cluster is permanently tracked, and the particles are not allowed to cross solid (cell) boundaries. The random walk of particles may then be confined inside the ECS, as shown in Figure 3.34. If the time allowed for the calculation is sufficiently large, the ECS is explored by the diffusing particles, as shown in Figure 3.35.

In a porous media, the distance between two points may be defined as the length of the shortest line in the fluid domain between two points (Figure 3.36), and tortuosity is then defined as the ratio between the distance in the liquid and the straight line distance.

It may be theoretically shown [19] that for any two-dimensional regular isotropic lattice of convex cells, tortuosity has a unique value

$$\tau = \sqrt{2} \tag{3.50}$$

and for three-dimensional lattices, the value of the tortuosity is

$$\tau = \sqrt{3} \tag{3.51}$$

It is relatively easy to be convinced of the validity of (3.50) and (3.51). Because the media is isotropic, we can estimate the tortuosity on a diagonal direction (Figure 3.37). We can approximate the length L_{AB} by a pixel discretization: By projection on the horizontal and vertical axis, assuming a discretization $\Delta x = \Delta y$, we obtain

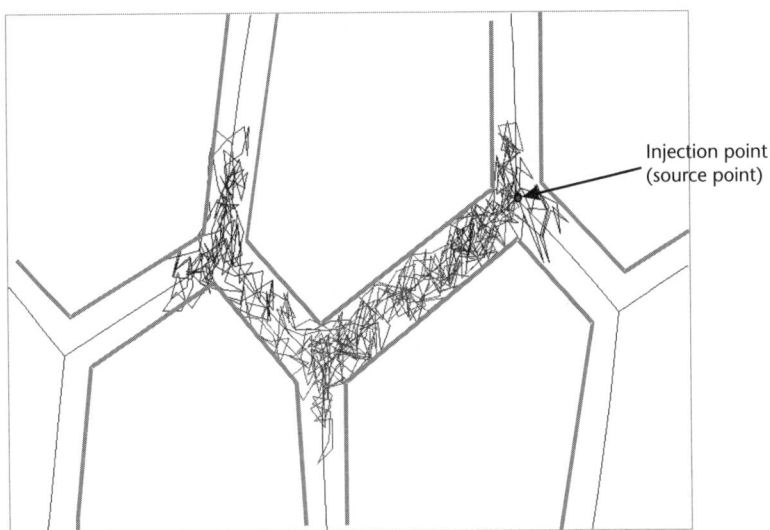

Injection point (source point)

Figure 3.34 Random walk of two particles inside an ECS calculated by a Monte Carlo method and constrained by ECS boundaries.

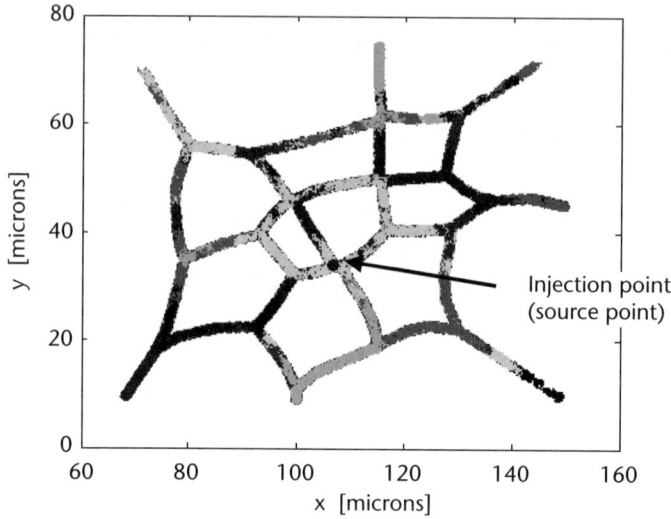

Figure 3.35   Random walk of 200 particles inside the ECS of the cluster.

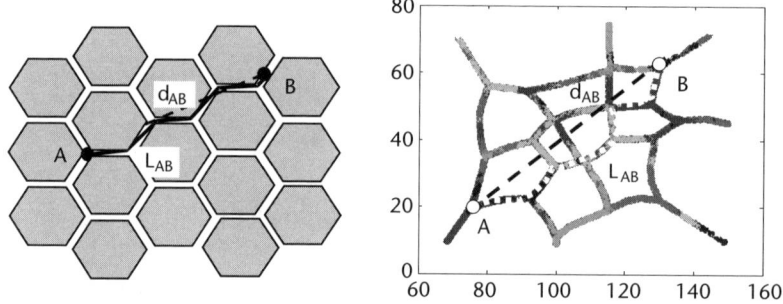

Figure 3.36   Definition of tortuosity in regular and irregular lattices. The tortuosity τ is equal to the ratio L_{AB}/d_{AB}; in a free media, $\tau = 1$.

$$L_{AB} = n\Delta x + n\Delta y = 2n\Delta x$$

On the other hand, the Pythagore relation is

$$d_{AB}^{2} = \left(n\Delta x\right)^{2} + \left(n\Delta y\right)^{2} = 2n^{2}\Delta x^{2}$$

Combining the two preceding equations yields

$$\tau^{2} = \frac{L_{AB}^{2}}{d_{AB}^{2}} = \frac{4n^{2}\Delta x^{2}}{2n^{2}\Delta x^{2}} = 2$$

Finally,

$$\tau = \frac{L_{AB}}{d_{AB}} = \sqrt{2}$$

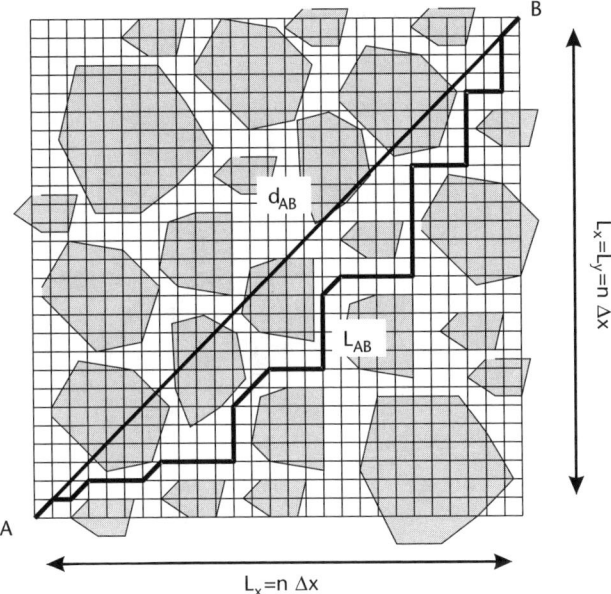

Figure 3.37 Estimation of the tortuosity in the case of convex, isotropic porous media.

The same reasoning also applies to the three-dimensional case.

It has been also shown that there is a very simple relation between the effective and the free diffusion coefficient involving the tortuosity

$$\frac{D_{eff}}{D} = \frac{1}{\tau^2} \tag{3.52}$$

Thus, if the width of the ECS is narrow and remains uniformly constant in the two-dimensional cluster, the value of the effective diffusion coefficient is half that of the free diffusion coefficient

$$\frac{D_{eff}}{D} = \frac{1}{2}$$

Assuming the same conditions for the ECS (isotropy and convexity), the Monte Carlo numerical model shows that (3.50) and (3.52) also apply for any isotropic cluster of irregular cells. Figure 3.38 shows the location of the diffusing particles initially starting from the injection point after a time interval of 15 seconds ($D = 10^{-10}$ m^2/s).

Let's introduce the normalized diffusion length β by

$$\beta = \frac{L}{\sqrt{4Dt}} \tag{3.53}$$

where L is obtained by averaging the distance of each particle between their location at time t and at time $t = 0$.

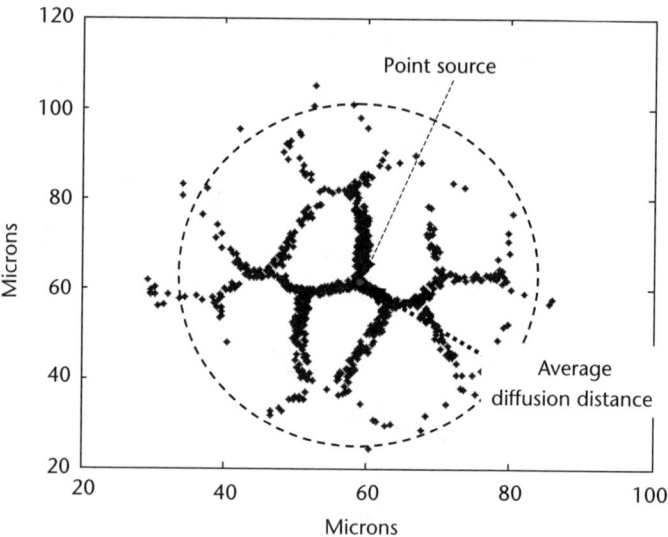

Figure 3.38 Diffusion distance from point source in an isotropic cell cluster.

In Figure 3.39, we plotted the normalized diffusion length versus time for different cluster morphologies: ordered (square and hexagonal cells) and disordered, with narrow ECS. A narrow ECS is defined here by a constant width less than one-tenth of the average size of the cells but larger than about three times the mean free path of the particles.

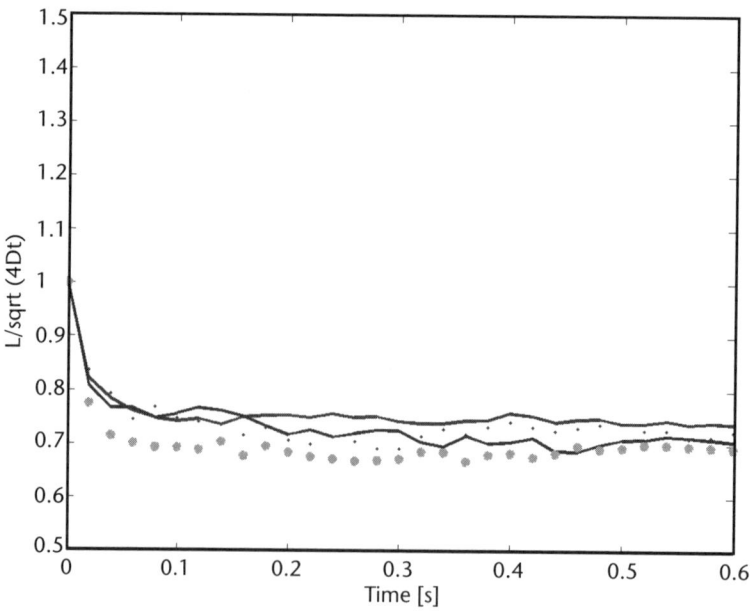

Figure 3.39 Normalized diffusion length versus time for different cluster morphologies (ordered and disordered) with narrow ECSs. The curves are similar to the experimental results of de Sousa et al. [26].

At very small times, the diffusing particles execute random walk inside a small region of the ECS where the particles are initially placed, so that the value of β is that of free diffusion: $\beta = 1$ at $t = 0$. After a short time, the particles have explored all of the available initial space and start diffusing inside the ECS. They are now constrained inside the ECS by the cells' boundaries, and β reaches a nearly constant value. The numerical results of Figure 3.39 are very similar to that of de Sousa et al. [26], obtained experimentally for regular triangle and square lattices.

The asymptotic value of β is 0.7, thus

$$\beta = \frac{L}{\sqrt{4Dt}} \approx 0.7 \approx \frac{1}{\sqrt{2}} \tag{3.54}$$

By definition, the apparent diffusion coefficient satisfies

$$\frac{L}{\sqrt{4D_{eff}t}} \approx 1 \tag{3.55}$$

From (3.54) and (3.55), we find the relation

$$\frac{D_{eff}}{D} = \beta^2 = \frac{1}{\tau^2} \approx \frac{1}{2} \tag{3.56}$$

leading to the value $\tau = \sqrt{2}$.

3.4.2.5   Intercleft Spaces and Channel Restrictions

Real ECSs in the human body are often more complex than that of a uniformly narrow gap between the cells, which we analyzed in the preceding section (see Figure 3.30). Very often, the spacing of the cells' lattice is not uniform, and there are intercleft spaces. It is shown that diffusion speed may be reduced by entrapment when the dimensions of the residual spaces are large and the connecting exits are sufficiently small. By "sufficiently small," we mean that the mean free path of the particle inside the carrier fluid is of the order of the cross dimensions of the ECS.

An idealized example is that of a cluster of round cells. If the dimension of the gaps between the cells is decreased, the apparent diffusion coefficient becomes smaller than the value predicted by (3.6). In the case defined in Figure 3.40, we obtain $\frac{D_{eff}}{D} \approx \frac{1}{4}$.

A limiting case is that of a gap width of the order of the mean free path of the particle, in such a case that the particles are trapped inside the intercleft space. In such a case, the relevant theory is the *percolation theory*, and there have been considerable efforts in this domain for biological applications.

3.4.2.6   Discussion

We have modeled the diffusion of biochemical species in a cluster of cells by a three-step algorithm: (1) generation of a cluster arrangement using the Evolver

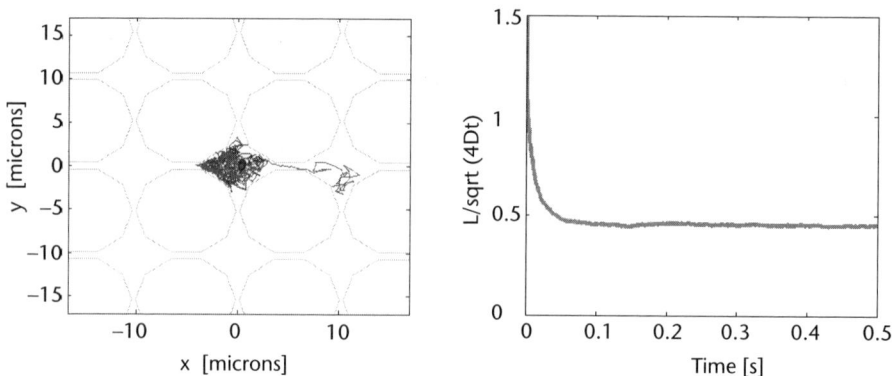

Figure 3.40 Normalized diffusion length versus time for a cluster of round cells with small gaps.

numerical software, (2) Monte Carlo random walk of the diffusing species, and (3) particle tracking to constrain the diffusing species inside the ECS.

The results of the model show that the ratio between the apparent diffusion coefficient and the free diffusion coefficient in dense cell clusters with small ECS is always the same, no matter what the morphology of the cluster (ordered or disordered) is. In a two-dimensional cluster

$$\frac{D_{eff}}{D} = \frac{1}{\tau^2} \approx \frac{1}{2}$$

where τ is the tortuosity of the porous media. However the situation is much more complex in the ECS of irregular and anisotropic clusters of cells, especially if there exists intercleft spaces. There are two cases where speed of diffusion can be considerably reduced: first, by particle entrapment in the intercleft spaces of the ECS, and second, in the case of an anisotropic medium, diffusion in the direction perpendicular to the preferred direction of the medium is delayed.

3.5 Conclusion

Diffusion is probably the main phenomenon concerning micro- and nanoparticles and target macromolecules in biotechnological applications. Estimation of diffusion time may be performed by solving the partial differential equation for the diffusion of concentration or by a discrete approach. The advantage of the *continuum* approach is the availability of numerical software—finite element method is recommended because it adapts best to the shape of the boundaries—and the relative fast computational time—at least in a two-dimensional case. A discrete approach—like the Monte Carlo method—is perhaps more demonstrative because it mimics the behavior of the particles and is well adapted to very complicated geometries. The drawback of the method is the computational time.

For the technological applications, diffusion is at the same time advantageous and not: for example, it takes advantage of the Brownian motion to make molecules recognize each other leading to the desired hybridization; on the other side, diffusion may disperse the target molecules or mix these molecules with other undesirable molecules. The art of the design of biotechnological components resides in part in the clever uses of molecular diffusion.

References

[1]   http://scienceworld.wolfram.com/physics/BrownianMotion.html.

[2]   Tabeling, P., *Introduction à la Microfluidique*, Paris: Belin, 2003.

[3]   Hiementz, P. C., and R. Rajagopalan, *Principles of Colloid and Surface Chemistry*, New York: Marcel Dekker, 1997.

[4]   Crank, J., *The Mathematics of Diffusion*, 2nd ed., Oxford, England: Oxford University Press, 1975.

[5]   Faucheux, L. P., and A. J. Libchaber, "Confined Brownian Motion," *Physical Review E.*, Vol. 49, No. 6, 1994.

[6]   Press, W. H., et al., *Numerical Recipes*, Cambridge, England: Cambridge University Press, 1987.

[7]   Berthier, J., and F. Chatelain, "Dimensioning Simultaneous Polymerase Chain Reactions in Capillary Tubes," *Proc. of 2005 FEMLAB Conference*, Paris, France, November 15, 2005.

[8]   Buckingham, E., "On Physically Similar Systems: Illustrations of the Use of Dimensional Equations," *Phys. Rev.*, Vol. 4, 1914, pp. 345–376.

[9]   FEMLAB reference manual. Stockholm: COMSOL AB, http://www.comsol.com.

[10] de Gennes, G., "Polymers at an Interface: A Simplified View," *Adv. Colloid Interface Sci.*, Vol. 27, No. 5, 1987, pp. 189–209.

[11] MATLAB software manual: http://www.mathworks.com.

[12] Johnson, E., et al., "Nanoscale Lead-Tin Inclusions in Aluminium," *J. Electron Microscopy*, Vol. 51, 2002, pp. S201–S209.

[13] Pollack, G. H., *Cells, Gels and the Engines of Life*, Seattle, WA: Ebner and Sons, 2001.

[14] Rumanian, S., et al., "Diffusion and Convection in Collagen Gels: Implications for Transport in the Tumor Interstitium," *Biophys. J.*, Vol. 83, 2002, pp. 1650–1660.

[15] Lankelma, J., et al., "A Mathematical Model of Drug Transport in Human Breast Cancer," *Microvascular Research*, Vol. 59, 2000, pp. 149–161.

[16] El-Kareh, A. W., S. L. Braunstein, and T. W. Secomb, "Effect of Cell Arrangement and Interstitial Volume Fraction on the Diffusivity of Monoclonal Antibodies in Tissue," *Biophys. J.*, Vol. 64, 1993, pp. 1638–1646.

[17] Herneth, A. M., S. Guccione, and M. Bednarski, "Apparent Diffusion Coefficient: A Quantitative Parameter for In Vivo Tumor Characterization," *European J. of Radiology*, Vol. 45, 2003, pp. 208–213.

[18] Nicholson, C., and E. Sykova, "Extracellular Space Structure Revealed by Diffusion Analysis," *TINS*, Vol. 21 No. 5, 1998, pp. 207–215.

[19] Chen, K. C., and C. Nicholson, "Changes in Brain Cell Shape Create Residual Extracellular Space Volume and Explain Tortuosity Behavior During Osmotic Challenge," *Proc. Natl. Acad. Sci. USA*, Vol. 97, No. 15, 1999, pp. 8306–8311.

[20] Saxton, M. J., "Lateral Diffusion in an Archipelago, the Effect of Mobile Obstacles," *Biophys. J.*, Vol. 52, 1987, pp. 989–997.

[21] Blum, J. J., et al., "Effect of Cytoskeletal Geometry on Intracellular Diffusion," *Biophys. J.*, Vol. 56, 1989, pp. 995–1005.

[22] Szafer, A., et al., "Theoretical Model for Water Diffusion in Tissues," *Magnetics Resonance in Medicine*, Vol. 33, No. 5, 1995, pp. 697–712.

[23] Berthier, J., F. Rivera, and P. Caillat, "Numerical Modeling of Diffusion in Extracellular Space of Biological Cell Clusters and Tumors," *Nanotech 2004*, Boston, MA, March 7–11, 2004.

[24] Martin, M., "Conséquences D'une Irradiation Ionisante Sur la Peau Humaine," *Clefs CEA*, Vol. 48, 2003, pp. 53–55.

[25] Brakke, K. A., "The Surface Evolver," *Experimental Mathematics*, Vol. 1, No. 2, 1992, pp. 141–165.

[26] de Sousa, P. L., D. Abergel, and J.-Y. Lallemand, "Experimental Time Saving in NMR Measurement of Time Dependent Diffusion Coefficients," *Chemical Physics Letters*, Vol. 342, 2001.

CHAPTER 4

Transport of Nanoparticles and Biochemical Species

4.1 Introduction

Biotechnology in general deals with the manipulation of biological targets. The ultimate goal is the manipulation of a single target (e.g., a single DNA strand or a single cell). To do so, different methods are used successively to allow for more and more selectivity. Figure 4.1 schematizes the different methods from the least selective to the most selective.

The first step is then the transport by microfluidic (continuous or digital) means. For example, the targets are to be extracted and concentrated from a liquid sample, or they have to be guided toward a reactive surface, or mixed with a reagent, or dispersed in another liquid, or transported to a mass spectrometer, and so forth. In any case, the knowledge of transport mechanism is mandatory. The next steps in selectivity depend on the particular application.

We have presented the behavior of the buffer liquid as a carrier fluid in Chapters 1 and 2, and the diffusion of the molecules of interest in Chapter 3. In this chapter, we deal with their transport by the buffer fluid. Transport by magnetic beads will be presented in Chapter 7; transport by electric field will be presented in Chapter 8.

We present first the governing equations of transport (advection-diffusion equation) under the continuum assumption and their nondimensional form, which introduces the characteristic Peclet number, and then we analyze some characteristic cases, like the flow in a microchannel, and present the Taylor-Aris model. This model will lead us to the major problem of mixing in microfluidics. To complete the approach, Langevin's equation is introduced for particles experiencing strong Brownian motion, and the particle trajectory approach is introduced for larger particles less affected by the Brownian motion. Applications of particle trajectory to field flow fractionation and chromatography columns are presented at the end of the chapter.

4.2 Advection-Diffusion Equation

4.2.1 Governing Equation for Transport

As we have done for the mass conservation equation and for the momentum equation, we write the concentration balance in an elementary volume $(\Delta x, \Delta y)$. For simplicity, we consider a two-dimensional element, but the reasoning is the same for

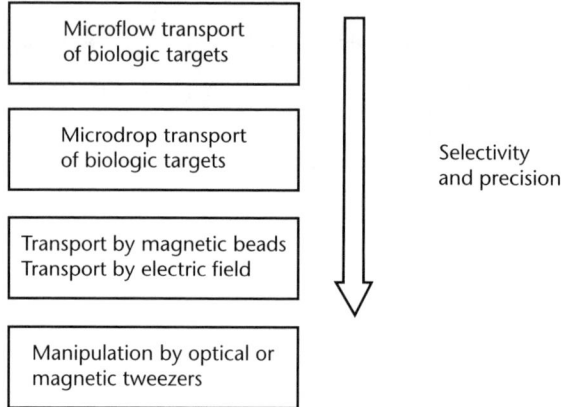

Figure 4.1   Schematic approach to micromanipulation of biologic targets.

a three-dimensional volume. The change in the mass of species in the volume is equal to the convective flux balance plus the diffusion flux balance (Figure 4.2).

Thus, we can write

$$\frac{\partial c}{\partial t}\Delta x \Delta y + u \frac{\partial c}{\partial x}\Delta x \Delta y + v \frac{\partial c}{\partial y}\Delta x \Delta y + \frac{\partial J_x}{\partial x}\Delta x \Delta y + \frac{\partial J_y}{\partial y}\Delta x \Delta y = 0 \qquad (4.1)$$

where J_x and J_y are the diffusion fluxes given by Fick's law

$$J_x = -D\frac{\partial c}{\partial x}$$
$$J_y = -D\frac{\partial c}{\partial y} \qquad (4.2)$$

In (4.2) D is the diffusion constant. Dividing (4.1) by $\Delta x\ \Delta y$ and substituting (4.2) yields

$$\frac{\partial c}{\partial t} + u\frac{\partial c}{\partial x} + v\frac{\partial c}{\partial y} = \nabla \cdot (D\nabla c) \qquad (4.3)$$

or

$$\frac{\partial c}{\partial t} + \vec{U}\cdot\nabla c = \nabla \cdot (D\nabla c) \qquad (4.4)$$

where U is the vector (u,v,w). Recall that the material derivative notation is

$$\frac{D}{Dt} = \frac{\partial}{\partial t} + u\frac{\partial}{\partial x} + v\frac{\partial}{\partial y} + w\frac{\partial}{\partial z} = \frac{\partial}{\partial t} + \vec{U}\cdot\nabla \qquad (4.5)$$

and then (4.3) can be cast under the form

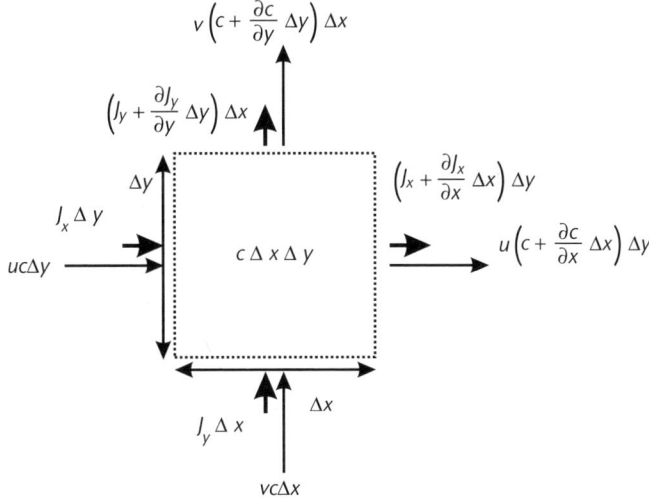

Figure 4.2   Concentration balance in an elementary volume.

$$\frac{Dc}{Dt} = \nabla \cdot (D\nabla c) \tag{4.6}$$

Assuming that D is constant, the advection-diffusion equation becomes

$$\frac{\partial c}{\partial t} + \vec{U} \cdot \nabla c = D\Delta c \tag{4.7}$$

or, in a Cartesian coordinate system,

$$\frac{\partial c}{\partial t} + u\frac{\partial c}{\partial x} + v\frac{\partial c}{\partial y} + w\frac{\partial c}{\partial z} = D\left[\frac{\partial^2 c}{\partial x^2} + \frac{\partial^2 c}{\partial y^2} + \frac{\partial^2 c}{\partial z^2}\right] \tag{4.8}$$

Suppose for an instant that the diffusion coefficient $D = 0$. In such a case

$$\frac{Dc}{Dt} = 0 \tag{4.9}$$

so that the concentration c remains the same along a trajectory (Figure 4.3). This propriety is valid for short times, where diffusion process has not had time to smear out the concentration. It is well known that very laminar flows, usually associated with biotechnological devices, are very unfavorable to diffusion, and "short times" are often rather long and special mixing devices promoting mixing have been developed to enhance diffusion [1].

Now, let us come back to the general case where the diffusion coefficient is not zero. Let's define a concentration norm in the whole domain occupied by the fluid by the mathematical function

$$\Theta = \int c^2 \, dx dy dz$$

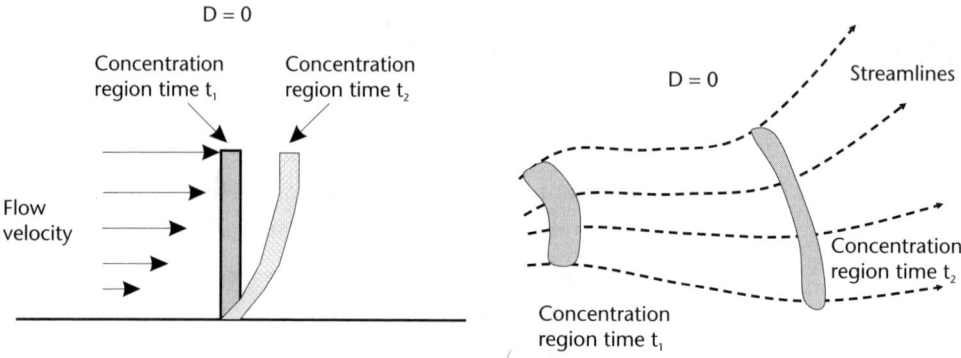

Figure 4.3 Sketches of mass transport in the absence of diffusion. Left: near a solid wall; right: in the bulk of the flow.

Mathematically, this function represents a measure of the concentration. In the absence of source or sink of substance, it is possible to derive [2]

$$\frac{\partial \Theta}{\partial t} = -D \int (\nabla c)^2 \, dxdydz$$

showing that Θ always decreases with time. Thus, the concentration spreads with time (Figure 4.4). If D were negative—which does not happen—the particles would concentrate and there would be *antidiffusion*—which of course does not exist.

4.2.2 Source Terms

If there are concentration source or sink terms, (4.7) becomes

$$\frac{\partial c}{\partial t} + \vec{U} \cdot \nabla c = D\Delta c + S \tag{4.10}$$

where S is the source term. The terms S in the advection-diffusion equation is a source or sink term, depending on its sign. The unit of S in the International Unit System is mole/m^3/sec or particles/m^3/sec.

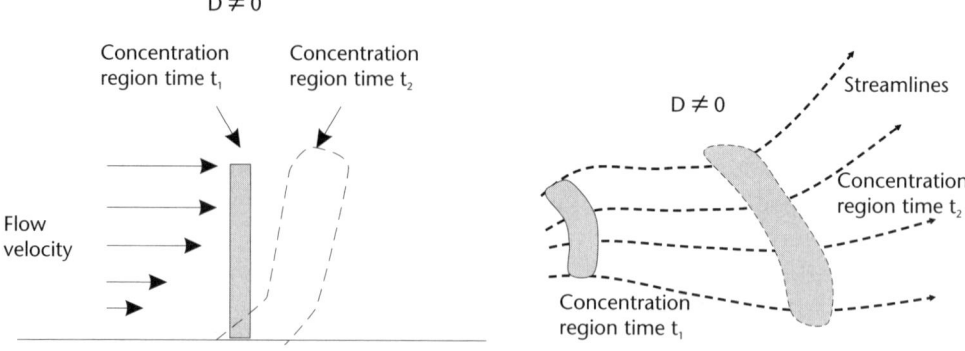

Figure 4.4 Sketch of transport of concentration in the real case where $D \neq 0$.

S may be a function depending on a volume, on a surface, on a contour, or on a point. Usually creation or removal of concentration of a constituent is linked to chemical or biochemical reactions. In the following chapter dedicated to biochemical reactions, we shall see some examples of source or sink terms.

4.2.3   Boundary Conditions

Many different forms of boundary conditions can exist for the advection-diffusion equation. However, two boundary conditions are remarkable.

The first one is the Dirichlet condition $c = 0$ at a solid wall. This condition means that the concentration of the studied macromolecules or nanoparticles vanishes at the solid wall. This is the case of total *adhesion*, when any particle of the concentration field that contacts the wall is immobilized and removed from the ensemble of the transported particles (Figure 4.5).

The second one is the homogeneous Neumann condition $\frac{\partial c}{\partial n} = 0$, where n is the unit vector defining the normal direction to the wall. In this case, there is no mass flux from the concentration field to (or through) the wall, and it corresponds to a situation of *no adherence* of the transported species and the wall (Figure 4.6).

Evidently, there exist more complicated boundary conditions, especially when the mass flux to the wall is governed by a biochemical reaction. In such a case, the mass flux at the wall is determined by the chemical reaction rate

$$J_n = -D\frac{\partial c}{\partial n} = \frac{d\Gamma}{dt} \tag{4.11}$$

where $\frac{d\Gamma}{dt}$ is the reaction kinetics. The boundary condition is the Neumann condition

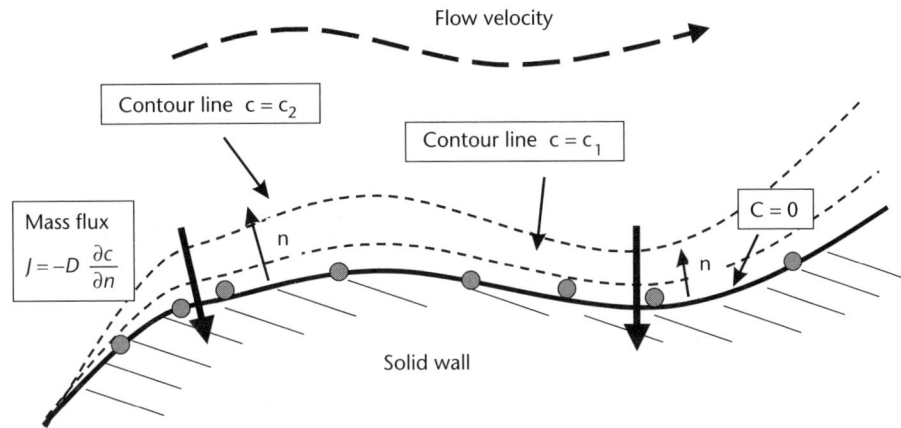

Figure 4.5   Dirichlet condition at a solid wall. Contour plot of concentration and mass flux. The particles are immobilized on the wall upon contact. A boundary layer of concentration develops along the solid wall.

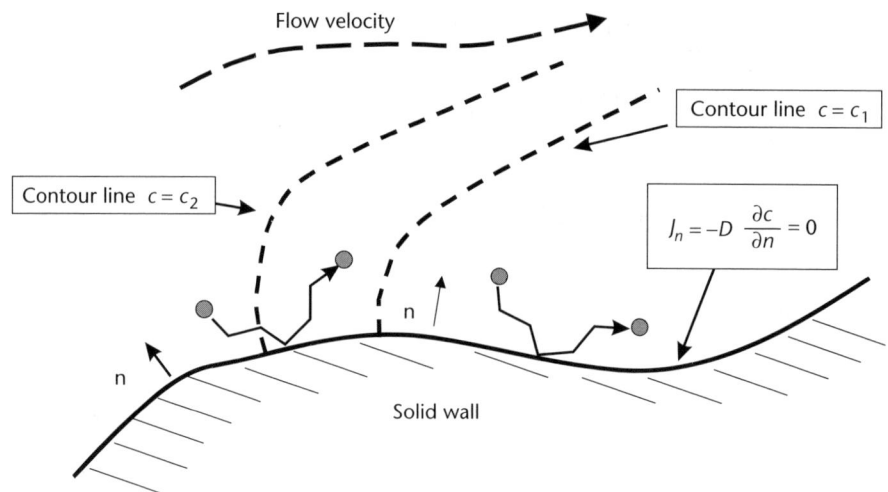

Figure 4.6 Homogeneous Neumann condition at the solid wall. Particles do not adhere on the contact surface.

$$\frac{\partial c}{\partial n} = -\frac{1}{D}\frac{d\Gamma}{dt} \qquad (4.12)$$

and we obtain the scheme of Figure 4.7, where a boundary layer of concentration develops along the wall. This type of problem will be treated in Chapter 5 in the section concerning concentration transport coupled to biochemical reaction.

4.2.4 Coupling with Hydrodynamics

The advection-diffusion equation is not sufficient to solve the complete problem by itself. The velocity field must be known. So the complete formulation is

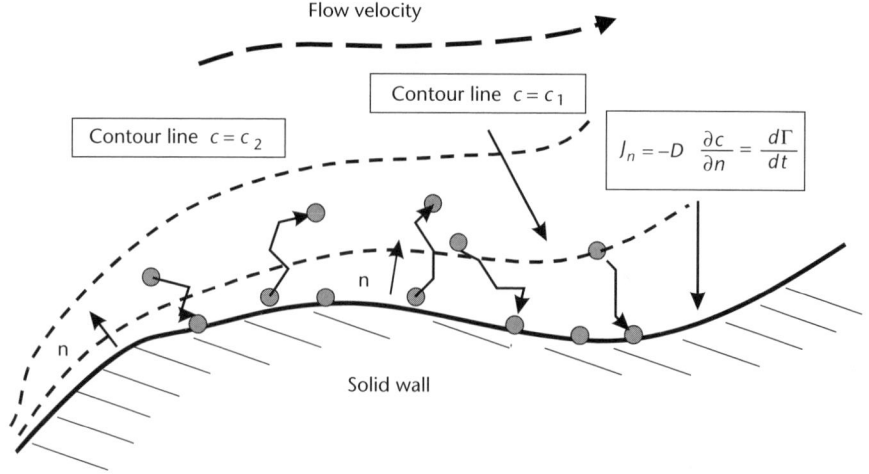

Figure 4.7 During some biochemical reactions, micro- and nanoparticles can be temporarily immobilized at the wall, until equilibrium is found. Depletion layers similar to concentration boundary layers form in the vicinity of the solid wall.

$$\frac{\partial \rho}{\partial t} + \nabla\left(\rho \vec{U}\right) = 0$$

$$\rho\frac{\partial \vec{U}}{\partial t} + \rho \vec{U} \cdot \nabla \vec{U} = -\nabla P + \eta \Delta \vec{U} + F \tag{4.13}$$

$$\frac{\partial c}{\partial t} + \vec{U} \cdot \nabla c = \nabla \cdot \left(D \nabla c\right) + S$$

The system (4.13) is a system of five scalar equations (in the three-dimensional case). There are five unknowns (u, v, w, P, c), two fluid properties (ρ and η), the diffusion constant of the species (D), and two external actions on the fluid (the body force par unit volume F and the concentration source or sink per unit volume S).

Note that (4.13) is only a weakly coupled system under the condition that the concentration is sufficiently small not to affect the buffer fluid viscosity and density. In the next section, we treat the problem of the variation of the fluid properties with concentration.

Usually, in a microfluidics microsystems, the flow of the buffer fluid is permanent (steady state), and only the concentration changes with time. Moreover, we can assume that ρ, η, and D are constant and that there are no body forces (gravity is usually negligible in very small systems). In such a case, (4.13) collapses to

$$\nabla \cdot \vec{U} = 0$$

$$\vec{U} \cdot \nabla \vec{U} = -\frac{1}{\rho}\nabla P + \nu \Delta \vec{U} \tag{4.14}$$

$$\frac{\partial c}{\partial t} + \vec{U} \cdot \nabla c = D \Delta c + S$$

and if the hypothesis of a creeping flow is valid (i.e., Stokes approximation is justified), (4.14) collapses to the linear system, under the condition that the function S is well behaved

$$\nabla \cdot \vec{U} = 0$$

$$\nabla P = \eta \Delta \vec{U} \tag{4.15}$$

$$\frac{\partial c}{\partial t} + \vec{U} \cdot \nabla c = D \Delta c + S$$

In the next section, we analyze the case where viscosity and density are not constant.

4.2.5   Physical Properties as a Function of Concentration of Species

When the concentration in transported species becomes substantial, the buffer fluid properties are modified. We analyze next the influence of concentration on viscosity, density, and diffusion.

4.2.5.1 Viscosity

The viscosity of the buffer fluid is a function of the concentration of the suspension. In the buffer liquid flow, micro- and nanoparticles are each animated with a rotation motion, so that molecular vortices form inside the buffer fluid by entrainment of the surrounding fluid. The result is an increase in viscosity of the fluid. A relative viscosity may be defined as

$$\eta_r = \frac{\eta}{\eta_0} \tag{4.16}$$

where η_0 is the viscosity of the buffer fluid (with no particles in it), and η is apparent (real) viscosity and a specific viscosity by

$$\eta_{sp} = \frac{\eta - \eta_0}{\eta_0} \tag{4.17}$$

The specific viscosity changes with the volume fraction of particles defined by

$$\phi = \frac{volume\ of\ particles}{volume\ of\ the\ fluid} \tag{4.18}$$

For very dilute solutions in which it can be assumed that the transported particles are independent (i.e., do not interact), the specific viscosity is given by Huggins's law [3, 4]

$$\eta_{sp} = [\eta]\varphi \tag{4.19}$$

and the apparent viscosity is then

$$\eta = \eta_0\left(1 + [\eta]\varphi\right) \tag{4.20}$$

In (4.19) and (4.20), $[\eta]$ is the intrinsic viscosity, which depends on the type of the particles. For spherical particles, the value of $[\eta]$ is approximately $[\eta] = 5/2$ (Figure 4.8).

Relation (4.20) is valid only for relatively small volume fraction. It is well known that there is a packing fraction—of the order of $\phi = 0.65$—at which the viscosity becomes infinite and the carrier liquid cannot flow anymore. In such case, (4.20) is just the linear part of the more complete relation

$$\eta_{sp} = [\eta]\phi + k[\eta]^2\phi^2 + \dots \tag{4.21}$$

In (4.21) the constant k is called the Huggins's constant. Relation (4.21) is plotted in Figure 4.9.

A rapid and approximate calculation shows that for most applications in biotechnology, the volume fraction of target macromolecules or nanoparticles is small. Suppose a concentration of substance c_0 expressed in M (mole/liters). Its value in mole per cubic meters is $10^3 c_0$. If we note R_H the hydraulic radius of a single

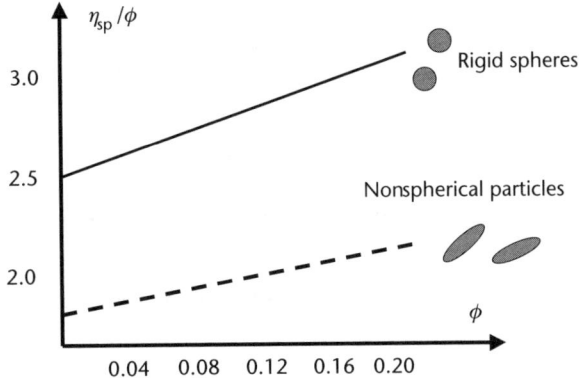

Figure 4.8 Relation between specific viscosity and volume fraction.

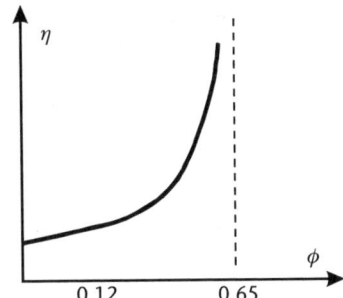

Figure 4.9 Apparent viscosity versus volume fraction of particles.

element of the substance, then the volume of this element is $V = 4/3\pi R_H^3$ and the volume fraction of the substance is

$$\varphi = 10^3 c_0 A V_{DNA} \tag{4.22}$$

where A is the Avogadro number ($A = 6.02 \cdot 10^{23}$). Typically for DNA analysis, the maximum concentration is $1\,\mu M$, and by taking an approximate hydraulic radius of $R_H = 20$ nm $= 20 \cdot 10^{-9}$ m, (4.22) gives the maximum volume fraction of $\phi = 0.02$. The value of the specific viscosity is then only 5%.

4.2.5.2 Diffusion Constant

In the preceding chapter, we introduced the Einstein relation for the diffusion of the particles

$$D = \frac{k_B T}{6\pi R_H \eta} \tag{4.23}$$

Substituting (4.19) in (4.23) yields

$$D(\phi) = \frac{k_B T}{6\pi R_H} \frac{1}{\eta_0 \left[1 + \dfrac{5}{2}\phi\right]} = \frac{D_0}{1 + \dfrac{5}{2}\phi} \tag{4.24}$$

The relative change in D is

$$\frac{D(\phi) - D_0}{D_0} \approx -\frac{5}{2}\phi$$

For the typical value $\phi = 0.02$, the relative change of D is -5%.

4.2.5.3 Density

The density of the buffer fluid is a function of the concentration. Using the definition of the volume fraction, the density is

$$\rho = \phi\rho_P + (1 - \phi)\rho_L \tag{4.25}$$

The relative change in ρ is

$$\frac{\rho - \rho_L}{\rho_L} = \phi\frac{\rho_P - \rho_L}{\rho_L}$$

For the typical value $\phi = 0.02$, $\rho_P = 2,000$ and $\rho_L = 1,000$ kg/m^3, the relative change is 2%.

4.2.5.4 Equations for Transport

In the case where the concentration effect on the fluid properties is not negligible, the advection-diffusion equation is not decoupled anymore. It has been seen in Section 4.2.5.1 that concentration is proportional to volume fraction. Taking advantage of the linearity of the transport equation, we obtain the following system for a creeping flow (Stokes' hypothesis)

$$\nabla \cdot \vec{U} = 0$$
$$\nabla P = \eta\Delta\vec{U} \tag{4.26}$$
$$\frac{\partial \phi}{\partial t} + \vec{U} \cdot \nabla\phi = D\Delta\phi + S$$

In the case where the concentration of species is sufficient to affect the properties of the liquid, we have to solve (4.26) using the constitutive relations

$$\eta(\phi) = \eta_0\left[1 + \frac{5}{2}\phi\right]$$
$$D(\phi) = \frac{k_B T}{6\pi R_H} \frac{1}{\eta_0\left[1 + \dfrac{5}{2}\phi\right]} = \frac{D_0}{1 + \dfrac{5}{2}\phi} \tag{4.27}$$

The system is now strongly coupled and no more linear due to terms of the form $\frac{1}{\phi}\Delta\phi$. The numerical solution requires a coupled multiphysics approach, where the unknowns are the vectors (u, v, w, P, ϕ) at each node of the computational domain. Also, because the system of equations is nonlinear, the use of a nonlinear solver is required.

4.2.6 Dimensional Analysis and Peclet Number

Let us start from the usual form of the diffusion-advection equation without source or sink.

$$\frac{\partial c}{\partial t}+u\frac{\partial c}{\partial x}+v\frac{\partial c}{\partial y}+w\frac{\partial c}{\partial z}=D\left[\frac{\partial^2 c}{\partial x^2}+\frac{\partial^2 c}{\partial y^2}+\frac{\partial^2 c}{\partial z^2}\right] \tag{4.28}$$

In this problem, there are four parameters: the velocity U_∞, the length scale L, the incoming concentration c_0, and the diffusion coefficient D. These four parameters contain three different units: m, s, kg (or mole). In such a case, Buckingham's Pi theorem implies that there is $4 - 3 = 1$ nondimensional number that governs the nondimensional equation and characterizes the phenomenon.

Suppose that we take for reference a velocity U_∞, a length scale L, and a concentration c_0. Relevant dimensionless variables may be defined as

$$c^* = \frac{c}{c_0}, x^* = \frac{x}{L}, y^* = \frac{y}{L}, z^* = \frac{z}{L}, u^* = \frac{u}{U_\infty}, v^* = \frac{v}{U_\infty}, w^* = \frac{w}{U_\infty}, t^* = \frac{t}{\frac{L}{U_\infty}} \tag{4.29}$$

The substitution of (4.29) in (4.28) yields

$$\frac{\partial c^*}{\partial t^*}+u^*\frac{\partial c^*}{\partial x^*}+v^*\frac{\partial c^*}{\partial y^*}+w^*\frac{\partial c^*}{\partial z^*}=\frac{D}{U_\infty L}\left[\frac{\partial^2 c^*}{\partial x^{*2}}+\frac{\partial^2 c^*}{\partial y^{*2}}+\frac{\partial^2 c^*}{\partial z^{*2}}\right] \tag{4.30}$$

As was expected from the Buckingham's theorem, only one dimensionless parameter appears in (4.30). This parameter represents the ratio of inertia to diffusion and is referenced by

$$Pe = \frac{U_\infty L}{D} \tag{4.31a}$$

The Peclet number is a key feature in the problems of dispersion under the action of diffusion and advection. We shall see in the following sections many examples where the Peclet number determines the solution of the advection-diffusion problem.

Note that the Peclet number may be written as a function of the Reynolds number

$$P_e = \frac{U_\infty L}{v}\frac{v}{D} = R_e S_c \tag{4.31b}$$

where S_c is the nondimensional Schmitt number.

4.2.7 Concentration Boundary Layer

In Chapter 1, we saw that the entrance length of microflows in capillary tubes is very short, because the hydrodynamic boundary layer develops and reaches very quickly the middle of the tube. It would be wrong to conclude that the same will happen to the mass transfer boundary layer. In fact, the picture looks like that of Figure 4.10, where the hydrodynamic flow is established but not the concentration field.

Figure 4.11 shows the calculated mass transfer boundary layer inside a typical detection chamber for a buffer fluid carrying DNA strands. The results are obtained by solving the advection-diffusion equation using a finite difference numerical scheme. It appears immediately that the vertical distance (1 mm) is too important because the compounds carried by the buffer flow (DNA strands) are mostly unaffected by the labeled wall and keep flowing through the chamber.

An estimate of the boundary layer thickness may be found by a dimensional analysis. The starting point is the advection-diffusion equation, assuming a steady-state concentration field, no source terms, and a two-dimensional problem

$$u\frac{\partial c}{\partial x} + v\frac{\partial c}{\partial y} = D\left[\frac{\partial^2 c}{\partial x^2} + \frac{\partial^2 c}{\partial y^2}\right] \tag{4.32}$$

Boundary layer hypothesis assumes that the vertical convection term is negligible as well as the axial diffusion term, so we are left with

$$u\frac{\partial c}{\partial x} = D\frac{\partial^2 c}{\partial y^2} \tag{4.33}$$

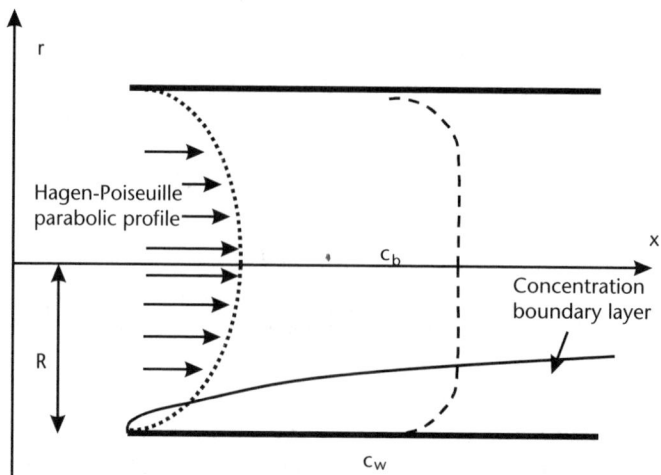

Figure 4.10   Schematic view of the flow and concentration boundary layer in a tube.

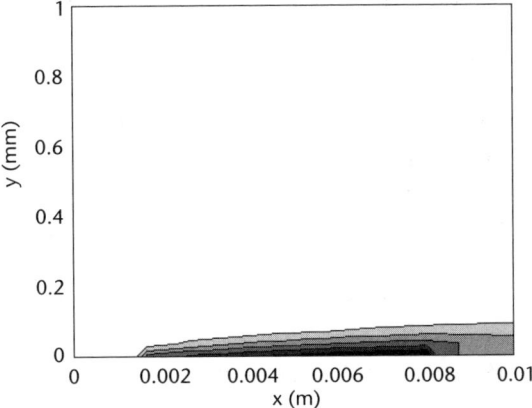

Figure 4.11 Typical mass transfer boundary layer on a partially labeled solid surface in an analysis chamber. Results obtained by solving the advection-diffusion equation with a finite difference algorithm (average velocity 1 mm/sec, labeled distance 6 mm).

After substitution of the Hagen-Poiseuille flow profile in (4.33), we obtain

$$\frac{3}{2}U\left[1-\frac{y^2}{d^2}\right]\frac{\partial c}{\partial x} = D\frac{\partial^2 c}{\partial y^2} \qquad (4.34)$$

where U is the average velocity, and d the half vertical distance between the plates (Figure 4.10). Now, we introduce the following scaling

$$x^* = \frac{x}{L}, y^* = \frac{y}{d}, c^* = \frac{c}{c_0} \qquad (4.35)$$

where L is a reference axial distance and c_0 is a reference concentration. Introducing the Peclet number, and taking into account (4.35), (4.34) can then be rewritten as

$$\frac{3}{2}P_e\frac{d}{L}\left[1-y^{*2}\right]\frac{\partial c^*}{\partial x^*} = \frac{\partial^2 c^*}{\partial y^{*2}} \qquad (4.36)$$

Now, we follow Levêques's approach and change the vertical origin

$$\tilde{y} = 1 - y^*$$

so that $\tilde{y}$ is zero at the wall. Equation (4.36) becomes

$$\frac{3}{2}P_e\frac{d}{L}\left[2\tilde{y}-\tilde{y}^2\right]\frac{\partial c^*}{\partial x^*} = \frac{\partial^2 c^*}{\partial \tilde{y}^2}$$

In the boundary layer, the distance $\tilde{y}$ is small, and we can assume $[2\tilde{y}-\tilde{y}^2] \approx 2\tilde{y}$. If we note $\tilde{\delta}$ the reduced boundary layer thickness, and note that at a distance from the wall $\tilde{\delta}$, the advection and diffusion terms are of the same order, and we obtain the scaling

$$\frac{\partial c^*}{\partial x^*} \approx \frac{c_w - c_0}{c_0 x^*}$$

$$\frac{\partial^2 c^*}{\partial \tilde{y}^2} \approx \frac{c_w - c_0}{c_0 \tilde{\delta}^2}$$

and finally

$$3P_e \frac{d}{x} \tilde{\delta}^3 \approx 1$$

Thus,

$$\frac{\delta}{d} \approx \left(\frac{x}{3d}\right)^{\frac{1}{3}} \frac{1}{P_e^{\frac{1}{3}}} \tag{4.37}$$

Take the case of Figure 4.11: $d = 1$ mm and $P_e = 10,000$. Equation (4.37) reduces to

$$\frac{\delta}{d} \approx 0.32 x^{\frac{1}{3}}$$

and for $x = 1$ cm, $\frac{\delta}{d} \approx 0.08$. This result agrees with the numerical result of Figure 4.11 and confirms the sketch of Figure 4.10.

To conclude, concentration boundary layers are often present in microfluidics, and they are to be taken into account for the comprehension and calculation of the transport phenomena.

4.2.8   Numerical Considerations

4.2.8.1   Boundary Layer

The small thickness of the concentration boundary layer has consequences on the numerical computation. It is essential to model precisely the mass transfer in the boundary layer because it determines the mass transfer to the solid wall. To do so, the computational grid needs to have at least two meshes in the boundary layer. It is then necessary to reduce the side of the geometrical elements in the boundary layer, especially where the boundary layer starts (Figure 4.12).

This reasoning often leads to very small values of the mesh size near the wall. Mesh size of the order of a few microns is usual.

4.2.8.2   The Question of the Mesh Size

A numerical Peclet number is defined, associated with the size of the mesh

$$P_{em} = \max\left(\frac{u\Delta x}{D}, \frac{v\Delta y}{D}\right) \tag{4.38}$$

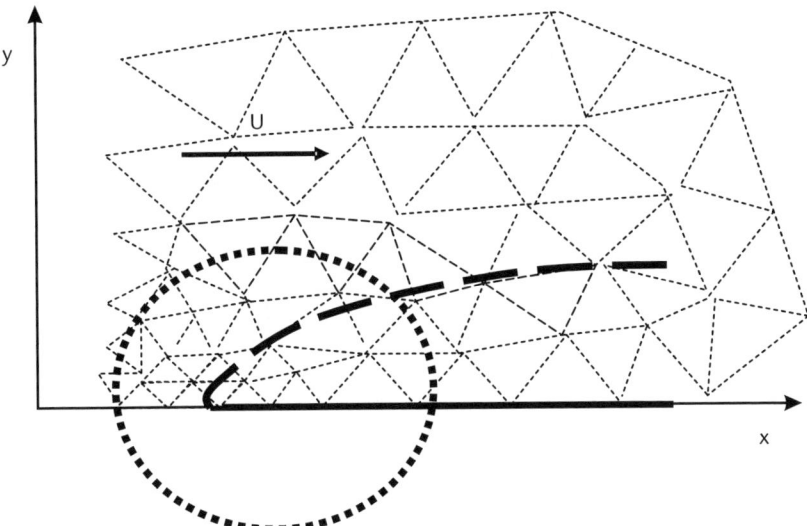

Figure 4.12 Nonuniform computational grid at the vicinity of the wall, especially at the beginning of the boundary layer.

For the numerical calculation to be stable, the numerical Peclet number must be smaller than 4

$$P_{em} = \max\left(\frac{u\Delta x}{D}, \frac{v\Delta y}{D}\right) < 4 \tag{4.39}$$

This condition requires the following limitations on the mesh size

$$\Delta x < \frac{4D}{u}$$
$$\Delta y < \frac{4D}{v} \tag{4.40}$$

Thus, the mesh size must be quite small. For $D = 10^{-10}$ m²/s and $u = 1$ mm/sec, one finds $\Delta x = 4\,\mu$m. This is approximately the same size as that of the meshes in the boundary layer. The problem could be rapidly intractable, especially if it is a three-dimension problem. In order to avoid this difficulty and relax the size of the mesh outside the boundary layer, an artificial diffusion is added to the diffusion constant in the meshes outside the boundary layers. If this additional diffusion is correctly chosen, it reduces the value of the mesh Peclet number—and allows for larger meshes—and does not affect the solution very much since it applies in a domain where the gradient of concentration is small. It is typical to choose the value of the added diffusion by

$$D_{add,x} = \beta\frac{u\Delta x}{2} \tag{4.41}$$

where β is a coefficient depending on the problem. Such a formulation respects the condition imposed by the boundary layer because u and Δx are small in the boundary layer. Note that the equivalent numerical diffusion coefficient is anisotropic

$$
D_{num} = \begin{bmatrix} D + \beta \dfrac{u\Delta x}{2} & 0 & 0 \\ 0 & D + \beta \dfrac{v\Delta y}{2} & 0 \\ 0 & 0 & D + \beta \dfrac{w\Delta z}{2} \end{bmatrix} \tag{4.42}
$$

4.2.8.3 Time Step

Time step and mesh size are not independent. If an explicit formulation is chosen, the Courant condition yields

$$
u\Delta t < \Delta x \tag{4.43}
$$

and, introducing the limitations on the mesh size,

$$
u\Delta t < \Delta x < \min\left(\frac{4D_{num,x}}{u}, \frac{\delta(x)}{2} \right) \tag{4.44}
$$

because we want at least two meshes in the boundary layer $\delta(x)$.

Thus,

$$
\Delta t < \min\left(\frac{4D_{num,x}}{u^2}, \frac{\delta(x)}{2u} \right) \tag{4.45}
$$

This condition is usually very restrictive. Typically it yields values of the time step of the order of 10^{-2} seconds. Keeping in mind that the duration of a biological reaction is of the order of 10 minutes to 10 hours, the number of the time step is quite large. This is why semi-implicit or implicit algorithms are preferred for solving transient concentration problems in microsystems.

To conclude, it is important for the numerical modeling of mass transfer in microsystems to have very small meshes in the boundary layers, to use an added diffusion coefficient to avoid numerical instabilities, and to choose a semi-implicit or implicit solution scheme.

4.2.9 Taylor-Aris Approach

4.2.9.1 Taylor-Aris Model

At a time when flow velocimetry methods were developing, it was found that the measurement of flow velocity inside tubes may be achieved using small particles. The principle was to introduce a radioactive substance or an electrolyte in the flow at a certain cross section and to follow its translation inside the pipe. However, it appeared immediately that the substance introduced into the stream diffused at the same time as it moved with the fluid. A model was then developed by Taylor [5]—and completed by Aris—to take into account in a simple manner both diffusion and advection. This model, although it was developed in the 1950s, has found a recent renewal in biotechnology where advection and diffusion of particles and macromolecules carried by a fluid is an everyday concern.

The basic idea behind Taylor's approach is that it is not possible to superpose advection and diffusion for a liquid flowing in a pipe under the action of pressure gradient: it is not correct to assume a mere translation where the particles would move as a whole with the fluid and diffuse with their usual diffusion coefficient. In a Lagrangian coordinate system moving at the average velocity of the fluid, the particles diffuse much more than is predicted by the usual diffusion theory. As will be seen, this is due to the radial gradient of velocity in the flow. But the strength of Taylor's approach is to show that a superposition may be done by allowing the particles to move at the average velocity of the fluid and using an effective or apparent diffusion coefficient for the particle—larger than the usual coefficient derived from Einstein's formula.

Suppose a capillary tube of radius R in which a liquid flows at a mean velocity U carrying particles with a concentration c (Figure 4.13).

The advection-diffusion equation written under an axisymmetric (x,r) form is

$$\frac{\partial c}{\partial t} + u(r)\frac{\partial c}{\partial x} = D\left[\frac{\partial^2 c}{\partial x^2} + \frac{1}{r}\frac{\partial c}{\partial r} + \frac{\partial^2 c}{\partial r^2}\right] \tag{4.46}$$

The local velocity $u(r)$ is given by the Hagen-Poiseuille solution

$$u(r) = 2U\left(1 - \frac{r^2}{R^2}\right) \tag{4.47}$$

Note that the velocity is zero at the wall and is maximum and equal to $2U$ on the central axis. The term $\frac{\partial^2 c}{\partial x^2}$ in (4.46) may be omitted if it is assumed that the axial change in concentration is much less than the radial change. After substituting (4.47) in (4.46), we obtain

$$\frac{\partial c}{\partial t} + 2U\left(1 - \frac{r^2}{R^2}\right)\frac{\partial c}{\partial x} = D\left[\frac{1}{r}\frac{\partial c}{\partial r} + \frac{\partial^2 c}{\partial r^2}\right] \tag{4.48}$$

In a Lagrangian system of coordinates moving with the average velocity of the fluid, (4.48) becomes

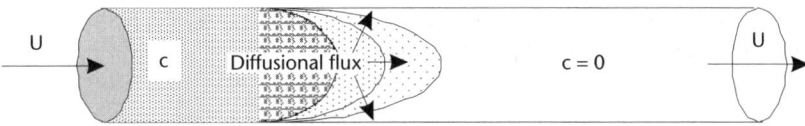

Figure 4.13   Schematic view a cylindrical capillary tube and the diffusion front of concentration.

$$\left[\frac{\partial c}{\partial t} + U\frac{\partial c}{\partial x}\right] + U\left(1 - \frac{2r^2}{R^2}\right)\frac{\partial c}{\partial x} = D\left[\frac{1}{r}\frac{\partial c}{\partial r} + \frac{\partial^2 c}{\partial r^2}\right] \tag{4.49}$$

The boundary condition at the capillary surface is

$$\frac{\partial c}{\partial r}\bigg|_{r=R} = 0$$

indicating that there is no flow of substance through the wall of the tube. The reasoning may be decomposed in two steps: First, assume temporarily that the concentration gradient along the x-axis is constant

$$\frac{\partial c}{\partial x} = \frac{\partial \bar{c}}{\partial x} = cste \tag{4.50}$$

Here $\bar{c}$ is the average concentration over the cross section of the tube, defined by

$$\bar{c} = \frac{1}{S}\int_0^R c2\pi r dr = \frac{2}{R^2}\int_0^R c r dr \tag{4.51}$$

Along the axis of the capillary, the concentration must have a finite value. Then, the solution of (4.49) may be written as

$$c = c_0 + \frac{UR^2}{4D}\frac{\partial \bar{c}}{\partial x}\left(\frac{r^2}{R^2} - \frac{1}{2}\frac{r^4}{R^4}\right) \tag{4.52}$$

$$\frac{\partial c}{\partial t} = 0$$

where c_0 is the concentration on the capillary axis ($r = 0$). In all of the following, remember that the time derivative is taken in the moving coordinate system. By using (4.51) and (4.52), we find the following relation between $\bar{c}$ and c

$$c = \bar{c} + \frac{UR^2}{4D}\frac{\partial \bar{c}}{\partial x}\left(-\frac{1}{3} + \frac{r^2}{R^2} - \frac{1}{2}\frac{r^4}{R^4}\right) \tag{4.53}$$

$$\frac{\partial c}{\partial t} = 0$$

The total mass flow of species through any cross section of the capillary is given by

$$Q = \int_0^R uc2\pi r dr = -\pi R^2\left(\frac{R^2 U^2}{48D}\right)\frac{\partial \bar{c}}{\partial x}$$

The corresponding flux is

$$j = \frac{Q}{\pi R^2} = -\left(\frac{R^2 U^2}{48D}\right)\frac{\partial \bar{c}}{\partial x} \tag{4.54}$$

Relation (4.54) shows that the mass flux of the concentration $\bar{c}$ has the form of Fick's law with the effective diffusion coefficient

$$D_{eff} = \frac{R^2 U^2}{48D} \qquad (4.55)$$

Now, if we assume that the axial concentration gradient is no more constant $\frac{\partial \bar{c}}{\partial x} \neq cste$, we are entitled to write

$$\frac{\partial \bar{c}}{\partial t} = -\frac{\partial j}{\partial x} \qquad (4.56)$$

and we obtain

$$\frac{\partial \bar{c}}{\partial t} = D_{eff} \frac{\partial^2 \bar{c}}{\partial x^2} \qquad (4.57)$$

This last equation shows that the average concentration is governed—in the moving coordinate system—by the usual diffusion equation for a stationary medium with the effective diffusion coefficient D_{eff} defined by (4.57).

A numerical example of the Taylor model is shown in Figure 4.14. The diffusion front progresses with the flow and, at the same time, smears out due to molecular diffusion.

One can make a simplified picture of the situation (Figure 4.15). In the case of the Taylor-Aris method, the concentration front has a parabolic shape; thus, the diffusional surface is much larger than if the front were flat. For this reason, the

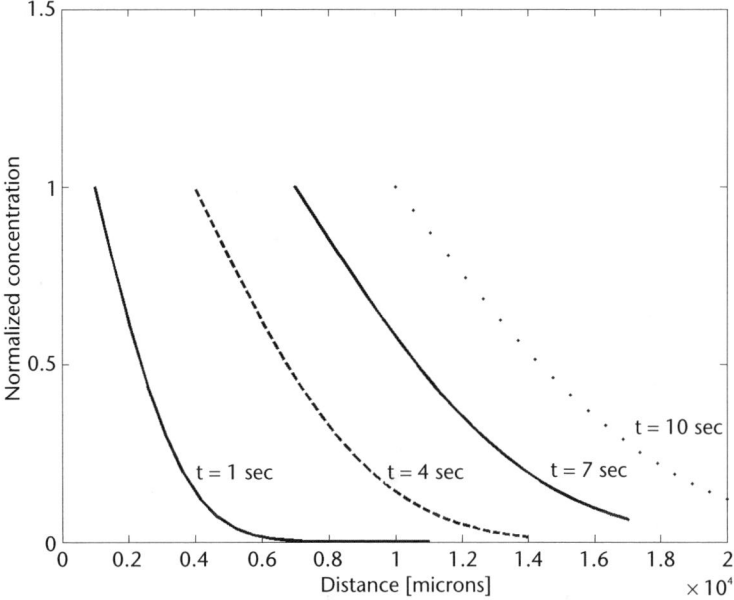

Figure 4.14   Concentration profiles obtained with (5.57) for a velocity of 1 mm/sec in a channel of radius $R = 100\ \mu m$.

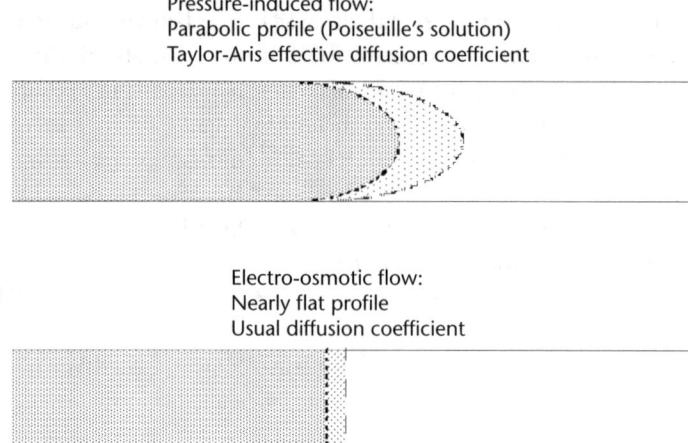

Figure 4.15 Comparison of advection/diffusion between a parabolic profile (pressure-induced flow) and a flat profile (electro-osmotic flow). In the moving coordinate system, diffusion is more important for the parabolic profile than for the flat profile.

same situation for a flow induced by electro-osmosis presents less diffusion because the diffusional front is nearly flat (Figures 4.15 and 4.16).

4.2.9.2 Conditions of Applicability of the Method

One question remains: What are the conditions for the Taylor-Aris approach to be valid?

First, we have neglected the axial diffusional $\dfrac{\partial^2 c}{\partial x^2}$ in (4.46) in front of the radial

diffusional term $\dfrac{1}{r}\dfrac{\partial c}{\partial r} + \dfrac{\partial^2 c}{\partial x^2}$. This radial term is important if the radial velocity

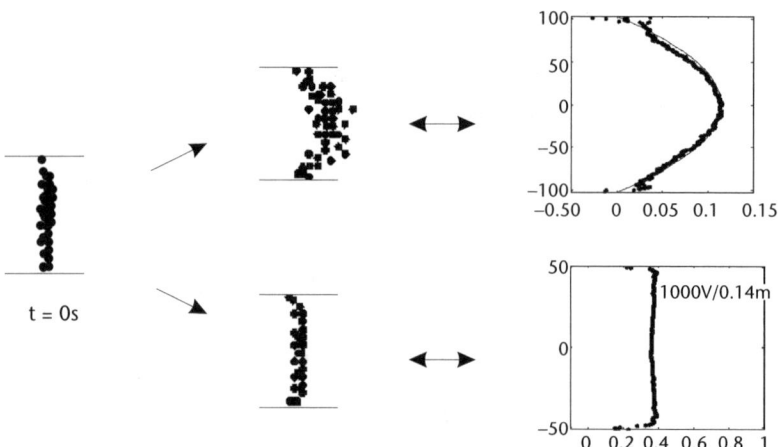

Figure 4.16 Experimental view of diffusing particles in a Poiseuille flow (top) and in an electro-osmotic flow (bottom). Experimental velocity profile in both cases. Dispersion is reduced if the diffusion front is flat.

gradient is large (the velocity profile varies from $2U$ to 0 along the radius)—that is, if the average velocity is sufficiently important. In such a case, we have

$$D << D_{eff}$$

which is equivalent to

$$\frac{UR}{D} >> 7$$

or

$$Pe = \frac{U\Phi}{D} >> 14 \tag{4.58}$$

Relation (4.58) requires that the Peclet number should be larger than 14.

Second, the concentration c must be a slowly changing function of x (we have supposed in a first approximation that $\dfrac{\partial c}{\partial x} = \dfrac{\partial \bar{c}}{\partial x} = cste$); from (4.53), we derive

$$\frac{\partial c}{\partial x} = \frac{\partial \bar{c}}{\partial x} + \frac{UR^2}{4D}\frac{\partial^2 \bar{c}}{\partial x^2}\left(-\frac{1}{3} + \frac{r^2}{R^2} - \frac{1}{2}\frac{r^4}{R^4}\right) \approx \frac{\partial \bar{c}}{\partial x} \tag{4.59}$$

To be satisfied, (4.59) requires

$$\frac{\partial \bar{c}}{\partial x} >> \frac{UR^2}{4D}\frac{\partial^2 \bar{c}}{\partial x^2}$$

If L is the length over which a noticeable change in c can occur, we may approximate the gradients by

$$\frac{\partial \bar{c}}{\partial x} \approx \frac{\bar{c}}{L}$$

$$\frac{\partial^2 \bar{c}}{\partial x^2} \approx \frac{\bar{c}}{L^2}$$

then the preceding inequality may be written as

$$\frac{LD}{UR^2} >> 1 \tag{4.60}$$

and, taking into account (4.58),

$$\frac{L}{R} >> \frac{UR}{D} >> 7 \tag{4.61}$$

To these two conditions, we add the condition for a laminar flow

$$Re = \frac{U\Phi}{\nu} \ll 2,000 \qquad (4.62)$$

The three conditions (4.58), (4.60), and (4.62) give the limits of applicability of (4.57) for a cylindrical tube.

The value $D_{eff} = \dfrac{R^2 U^2}{48D}$ refers to cylindrical tubes only. Another value of D_{eff} can be obtained by the same reasoning for a flow limited by two parallel plates [2]

$$D_{eff} = \frac{H^2 U^2}{210\,D} \qquad (4.63)$$

where H is the half-distance between the two plates. The general form for the equivalent diffusion coefficient using the Peclet number is

$$\frac{D_{eff}}{D} = \frac{1}{\beta} Pe^2 \qquad (4.64)$$

where β is a geometric coefficient depending on the shape of the cross section. Note that the lower limit for the Peclet number (4.58) is not universal and depends on the cross section of the capillary tube. The general formulation would be

$$Pe > 2\sqrt{\beta} \qquad (4.65)$$

4.2.9.3 Applicability to Microflows

The conditions defined in the preceding section are very often satisfied by microflows in microsystems. Generally, tube diameters are in the range $100\ \mu m$–1 mm, and velocities vary from a few microns per second to a few millimeters per second. So, for a water-based flow, the Reynolds number is smaller than 10, and the flow is strongly laminar. Because diffusion coefficients are very small (seldom larger than $10^{-10}\ m^2/s$), the Peclet number is at least of the order of 10 and most of the time larger than that. Finally, the condition $L/R > 7$ requires the microchannel to be sufficiently long, which is very often the case.

When applicable, the Taylor-Aris approach is very simple and useful. It has many applications in chemistry [6, 7] and for immunoassyas, as we will see in Chapter 5.

4.2.10   Distance of Capture in a Capillary

In biotechnology, the capture of particles advected by a carrier fluid flowing inside a capillary tube is a fundamental question. For example, we may want to dimension an annular surface to capture a certain type of particle in the carrier fluid. In this section, we do not deal with the capture itself (this will be done in Chapter 5), but with the contact of the particles with the solid wall.

4.2.10.1 Analytical Approach

A very simple approximation may be done by comparing axial convection to radial diffusion. Particles near the wall are not going to have very long axial displacement before impacting the wall, whereas the particles initially located at the center of the capillary will follow the longest trajectory before impacting the wall (Figure 4.17).

The average maximum time necessary for a particle to diffuse radially to the wall is

$$\tau \approx \frac{R^2}{4D} \tag{4.66}$$

During this time τ of radial diffusion, the particle has moved along the axial direction on a distance of

$$L \approx 2V\tau \approx \frac{VR^2}{2D} \tag{4.67}$$

The coefficient 2 in (4.67) corresponds to the maximum Hagen-Poiseuille velocity $2V$ at the center of the tube.

With this reasoning, it can be deduced that after a distance L, statistically all of the particles will have impacted the wall at least one time. If the wall property is such that there is a capture upon contact, then (4.67) is a good approximation of the dimension of the surface of capture. Equation (4.67) may be rewritten in a nondimensional form

$$\frac{L}{R} \approx \frac{VR}{2D} = \frac{1}{2}Pe \tag{4.68}$$

where Pe is the Peclet number. We see here another significance of the Peclet number.

Suppose now that the flow rate is imposed by the experimental conditions (syringe pump for example), and using the relation between the average velocity and the flow rate

$$Q = SV = \pi R^2 V$$

The length L becomes

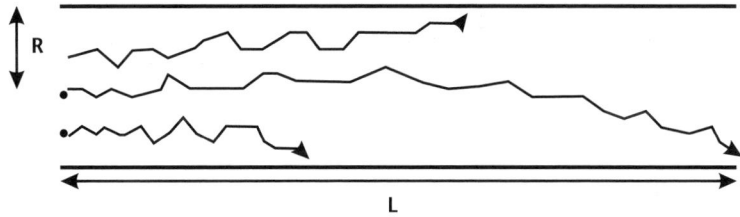

Figure 4.17 Sketch of particle trajectories depending on their starting point.

$$L \approx \frac{Q}{2\pi D} \qquad (4.69)$$

and the radius of the capillary does not appear in the equation anymore. For a given mass flow rate, if the radius is increased, the fluid velocity decreases: The convection distance is then shorter but the radial diffusion distance is longer. The two effects balance each other (Figure 4.18).

4.2.10.2 Numerical Approach

Confirmation of the preceding approach can be done by using numerical modeling. We have two choices to set up a numerical approach. Either, a Hagen-Poiseuille flow formulation can be introduced in the advection-diffusion equation and we have to solve in a cylindrical geometry (r,z) the equation

$$\frac{\partial c}{\partial t} + 2U\left(1 - \frac{r^2}{R^2}\right)\frac{\partial c}{\partial r} = D\left[\frac{\partial^2 c}{\partial z^2} + \frac{1}{r}\frac{\partial c}{\partial r} + \frac{\partial^2 c}{\partial r^2}\right] \qquad (4.70)$$

or we start from the Navier-Stokes equations and solve the system of equations

$$\frac{\partial u}{\partial z} + \frac{1}{r}\frac{\partial rv}{\partial r} = 0$$

$$u\frac{\partial u}{\partial z} + v\frac{\partial u}{\partial r} = -\frac{1}{\rho}\frac{\partial P}{\partial z} + \frac{\mu}{\rho}\left[\frac{\partial^2 u}{\partial z^2} + \frac{1}{r}\frac{\partial u}{\partial r} + \frac{\partial^2 u}{\partial r^2}\right]$$

$$u\frac{\partial v}{\partial z} + v\frac{\partial v}{\partial r} = -\frac{1}{\rho}\frac{\partial P}{\partial r} + \frac{\mu}{\rho}\left[\frac{\partial^2 v}{\partial z^2} + \frac{1}{r}\frac{\partial v}{\partial r} + \frac{\partial^2 v}{\partial r^2}\right] \qquad (4.71)$$

$$\frac{\partial c}{\partial t} + u\frac{\partial c}{\partial z} + v\frac{\partial c}{\partial r} = D\left[\frac{\partial^2 c}{\partial z^2} + \frac{1}{r}\frac{\partial c}{\partial r} + \frac{\partial^2 c}{\partial r^2}\right]$$

Equations (4.70) and (4.71) treat the case of a permanent buffer fluid flow moving inside a cylindrical tube of constant diameter, with a transient concentration advected by the fluid. Thus, there is no transient terms $(\frac{\partial u}{\partial t} = \frac{\partial v}{\partial t} = 0)$ in the

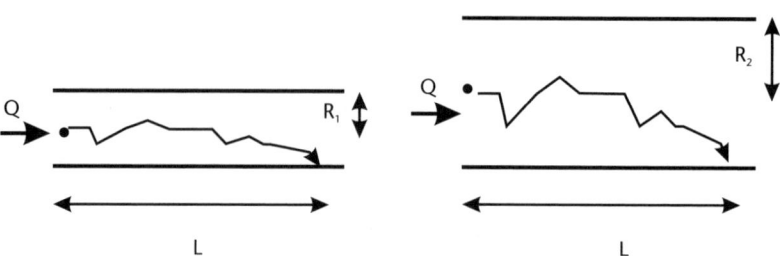

Figure 4.18 The length L is the same regardless R if the flow rate is identical in both cases.

Navier-Stokes equations, but the term $\dfrac{\partial c}{\partial t} \neq 0$ is present in the advection-diffusion equation.

This system can be decoupled if the velocity field does not depend on the concentration—this is usually the case since the concentration is assumed to be small and there are no considerable aggregation regions. In such a case, the system can be solved in two steps: (1) hydrodynamics; and (2) advection of particles. This process has the advantage of mobilizing less computational memory, but it requires loading a file with the results of the velocity field and transmitting it to step 2. If enough computational memory is available and if the same meshing of the computational domain is possible, it may be advantageous to solve the system as a totally coupled system—even if it is not the case—and have only one (large) matrix to invert in the numerical algorithm.

This last coupled approach was performed using the FEMLAB numerical software [8]. Figure 4.19 shows the results of the calculation: the velocity field has the parabolic shape of the Hagen-Poiseuille's solution except at the entrance where the flow is not yet established. Concentration flow lines have been plotted, confirming that the particles entering in the middle of the channel have the longest axial trajectory (a flow line is the line defined by the gradient of the concentration function and collinear to the Fick's concentration flux). The flow lines are perpendicular to the side walls because the boundary condition is $c = 0$ at the walls.

Equation (4.70) is much easier to solve. Of course most numerical software aimed at the resolution of PDEs will do the job, but a straightforward discretization can be performed and implemented with mathematical software like MATLAB if the geometry of the computational domain is simple (rectangle or cylinder). A finite volume method can be set up by using a semi-implicit Crank-Nicholson discretization scheme.

Figure 4.20 shows the indices for r and z; the discretized equation at the node (i,j) is

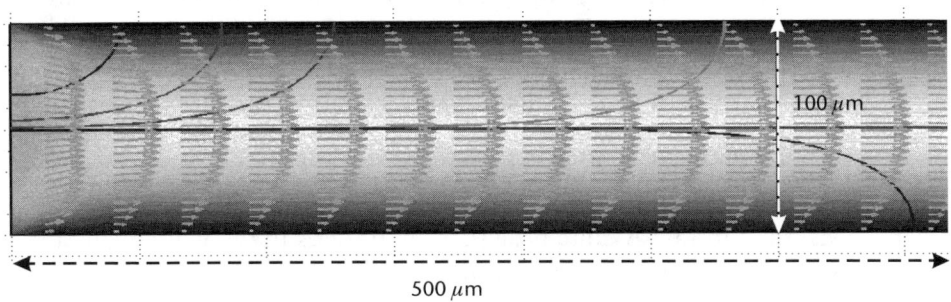

100 μm

500 μm

Figure 4.19   Results of the numerical modeling: velocities are indicated by the arrows and show a Poiseuille parabolic profile—except at the entrance of the channel where the flow is totally established. A few flow lines for concentration have been plotted, proving that the distance of capture depends on the initial position of the particle.

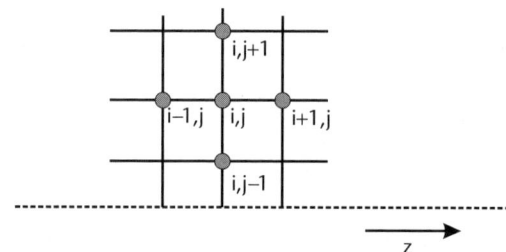

Figure 4.20 Schematic view of the computational nodes and grid.

$$\frac{c_{i,j}^{n+1} - c_{i,j}^{n}}{\Delta t} + U\left(1 - \frac{r_j^2}{R^2}\right)\left[\frac{c_{i,j+1}^{n+1} - c_{i,j}^{n+1}}{\Delta r} + \frac{c_{i,j+1}^{n} - c_{i,j}^{n}}{\Delta r}\right]$$

$$= \frac{D}{2}\left[\frac{c_{i+1,j}^{n+1} - 2c_{i,j}^{n+1} + c_{i-1,j}^{n+1}}{(\Delta z)^2} + \frac{c_{i,j+1}^{n+1} - 2c_{i,j}^{n+1} + c_{i,j-1}^{n+1}}{(\Delta r)^2} + \frac{1}{r_j}\frac{c_{i,j+1}^{n+1} - c_{i,j}^{n+1}}{\Delta r}\right] \qquad (4.72)$$

$$+ \frac{D}{2}\left[\frac{c_{i+1,j}^{n} - 2c_{i,j}^{n} + c_{i-1,j}^{n}}{(\Delta z)^2} + \frac{c_{i,j+1}^{n} - 2c_{i,j}^{n} + c_{i,j-1}^{n}}{(\Delta r)^2} + \frac{1}{r_j}\frac{c_{i,j+1}^{n} - c_{i,j}^{n}}{\Delta r}\right]$$

with the precaution that on the centerline $(r = 0)$, the terms in $1/r$ should be removed (because $\frac{\partial c}{\partial r} = 0$). More on the numerical algorithm for the solution of the advection-diffusion equation may be found in [9].

The results for a concentration "burst" of particles have been plotted in Figure 4.21.

4.2.11 Determination of the Diffusion Coefficient

Measurement of liquid phase diffusion coefficients is based on the observation of the spreading of the diffusing substance/solute. Diffusion coefficients are very small, and measurements in macroscopic systems are not reliable because of uncontrolled fluctuations of velocity. Because they are very laminar, easily controllable, and not distorted, microflows are well adapted to the measurements of liquid phase diffusion coefficients. The experimental principle is based on the mixing of the buffer liquid alone and the buffer liquid with a concentration of the targeted substance, as shown in Figure 4.22.

In the "diffusing zone," the streamlines are parallel and directed along the x-axis; the substance/solute progressively diffuses in the y-direction, and there is a growing distance $\delta(x)$ of concentration gradient. It can be shown that the concentration profile is given by [2, 10, 11]:

$$c(x,y) = \frac{1}{2}c_0\left(1 - erf\frac{y\sqrt{U}}{\sqrt{4Dx}}\right) \qquad (4.73)$$

where U is the mean flow velocity. A fit of (4.73) with the experimental concentration profile produces the value of the diffusion coefficient D (Figure 4.23).

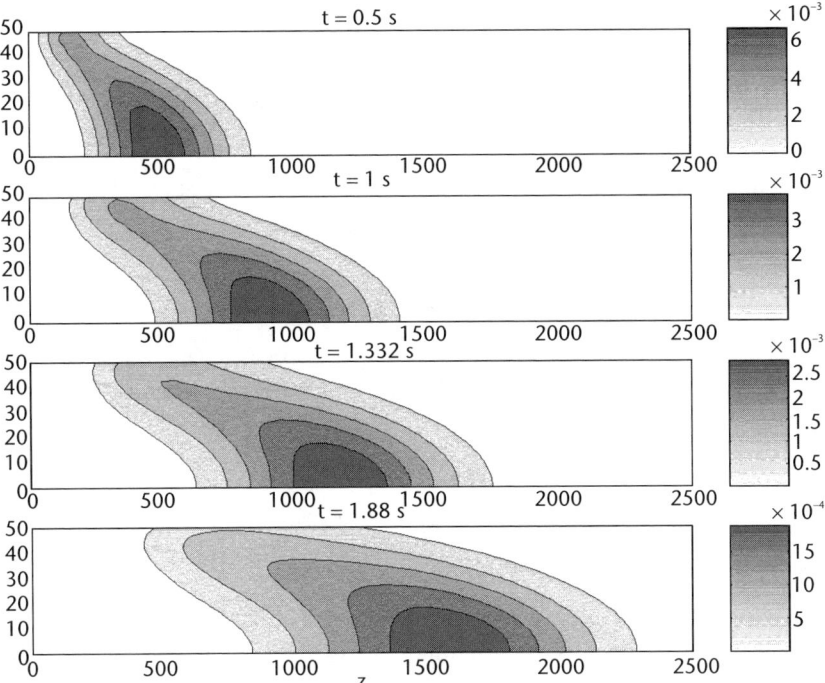

Figure 4.21 Contour plot of concentration at four different times after injection. The calculation has been performed in a (r,z) coordinate system and only half of the channel is represented. The solid wall is located at the top of each picture. The velocity field is given by the Hagen-Poiseuille solution.

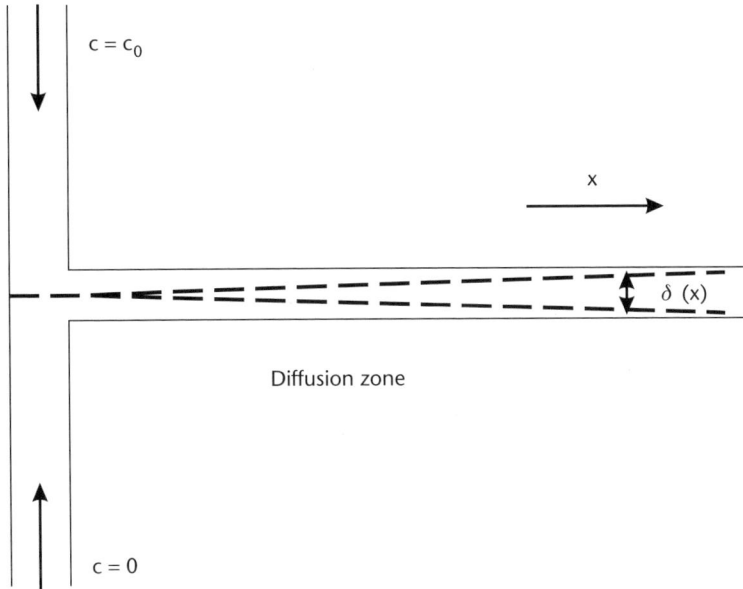

Figure 4.22 Experimental principle for the measurement of the diffusion coefficient.

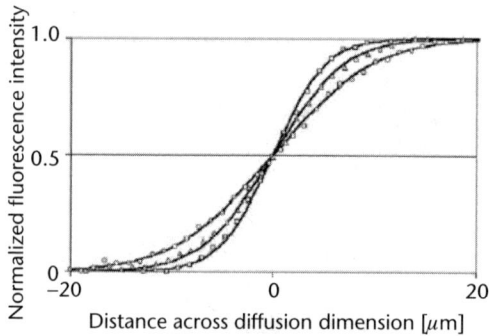

Figure 4.23 Concentration profile of diffusing species marked with fluorescent markers at three different locations in the channel. (*From:* [10]. © 2001 Biophysical Journal. Reprinted with permission.)

4.2.12 Mixing of Fluids

4.2.12.1 Introduction

Mixing of liquid constituents is a major problem in biochips and bioMEMS. The high degree of laminarity of microflows delays the mixing of constituents. Two different liquids can flow side by side for a rather long distance before complete mixing occurs. This is a real difficulty for the miniaturization and compactness expected from a biochip. It is always possible to design a fluidic system with zigzags to have more capillary length in a compact surface, as sketched in Figure 4.24 and according to the photograph of the microsystem of Figure 4.25.

However, another difficulty is linked to poor mixing in biochips: The time required to execute the different biological processes may be important. Besides miniaturization, another advantage expected from microsystems is the reduction of reaction time. It is then often necessary to accelerate the mixing process. Many different micromixers have been developed. They fall into two categories: active and passive. Active micromixers use actuated devices, like piezoelectric actuated

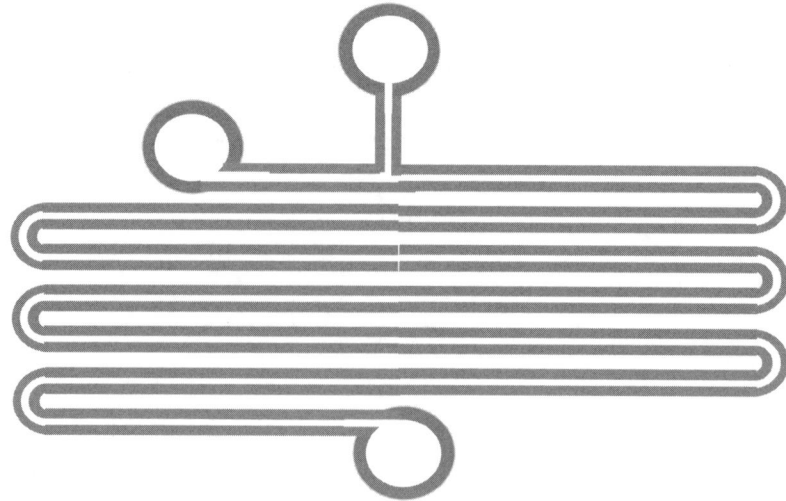

Figure 4.24 Sketch of compaction of a long channel on a biochip to realize micromixing.

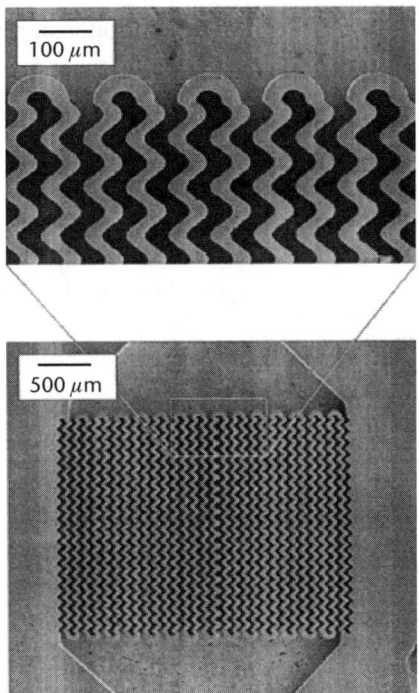

Figure 4.25   Detailed view of a micromixer based on a long mixing length. (*From*: [12]. © 2005 American Chemical Society. Reprinted with permission.)

membranes [13], whereas passive micromixers use only the energy of the flow and special morphological design promoting mixing of the compounds [2, 13].

4.2.12.2   Parallel Flows

In the preceding section, we established a relation for the concentration diffusion between parallel flows. Let us analyze this relation. Figures 4.26 and 4.27 show the concentration distribution in a half channel. The difficulty of mixing the flows is obvious: For typical dimensions, velocity, and liquids, the mixing length is very long and often not acceptable for compact microsystems.

It is interesting to estimate the length L at which a relative concentration of $c/(c_0/2) = 90\%$ at the wall is reached. Using the inverse *erf* function and (4.73), and $y = R$, we find

$$\frac{R\sqrt{U}}{\sqrt{4DL}} = erfinv(0.1) = 0.0889$$

where the function *erfinv* is the inverse of the *erf* function. Thus, the length L is

$$\frac{L}{R} = \frac{1}{4\left(erfinv(0.1)\right)^2}\frac{RU}{D} \approx 32\frac{UR}{D} \tag{4.74}$$

and we see that the mixing length is a function of the Peclet number

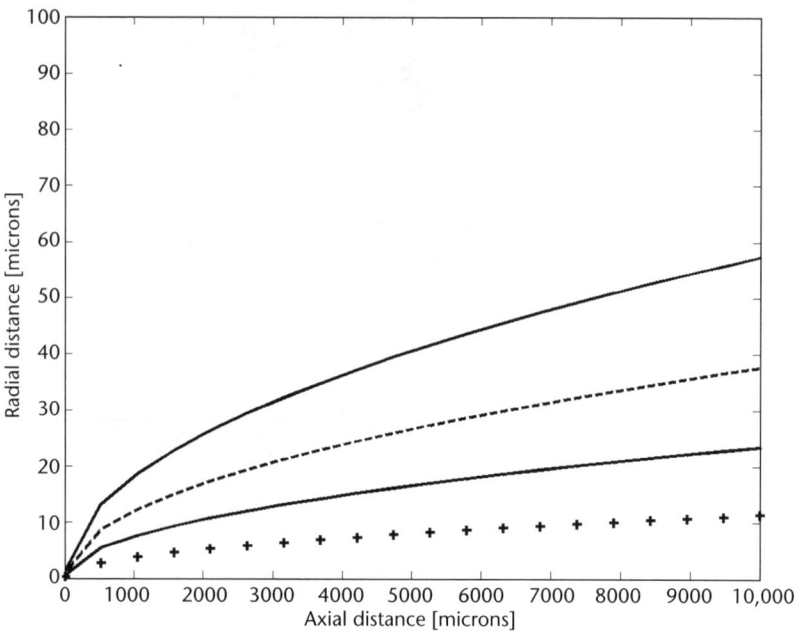

Figure 4.26 Contour plot of concentration c/c_0 from (4.73) for $R = 100$ mm, $L = 1$ cm, $D = 10^{-10}$ m^2/s, and $U = 1$ mm/sec. The four plots correspond to $c/c_0 = 0.1, 0.2, 0.3$, and 0.4.

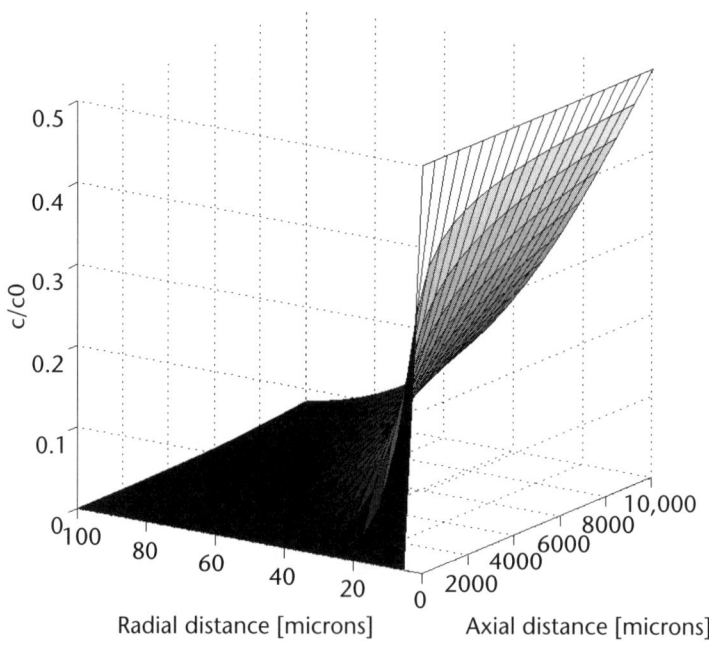

Figure 4.27 Perspective view of (4.73) showing the concentration profiles in the channel for $R = 100$ mm, $L = 1$ cm, $D = 10^{-10}$ m^2/s, and $U = 1$ mm/sec. The four plots correspond to $c/c_0 = 0.1, 0.2, 0.3$, and 0.4.

$$\frac{L}{R} \approx 32Pe \qquad (4.75)$$

It is worth comparing the mixing length from (4.75) with the entrance length in a channel (1.25) established in Chapter 1. There is an obvious similarity if we write the two relations as

$$L_{mix} \approx 32RPe_R \approx 8(2R)Pe_D$$
$$h \approx 0.04(2R)Re_D \qquad (4.76)$$

In the case of the hydrodynamic entrance, it is the action of the viscosity that homogenizes the flow to reach a fully developed flow. In the case of the mixing of microflows, it is the action of the molecular diffusion that homogenizes concentration. The physics of the two problems is similar: In both cases, the phenomenon is linked to the growth of a boundary layer. It is no wonder then that the form of the two equations (4.76) is similar. The major difference is that the cinematic viscosity is of the order of 10^{-6} m²/s, whereas the diffusion coefficient is only 10^{-10} m²/s. The hydrodynamic entrance is then very short, whereas the mixing length is very long.

4.2.12.3   Improving the Mixing of Parallel Flows

Usually in bio-MEMS, the flow rate is imposed and, from (4.75), the only way to reduce the mixing length for parallel flows is to reduce the radius R. This will have a considerable effect since the mixing length L_{mix} varies as the square of R. If the radius is reduced and if the flow rate is imposed, a micromixer design consists of dividing the flow in multiple branches. A typical design based on the reduction of the channel cross section is shown in Figure 4.28. Another design is that of Figure 4.29, showing a micromixer based on the principle of flow lamination and hydrodynamic focusing.

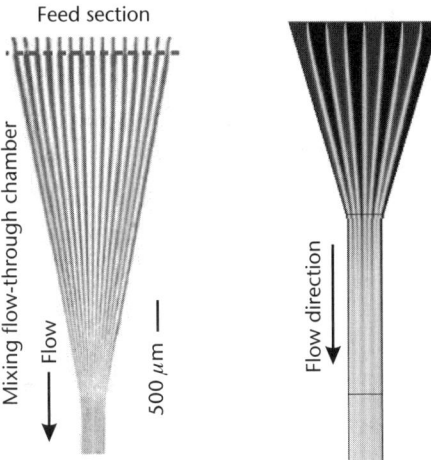

Figure 4.28   Schematic view of a parallel micromixer. The two fluids are mixed together in channels of reduced cross sections.

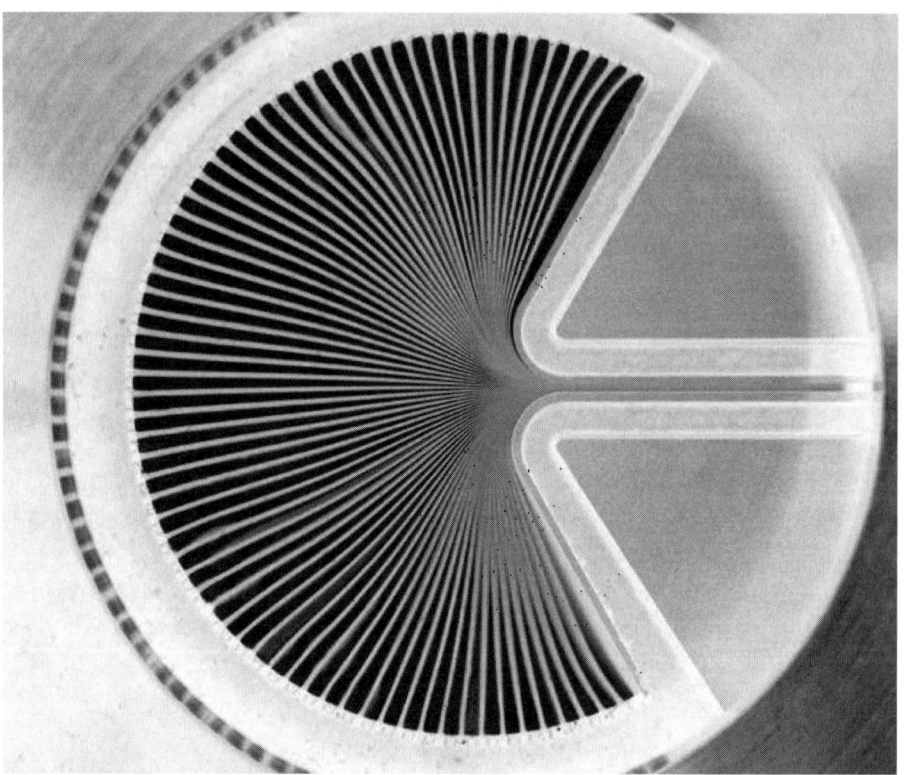

Figure 4.29 Institut Microtech (IMM) micromixer based on the multilamination principle combined with hydrodynamic focusing. In this picture, the flow enters by the many openings into the outer circle and converge in a very laminar flow toward a narrow channel outlet. Mixing of the fluid is realized in this narrow outlet channel. (*From:* [14]. © 2005 Wiley-VCH. Reprinted with permission from Chemical Engineering and Technology.)

4.2.12.4 Chaoting Mixing

The principle of chaoting mixing is based on successive stretching and bending of fluid streamlines. In Figure 4.30, we show how a domain of liquid 1 immersed in liquid 2 is deformed by chaoting mixing.

If the succession of stretching and folding is done rapidly, at a time scale much smaller than that of diffusion, the time interval for the stretching- folding process may be neglected, and it is possible to compare the corresponding mixing zones in Figure 4.31. Suppose a time scale τ, and then diffusion length is approximately

$$\lambda \approx \sqrt{4D\tau} \tag{4.77}$$

If we chose the value of τ so that the distance λ is approximately the distance between the folded regions, the mixing zones at the time τ corresponding to Figure 4.30 are shown in Figure 4.31. The mixing zone after chaoting mixing has a more important surface than the original one. This proves the efficiency of chaoting mixing. The important thing here is that the stretching/folding deformations are performed in a short time compared to diffusion time.

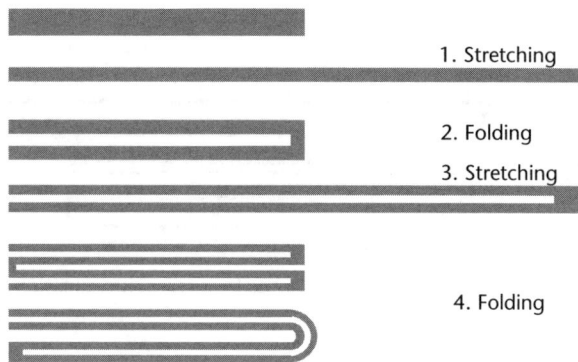

Figure 4.30   Principle of chaoting mixing by successive stretching and folding of flow domain. (*After:* [2].)

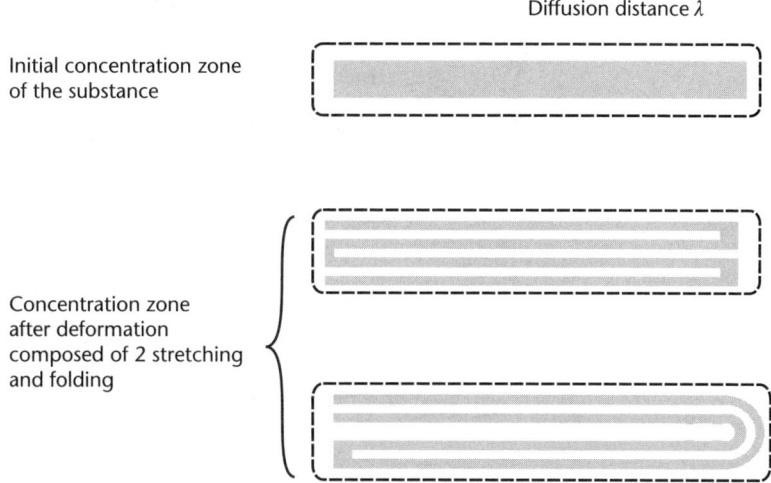

Figure 4.31   The mixing zone is enlarged by a succession of stretching and folding.

4.2.12.5   Mixing in Two-Phase Flows

Mixing is not a problem reserved to single-phase microflows. Obtaining short time for mixing is also important in two-phase flow and plug flow. Mixing of plug flows has recently been a subject of interest. Qualitatively, the internal convective motion in a liquid plug is that sketched in Figure 4.32. Usually—not always [15]—the plug keeps the same geometry during its motion because of the effect of surface tension. Due to the displacement of the plug and the friction at the solid walls, two convective cells forms in the plug.

Modeling the internal convection in the plug may be performed by considering the problem in the moving coordinates system. In this system, the plug has fixed boundaries (to the first order) and the solid walls are moving with a velocity $-V$. Specifying this value of the velocity on the solid walls, and symmetry conditions on the side surfaces with no contact of any wall, the numerical solution is straightforward. We show in Figure 4.33 the result obtained with the FEMLAB numerical software [8].

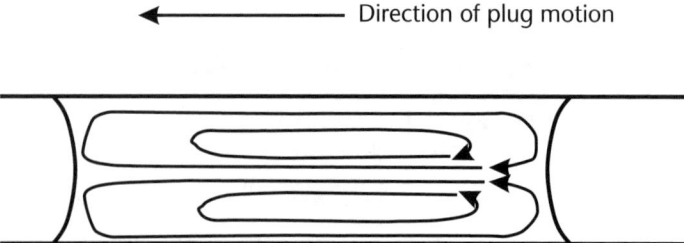

Figure 4.32   Mixing in a plug flow promoted by internal recirculation due to the friction at the walls.

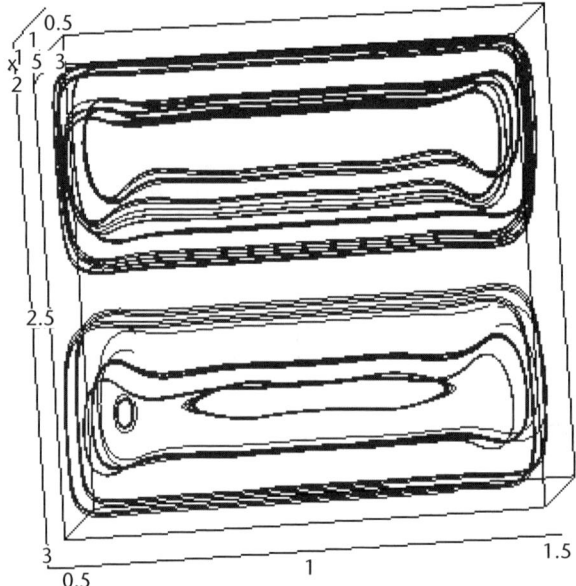

Figure 4.33   Modeling of the internal motion in a plug flow with FEMLAB.

4.2.12.6   Mixing in Digital Microfluidics

The preceding examples concerned mixing in microflows. With the fast development of digital microfluidics (see Chapter 2), mixing of fluids and substances in microdrops is now a growing subject of study. As of the writing of this book, the understanding of the mixing phenomena in microdrops is only at a qualitative stage. In order to illustrate this problem and to familiarize the reader with internal microdrop motion, we show in Figure 4.34 the principle of mixing two fluids in a microdrop by moving the drop along a designated path [16]. Figure 4.35 shows the mixing of two drops in a confined EWOD system (see Chapter 2). During the motion from one electrode to the other, mixing occurs by the motion of the concentration along the interfaces.

It appears that mixing in digital microfluidics shows very special patterns. Right now it is a new topic of investigations.

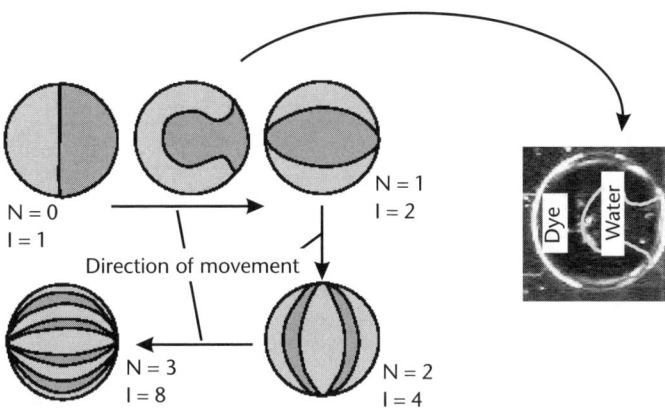

Figure 4.34 Mixing of two constituents in a drop by electrowetting (open EWOD) displacement. (*From:* [16]. © 2002 IEEE. Reprinted with permission.)

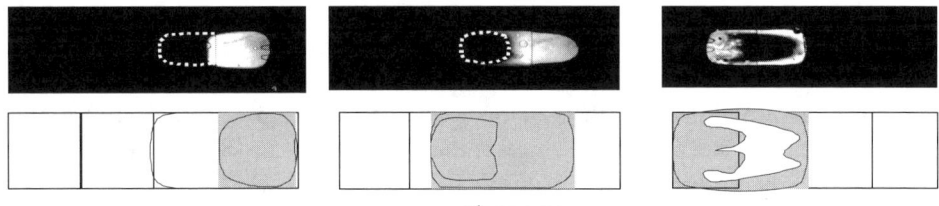

Figure 4.35 Mixing of two microdrops confined by electrodes in an EWOD microsystem. One can see on this figure that the mixing is initiated the motion of the interface. (Courtesy of C. Peponnet, CEA/LETI.)

4.3 Trajectory Calculation

Computation of transport of substance using the concentration equation requires that the molecules or particles comprising this substance have a small size. Because gravity is not taken into account in the advection-diffusion equation, the particles are not allowed to sediment and their size and weight is limited by the sedimentation size and weight.

If this is not the case—for example, when transporting large biologic objects like cells or proteins, or heavy particles like some magnetic beads (see Chapter 7)—gravity has an important influence on the transport. However, in such a case, the influence of Brownian motion on the particles is reduced. A first step in this approach is to calculate the trajectories of the particles in a deterministic (i.e., without taking account of the Brownian motion).

4.3.1 Trajectories of Particles in a Microflow

Larger particles experience less diffusion under the action of Brownian motion. In such a case, these particles follow trajectories determined by the forces acting upon them. At a macroscopic scale, kinematics theory relates the mass acceleration of a body to the resultant of the external forces that act upon it. This is the well-known Newton's theorem.

$$m\frac{d\vec{V}_p}{dt} = \sum \vec{F}_e \qquad (4.78)$$

where m is the mass of the particle, $\vec{V}_p$ is the velocity, and $\vec{F}_e$ is the external forces. We will treat here the case of particles submitted to gravity force and hydrodynamic drag force. Newton's equation can then be written under the form

$$m\frac{d\vec{V}_p}{dt} = \vec{F}_{hyd} + \vec{F}_{grav} \qquad (4.79)$$

The hydrodynamic drag is derived from the velocity field according to the equation

$$\vec{F}_{hyd} = -C_D\left(\vec{V}_p - \vec{V}_f\right) = -6\pi\eta r_h\left(\vec{V}_p - \vec{V}_f\right) \qquad (4.80)$$

where C_D is the drag coefficient, η is the dynamic viscosity of the carrier fluid, r_h is the hydrodynamic diameter of the particle, and V_f is the velocity of the carrier fluid. It is assumed here that the velocity field of the carrier fluid is not affected by the presence of the particles, which is the general case, except if the volume fraction of particles is important, leading to the formation of aggregates. Under this assumption, the velocity field of the carrier fluid must be calculated before attempting the calculation of the particles' trajectories, using classical hydrodynamics equations (i.e., the Navier-Stokes equations). A typical situation in microfluidics is the Hagen-Poiseuille flow between two plates or in a rounded capillary. The gravity term is given by

$$\vec{F}_{grav} = g\,vol_p\,\Delta\rho\hat{y} \qquad (4.81)$$

where g is the acceleration of gravity, vol_p is the volume of the particle, $\hat{y}$ is the vertical unit vector (oriented downwards), and $\Delta\rho$ is the difference between the volumic mass of the particle and that of the liquid. After substitution of (4.80) and (4.81) in (4.79), one obtains the equation for the particles' velocity

$$m\frac{d\vec{V}_p}{dt} = -6\pi\eta r_h\left(\vec{V}_p - \vec{V}_f\right) + gvol_p\Delta\rho\hat{y} \qquad (4.82)$$

This relation can be decomposed along each coordinate (here we choose a two-dimensional configuration and consider the central vertical plane)

$$m\frac{du_p}{dt} = -6\pi\eta r_h\left(u_p - u_f\right)$$
$$m\frac{dv_p}{dt} = -6\pi\eta r_h v_p + gvol_p\Delta\rho \qquad (4.83)$$

Using the notations

$$c_1 = \frac{6\pi\eta r_b}{m}$$

$$c_2 = \frac{g \, vol_p \Delta\rho}{m}$$

(4.84)

this system becomes

$$\frac{du_p}{dt} = -c_1\left(u_p - u_f\right)$$

$$\frac{dv_p}{dt} = -c_1 v_p + c_2$$

(4.85)

and this system can be solved analytically

$$u_p = u_{p,0}e^{-c_1 t} + u_f\left[1 - e^{-c_1 t}\right]$$

$$v_p = v_{p,0}e^{-c_1 t} + \frac{c_2}{c_1}\left[1 - e^{-c_1 t}\right]$$

(4.86)

By definition, x and r coordinates of the particle at a given time are linked to the velocity by the relations

$$\frac{dx_p}{dt} = u_p = u_{p,0}e^{-c_1 t} + u_f\left[1 - e^{-c_1 t}\right]$$

$$\frac{dr_p}{dt} = v_p = v_{p,0}e^{-c_1 t} + \frac{c_2}{c_1}\left[1 - e^{-c_1 t}\right]$$

(4.87)

If the starting velocity of the particle is zero, we obtain a simple relation between the coordinates of the particle

$$\frac{dx_P}{dr_P} = u_f \frac{c_1}{c_2}$$

(4.88)

where the ratio c_1/c_2 is

$$\frac{c_1}{c_2} = \frac{C_D}{g\Delta m}$$

Assuming a Hagen-Poiseuille flow in the duct, (4.88) becomes

$$\frac{dx_P}{dr_P} = \frac{V_0}{2}\left(1 - \frac{r_P^2}{R^2}\right)\frac{c_1}{c_2}$$

(4.89)

Integration of this relation gives the relation

$$x_P = \frac{V_0}{2}\frac{c_1}{c_2}\left[r - r_{p,0} + \frac{r_P^3}{3R^2} - \frac{r_{p,0}^3}{3R^2}\right]$$

(4.90)

where $(0, r_{p0})$ is the starting location of the particle. A particle starting from the middle of the duct $(r_{p0} = 0)$ will contact the wall at an axial distance of

$$L = \frac{C_D}{g\Delta m} \frac{2V_0}{3} R \qquad (4.91)$$

It is interesting to compare this result (4.91) with (4.67). The two results are quite different. In the present case, we have calculated the distance of travel of a particle submitted to hydrodynamic drag considering that the particle is supposed sufficiently large (or heavy) to neglect the Brownian motion in front of the gravity force. In the previous case, we have calculated the same distance for a particle submitted to hydrodynamic drag force but small enough to neglect gravity in front of Brownian motion. The ratio of the two calculated lengths is

$$\frac{L_{diff}}{L_{grav}} = \frac{3}{4} \frac{R}{D} \frac{g\Delta m}{C_D} = \frac{3}{4} \frac{g\Delta m}{k_B T} R \qquad (4.92)$$

Interestingly, the average buffer fluid velocity has disappeared from (4.92), which is just a balance between gravity forces and Brownian motion energy $(k_B T)$. Depending on the relative particle mass Δm, the travel distance will be either the gravity model distance defined by (4.91) or the diffusion model distance given by (4.67).

Note that it is very seldom that the trajectory equation can be solved analytically, like it is here; however, it is always interesting to spend some time investigating whether an analytical solution may exist—even with the price of some simplification (initial velocities set to zero). Most of the time, a numerical approach is required: Different methods can be used, like Runge Kutta or predictor-corrector. We give an example of predictor-corrector scheme in Chapter 7.

4.3.2 Trajectories and Brownian Motion

In this section, we combine particles' entrainment by the flow with Brownian motion. We have seen in Chapter 3 that discrete models, like the Monte Carlo model, are interesting because they bring new insight to the understanding of the effect of Brownian motion. With this in mind, a similar approach may be used for microparticles transport. The buffer fluid—or carrier fluid—flow being the continuous phase, its behavior is still obtained by solving the Navier-Stokes hydrodynamic equations. If the entrainment of the microparticles is strong enough, one can assume that the transported microparticles are following trajectories slightly modified by the effect of Brownian motion (Figure 4.36).

The real force balance on a particle is given by Langevin's equation [17]

$$m_e \frac{dV_c}{dt} = C_D \left(V_f - V_c \right) + F(t) \qquad (4.93)$$

where the function $F(t)$ represents the Brownian forces. Although this is not strictly correct, we approximate (4.93) by the superposition of a deterministic trajectory

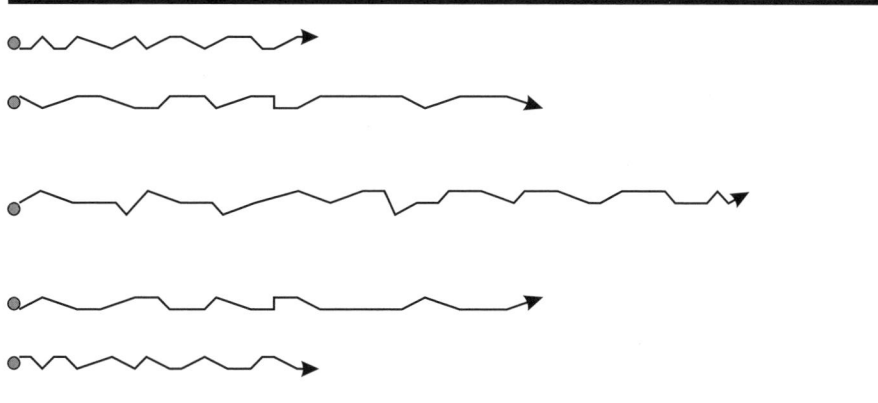

Figure 4.36 Sketch of the superposition of advection by the buffer fluid and Brownian motion.

(see Section 4.3.1), modified by the effect of the Brownian motion modeled by a Monte Carlo method. This is approximately correct if the particle trajectory is not affected too much by Brownian motion (i.e., if the entrainment is strong in front of the Brownian motion).

In this example, for simplicity we do not take into account gravity force. In such a case,

$$V_p = V_f \left(1 - e^{-\frac{C_D}{m} t} \right) \qquad (4.94)$$

For a very small particle, its velocity is that of the fluid. Now we account for Brownian motion by introducing the relations

$$V_{p,x} = V_{f,x} + \sqrt{\frac{4D}{\Delta t}} \cos(\alpha)$$

$$V_{p,y} = V_{f,y} + \sqrt{\frac{4D}{\Delta t}} \sin(\alpha) \qquad (4.95)$$

$$\alpha = random\,(0,2\pi)$$

and for a three-dimensional computation

$$V_{p,x} = V_{f,x} + \sqrt{\frac{4D}{\Delta t}} \cos(\alpha)\sin(\beta)$$

$$V_{p,y} = V_{f,y} + \sqrt{\frac{4D}{\Delta t}} \sin(\alpha)\sin(\beta)$$

$$V_{p,z} = V_{f,z} + \sqrt{\frac{4D}{\Delta t}} \cos(\beta) \qquad (4.96)$$

$$\alpha = random\,(0,2\pi)$$

$$\beta = a\cos\left(1 - 2\ random(0,1)\right)$$

More explanation about these equations can be found in Chapter 3. Suppose now that a uniform concentration of target particles arrives at the entrance of the duct, meaning that they are uniformly dispersed in the entrance cross section. An easy way to obtain a uniform concentration in a circular cross section is to generate two sets of values uniformly distributed:

$$x = R \; random \, (0,1)$$
$$y = R \; random \, (0,1)$$

and to reject the values (x, y) located outside the circle of radius R. If the particles are referenced by their polar coordinates (r, ϕ), then the distribution in ϕ is uniform and linear in r, as shown by Figure 4.37.

The buffer flow velocity profile is given by the Hagen-Poiseuille relation

$$V(r) = 2V_0 \left(1 - \left(\frac{r}{R} \right)^2 \right) \tag{4.97}$$

In the present model, we suppose that the wall is completely adherent (i.e., the particles contacting the wall are immobilized immediately), as shown in Figure 4.38. The model is then

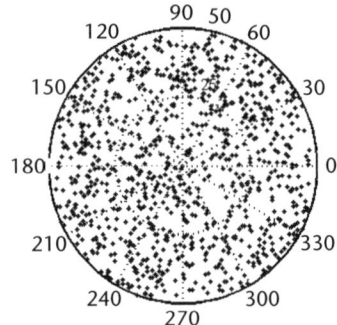

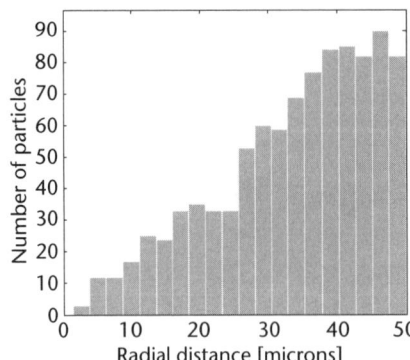
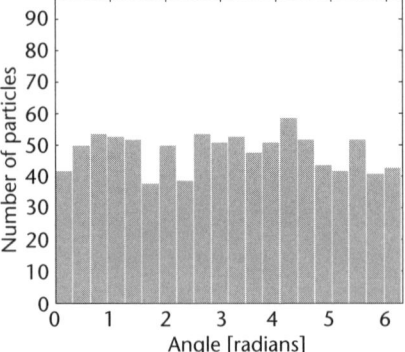

Figure 4.37   Initial distribution of particles in the entrance cross section.

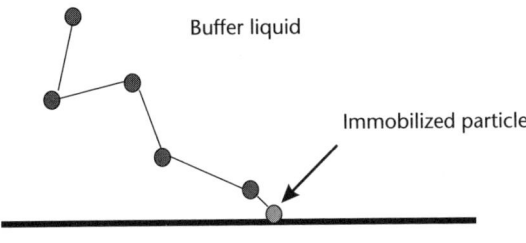

Figure 4.38   Sketch of a particle impacting the wall.

$$V_{p,x} = V_{f,x} + \sqrt{\frac{4D}{\Delta t}} \cos(\alpha)\sin(\beta)$$

$$V_{p,y} = V_{f,y} + \sqrt{\frac{4D}{\Delta t}} \sin(\alpha)\sin(\beta)$$

$$V_{p,z} = 2V_0\left(1-\left(\frac{r}{R}\right)^2\right) + \sqrt{\frac{4D}{\Delta t}} \cos(\beta) \tag{4.98}$$

$$\alpha = random\,(0,2\pi)$$

$$\beta = a\cos\big(1-2\,random(0,1)\big)$$

Equation (4.98) has been implemented in MATLAB [17]. Figure 4.39 shows the trajectories of the particles in the capillary tube.

The photograph of the particles at a given time is shown in Figure 4.40. The target particles are dispersed following the Hagen-Poiseuille parabolic profile of velocity.

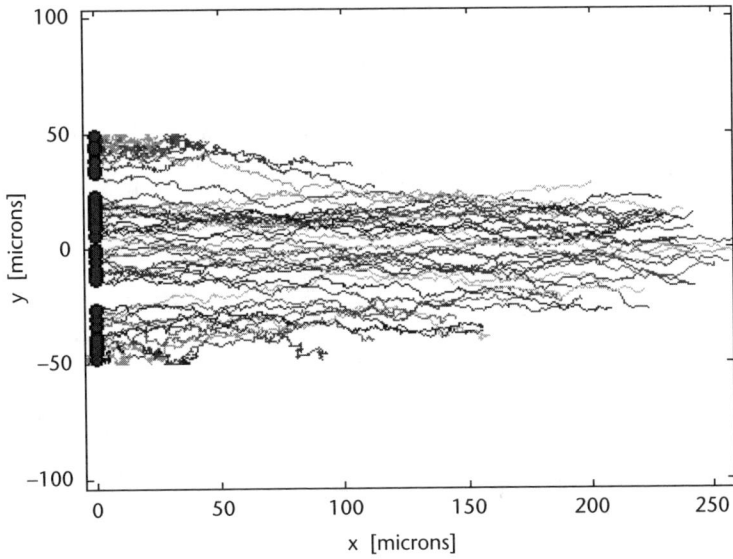

Figure 4.39   Calculated trajectories of particles (100-nm diameter) transported by a buffer fluid flow (500 μm/sec) in a 50-μm radius capillary tube.

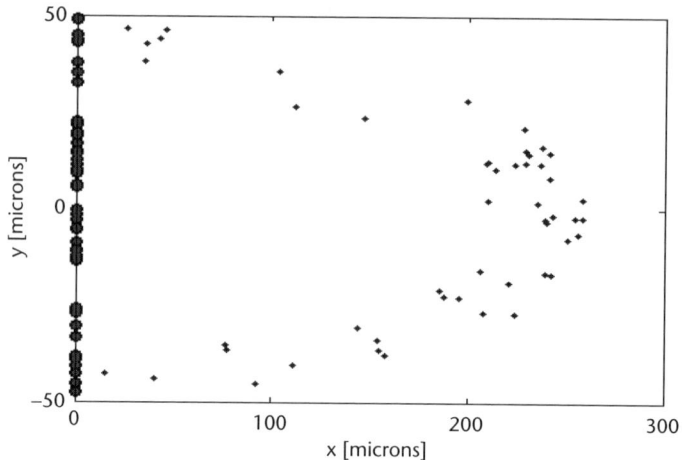

Figure 4.40   Calculated location of the particles at a given time. Note the similarity to Figure 4.17.

By superposition of images of location of the particles at different times, we understand the pattern of the transport of the microparticles (Figure 4.41). In a cross section, the location of the particles is given in Figure 4.42. The particles adhering to the wall are clearly seen on the periphery.

It is interesting to compare these results to that of Figure 4.21 obtained by solving the advection-diffusion equation for the concentration. In Figure 4.43 we have placed the particles on the concentration contour lines. The comparison is rather good; Monte Carlo computation for more particles would have been still more accurate.

4.4   Separation/Purification of Bioparticles

Separation and purification of bioparticles are required for many different applications and targets (purification of proteins, separation of DNA strands by length, and so on). Several techniques have been developed to perform these processes. We present here, mostly qualitatively, the principle of field flow fractionation (FFF) and chromatography columns.

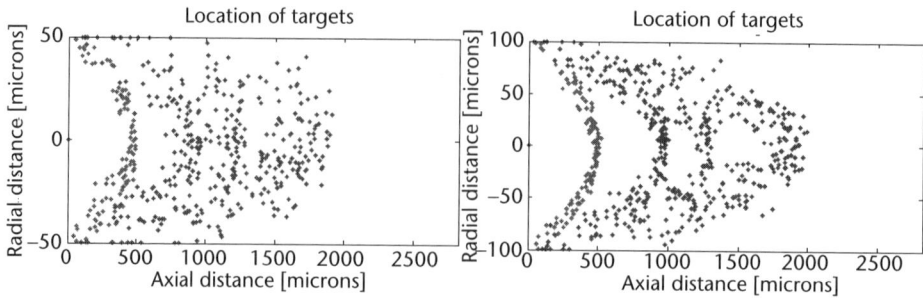

Figure 4.41   Superposition of the location of particles at different times.

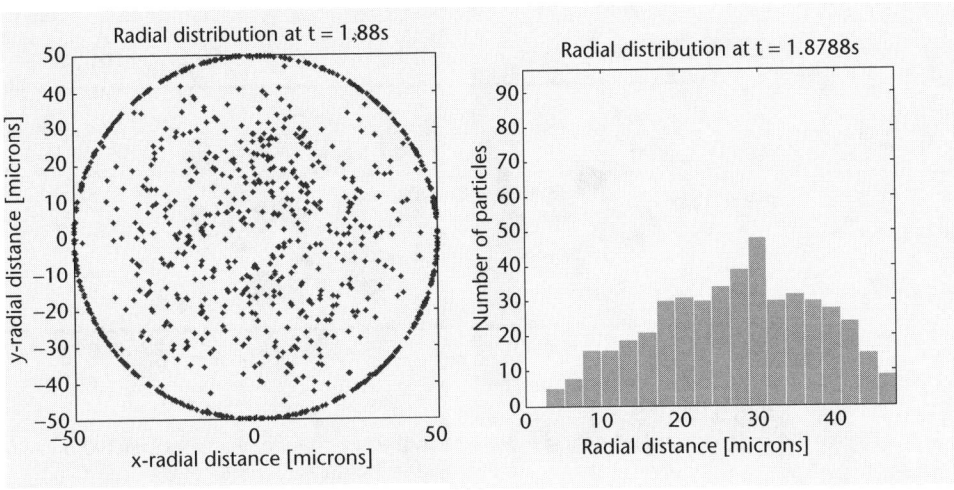

Figure 4.42 Image of the particles in a cross section (left), and radial distribution of particles corresponding to left picture (right).

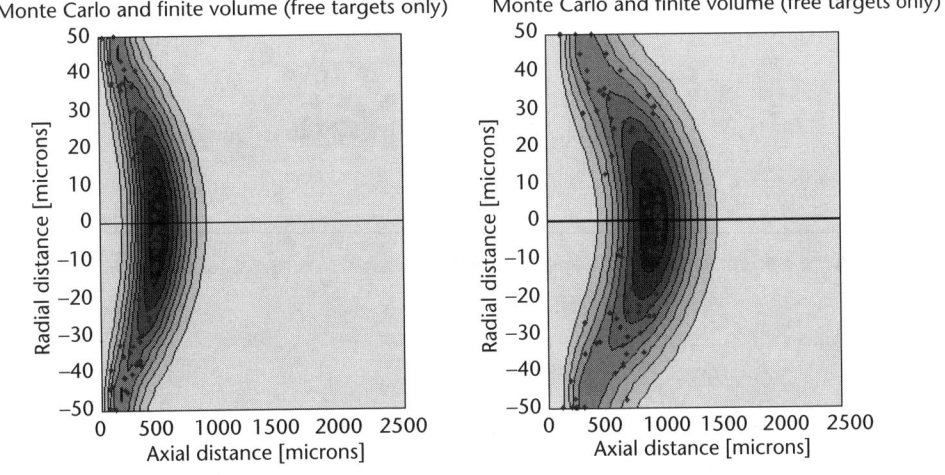

Figure 4.43 Comparison of the concentration contour lines (from Figure 4.21) with the results of the Monte Carlo simulation. Radius $R = 50\ \mu$m, diffusion coefficient $D = 2 \cdot 10^{-10}\ m^2$/s, and average velocity $V = 500\ \mu$m/sec.

4.4.1 The Principle of FFF

FFF is a group of techniques to separate different types of particles [18]. The principle is shown in Figures 4.44 through 4.46. The principle here is that particles in a liquid flow separate according to their physical properties like volume, mass, electric charge, or magnetic moment. Suppose a horizontal flow: Drag force depends on the size of the particle. If another force field is applied vertically (like gravitation), the particles will gather at different places on the lower solid wall.

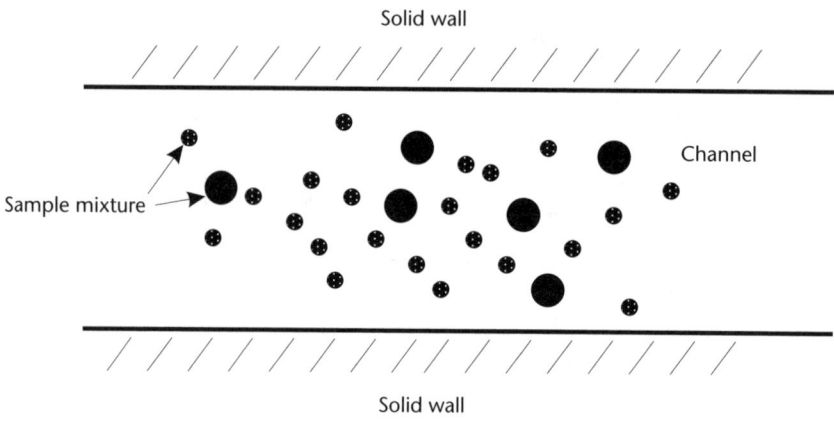

Figure 4.44   Injection of the sample mixture in the FFF channel. Particles injected onto the column without the field or flow turned on are evenly distributed across the column.

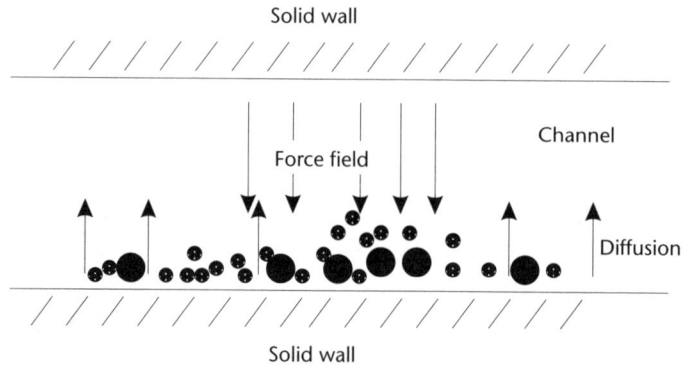

Figure 4.45   When a field is applied, the solute zone is compressed into a narrow layer against one wall.

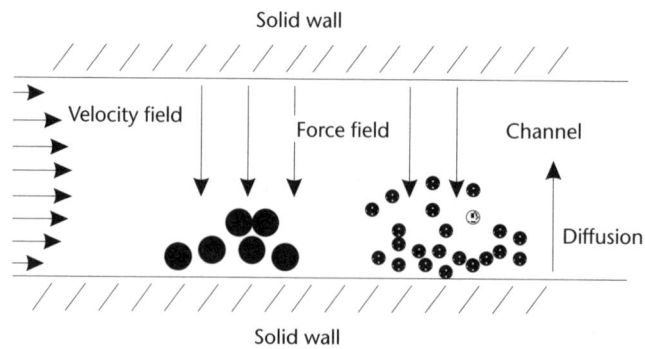

Figure 4.46   An applied velocity field in the channel exerts a different hydrodynamic drag on the two types of particles. The larger particles stay behind and are separated from the smaller particles.

When the liquid flow is initiated, the solute zone is carried downstream at a rate depending on the particle size and mass. Figure 4.47 shows a correct separation when the immobilized particles form two peaks completely disjointed.

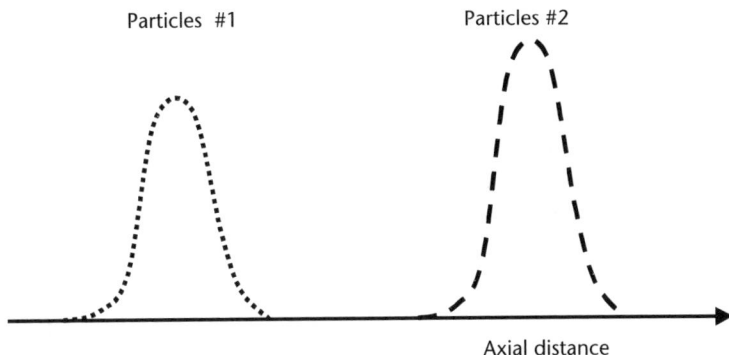

Figure 4.47   Particle distribution on the wall: separation is correct when the two aggregates are completely disjointed.

In Chapter 7, dedicated to the study of the behavior of magnetic particles, an example of magnetic FFF is given, with the calculation of trajectories inside the channel.

4.4.2   Chromatography Columns

Chromatography is a separation technique mostly employed in chemical and biochemical analysis [19, 20]. In a single-step process, it can separate a mixture (buffer fluid carrying different types of bioparticles) into its individual components and simultaneously provide an quantitative estimate of each constituent. In biotechnology, the analysis is usually carried out on a mass spectrometer (MS) placed behind the chromatography column.

The name *chromatography* may look strange at first sight. Color has nothing to do with modern chromatography, but the name was given to this method of separation by the Russian botanist Tswett, who used a simple form of liquid-solid chromatography to separate a number of plant pigments. The colored bands he produced on the adsorbent bed evoked the term chromatography for this type of separation.

In a chromatography device, there are two phases: a mobile phase and a stationary phase. The mobile phase transports the sample with the targets and other compounds, and the stationary phase is designed to retain longer in the column the compounds with which it associates best. Targets and compounds are then separated in zones or bands [21]. A typical design of chromatography column is that of the size exclusion chromatography sketched in Figure 4.48. In such a device, the small size proteins or peptides travel at a smaller speed through the column, because they can enter the gel beads, whereas large size proteins are excluded and travel faster in the connected porosities. The result is the separation of the particles into two disjointed bands (Figure 4.49).

An example of a different type of chromatography column is that of the proteomic reactor [22]. Proteins are first digested into peptides, then all of the peptides are immobilized in a microfabricated (Figures 4.50 and 4.51) chromatography column. A flow of acetonitril (CH_3CN) is then used at different concentrations to elute progressively the peptides. The eluted peptides are transported to a spray nozzle and sprayed into the mass spectrometer.

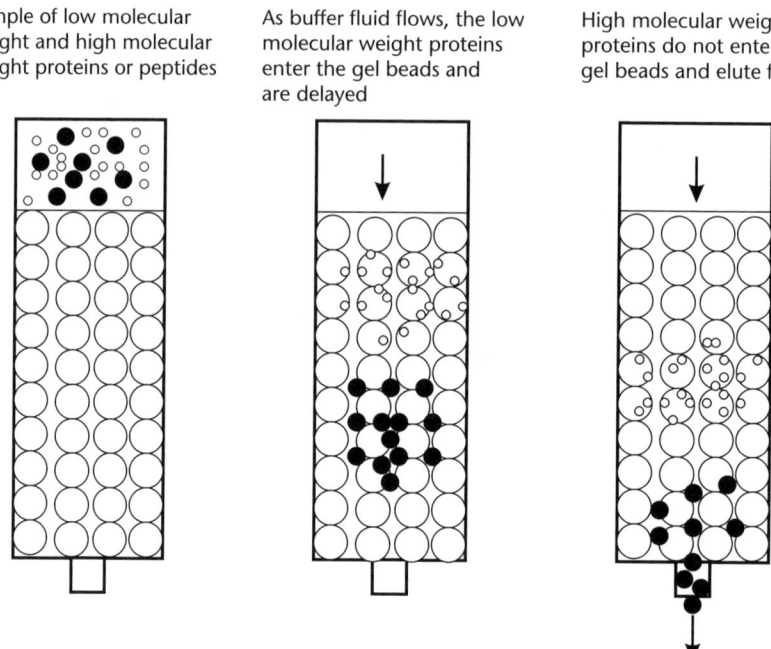

Figure 4.48 Proteins purification by size-exclusion chromatography.

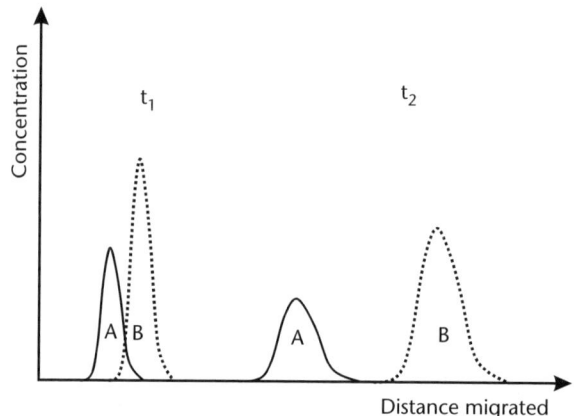

Figure 4.49 Separation principle of substance traveling through a chromatography column. Progressively the two peaks separate.

Figure 4.52 shows the detection by the mass spectrometer of the well-known peptide β-Galactosidase, with its three characteristic peaks.

4.5 Conclusion

In this chapter, we presented the governing equations for the transport of biological substance. Two different approaches are possible, depending on the size and mass of the molecules/particles of the substance. If they are sufficiently small, the

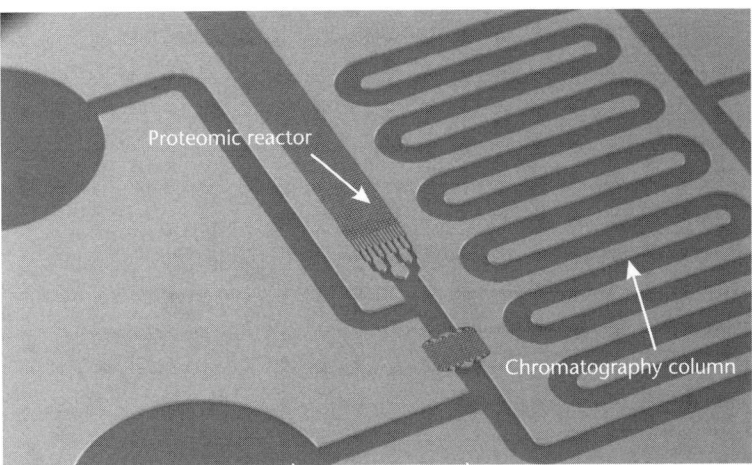

Figure 4.50 View of a proteomic reactor with the chromatography column at the right part of the picture. (Courtesy of N. Sarrut, CEA/LETI.)

Figure 4.51 Left: Detailed view of a microfabricated chromatography column in a proteomic reactor; right: microscope view of the spray nozzle. (Courtesy of N. Sarrut, CEA/LETI.)

advection-diffusion equation is the one to choose. If the particles are larger and submitted to nonnegligible gravity forces, it may be interesting to calculate individually their trajectory, with or without the introduction of a Brownian perturbation.

Transport mechanisms of micro- and nanoparticles or macromolecules or cells are at the heart of any biotechnological microdevice. As the ultimate goal is the handling or detection of the smallest possible number of these objects, it is necessary to have very precise control over the particles. This cannot be done in just one step. Transport by microflows and microdrops constitutes the first step to manipulating the particles. Other more specific steps are magnetic and electric methods and will be presented in the following chapters.

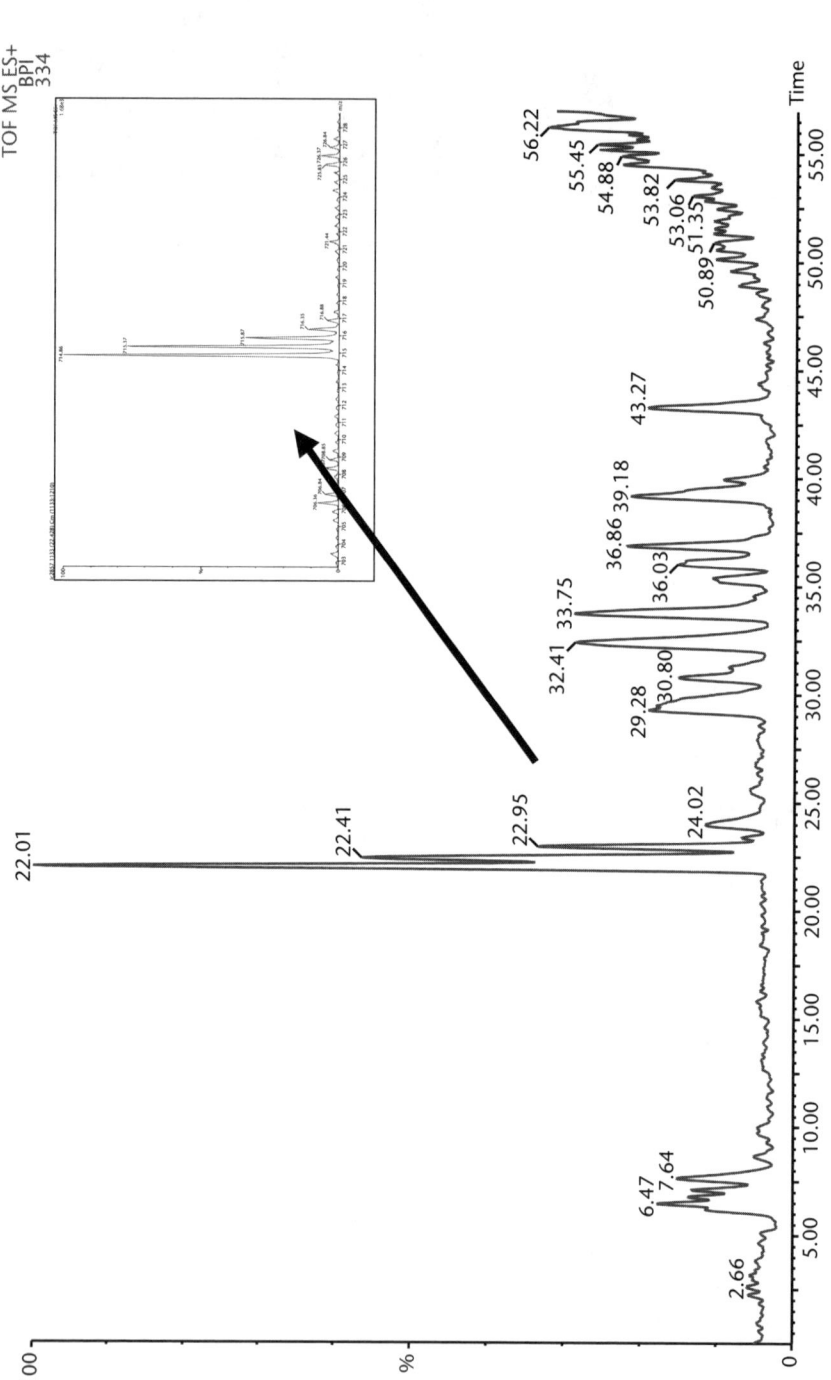

Figure 4.52 Experimental result of chromatography separation of peptides in a proteomic reactor. Mass spectrometry trace of separation in the microfabricated column of Figure 4.51. Experiment using 50 fento-mol of a protein tryptic digest (β-Galactosidase) and liquid chromatography flow rate 300 nanoliters/min. Reconstructed chromatograms and corresponding mass spectra for a tryptic peptide of β-Galactosidase.

References

[1] Nguyen, N. -T., and S. T. Wereley, *Fundamentals and Applications of Microfluidics*, Norwood, MA: Artech House, 2002.

[2] Tabeling, P., *Introduction à la Microfluidique*, Paris, France: Editions Belin, 2004.

[3] Starkey, T. V., "The Laminar Flow of Suspensions in Tubes," *British Journal of Applied Physics*, Vol. 6, January 1955, pp. 34–37.

[4] Tanner, R. I., *Engineering Rheology*, Oxford, England: Oxford University Press, Oxford Engineering Science Series, 2000, p. 52.

[5] Levich, V. G., *Physicochemical Hydrodynamics*, Upper Saddle River, NJ: Prentice Hall, 1962.

[6] Wissler, E. H., "On the Applicability of the Taylor-Aris Axial Diffusion Model to Tubular Reactor Calculations," *Chemical Engineering Science*, Vol. 24, 1969, pp. 527–539.

[7] Batycky, R. P., D. A. Edwards, and H. Brenner, "Thermal Taylor Dispersion in an Insulated Circular Cylinder—1. Theory," *Int. J. Heat Mass Transfer*, Vol. 36, No. 18, 1993, pp. 4317–4325.

[8] FEMLAB, Multiphysics modeling: http://www.comsol.com.

[9] Dehghan, M., "Numerical Solution of the Three-Dimensional Advection-Diffusion Equation," *Applied Mathematics and Computation*, Vol. 150, 2004, pp. 5–9.

[10] Kamholz, A. E., and P. Yager, "Theoretical Analysis of Molecular Diffusion in Pressure-Driven Laminar Flow in Microfluidic Channels," *Biophysical Journal*, Vol. 80, 2001, pp. 155–160.

[11] Kamholz, A. E., and P. Yager, "Molecular Diffusive Scaling Laws in Pressure-Driven Microfluidic Channels: Deviation from One-Dimensional Einstein Approximations," *Sensors and Actuators, B: Chemical*, Vol. 82, No. 1, 2002, pp. 117–121.

[12] Ehrfeld, W., et al. "Characterization of Mixing in Micromixers by a Test Reaction: Single Mixing Units and Mixer Arrays," *Ind. Eng. Res.*, Vol. 38, 1999, pp. 1075–1082.

[13] Hessel, V., H. Loewe, and F. Schönfeld, "Micromixers—A Review on Passive and Active Mixing Principles," *Chemical Engineering Science*, Vol. 60, 2005, pp. 2479–2501.

[14] Loeb, P., et al., "Steering of Liquid Mixing Speed in Interdigital Mixers—From Very Fast to Deliberately Slow Mixing," *Chemical Engineering and Technology*, Vol. 27, No. 3, 2004, pp. 340–345.

[15] Berthier, J., and F. Ricoul, "Numerical Modeling of Ferrofluid Flow Instabilities in a Capillary Tube at the Vicinity of a Magnet," *Proc. MSM 2002 Conference*, Puerto Rico, 2002.

[16] Fowler, J., H. Moon, and C. -J. Kim, "Enhancement of Mixing by Droplet-Based Microfluidics," *Proc. IEEE Conference MEMS*, Las Vegas, NV, January 2002, pp. 97–100.

[17] MATLAB, "The Language of Technical Computing," Version 6.2, The MathWorks, 2000.

[18] Giddings, J. C., "Field-Flow Fractionation," *C & E News*, Vol. 66, 1988, pp. 34–45.

[19] Scott, R. P. W., "Dispersion in Chromatography Columns," Chrom-Ed-series, http://www.chromatography-online.org/Dispersion/Rate-Theory/rs1.html.

[20] Martin, A.J.P., and R. L. M. Synge, "A New Form of Chromatogram Employing Two Liquid Phases. 1. A Theory of Chromatography. 2 Application of the Microdetermination of the Higher Monoaminoacids in Proteins," *Biochem. J.*, Vol. 35, 1941, p. 1358.

[21] Shuler, M. L., and F. Kargi, *Bioprocess Engineering: Basic Concepts*, Upper Saddle River, NJ: Prentice Hall, 2002.

[22] Sarrut, N., et al., "Enzymatic Digestion and Liquid Chromatography in Micro-Pillar Reactors—Hydrodynamic Versus Electro-Osmotic Flow," *SPIE, Photonics West—MOEMS-MEMS*, San Jose, CA, 2005.

CHAPTER 5

Biochemical Reactions in Biochips

5.1 Introduction

In this chapter, we come to the very purpose of biochips. So far, we have dealt with the principles of microfluidic transport of macromolecules or microparticles, and we have shown how these principles are used to displace and manipulate these objects inside microsystems. It is recalled here that the approach has been done in two steps: first, the study of the microfluidic flow as a carrier fluid; second, the study of the behavior of macromolecules and microparticles in such microfluidic flows. Up to now, we have dealt only with tools to perform a task. And the essential question is: What task are we going to perform with such tools and what have all these techniques been developed for? This brings us to the purpose of biochips or bioMEMS.

Basically, the main purpose of biochips is the analysis and recognition of macromolecules—DNA, proteins, and such. Ultimately, recognition processes should be fast, sensitive, and reliable, with the less possible false results, and largely parallelizable, allowing for simultaneous samples recognition.

We will see in this chapter that biorecognition is based on a mechanism of *key lock* [1], which is in reality a biochemical reaction. This leads us to present the kinetics of chemical and biochemical reactions, with a special attention to some key reactions, like enzyme-catalyst reactions for proteins and adsorption reactions for DNA hybridization. Because in biochip technology, the targets to analyze are immersed and carried by a buffer fluid, recognition times depend not only on the biochemical reaction kinetics but also on the presence of targets in the vicinity of the reagents. It is then necessary to treat the coupling between biochemical reactions and the advection-diffusion of targeted molecules. In conclusion, we point out that biorecognition is very dependent on detection sensitivity. The same care that is taken for developing an efficient bioreaction should be taken also to the detection process.

5.2 From the Principle of Biorecognition to the Development of Biochips

5.2.1 Introduction to Biorecognition

The discovery of the recognition potential by the immune system—sometimes called immune specificity—has been a major milestone in the development of biology and has been awarded many Nobel prizes. The first step was the discovery of the model key lock by Fisher in 1892, sketched in Figure 5.1. In such an approach, a macromolecule binds to a specific complementary macromolecule and with no other one—at least in theory—due to a sufficient interaction force.

The discovery of the key-lock model was followed by the development of synthetics analogs by Landsteiner in 1930, and by the principle of solid phase immunoassay by Langmuir and Shaefer in 1942, associated with the detection by immunofluorescence introduced by Coonsz the same year. With the determination of the structure of antibodies and their three-dimensional structure by Yalow and Berson in 1959 and Poljak in 1973, and the production of specific monoclonal antibodies by Köhler and Milstein in 1986, all of the pieces of the puzzle were present to give birth to antigen, antibody, or protein biorecognition by a process that is now called *immunoassay*.

5.2.2 Biorecognition

As we just have seen, biorecognition is a process based on the lock-and-key principle. We show here two examples of biorecognition that are the basis of DNA and proteins biochips.

Start with the case of DNA. DNA has a well-known double helix structure, as indicated in Figure 5.2. The two helices are linked by hydrogen bonding between two of the four groups of base: A with G and C with T.

It has been observed that, given some favorable conditions of temperature or pH, the double helix can dissociate (a process known as denaturing). Then a single DNA strand that has a known sequence can recognize a complementary sequence on another single strand of DNA. This process is called hybridization (Figure 5.3). So, if we can fabricate a given DNA sequence, and have many copies of this sequence, these strands will recognize specific DNA strands with complementary sequence. Note that the bonding is reversible, and it corresponds to an equilibrium reaction, as we will see later in this chapter.

A similar lock-and-key approach can be done for antigen (or proteins) [2]. Roughly speaking, an antibody is a very complex molecule having a Y shape, as symbolized in Figure 5.4. An antibody can recognize a specific antigen approaching one of the two binding sites located on both ends of the Y. The bond is made of multiple, noncovalent interactions, like hydrogen bonds, van der Waals forces, or Coulombic interactions. As for DNA, it is a reversible equilibrium reaction, called *immunoreaction*. Biologists use the term *affinity* to name the interaction force between antibody and antigen.

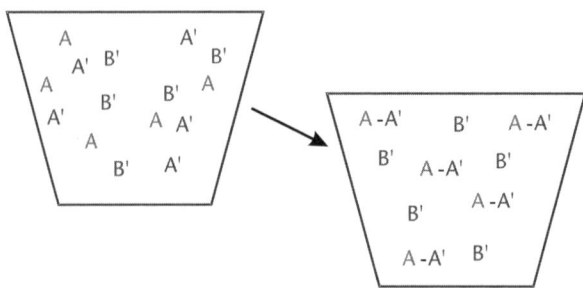

Figure 5.1 Principle of molecular recognition imagined by Fisher in 1892.

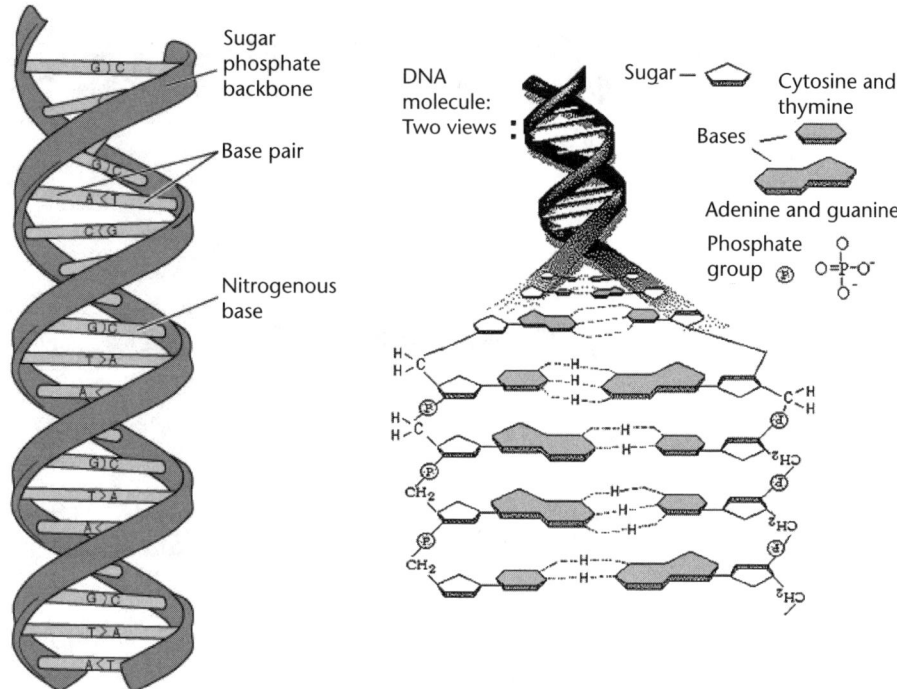

Figure 5.2   Left: Double-helix DNA structure. Right: Sketch of hydrogen bondings between A and T (adenine and thymine) and C and G (cytosine and guanine).

5.2.3   Biochip Technology

Biochips derived directly from the principle of biorecognition. First, let us consider two Eppendorf tubes, each one functionalized with a specific antibody (Figure 5.5). In such a case, we want to detect a precise antigen that can be recognized by the grafted antibodies. First, the samples are filled into the Eppendorf tubes, and after a sufficient time to allow for recognition, the tubes are washed. Only the couples of specific antibody-antigen resist to the washing process. Fluorescent markers are then introduced in the tubes and, again after washing, there remain only the marked specific couples. Detection is made by comparison of the emitted light between the positive and negative Eppendorf tubes.

Biochips or bio-MEMS are just a miniaturization the preceding principle (Figure 5.6), plus a systematic recognition due to the many different spots grafted (functionalized) with different types of antibodies.

At this stage, we can make a distinction between two types of biochips: a first category where the buffer liquid (the liquid containing the samples) is at rest (Figure 5.7) and another whereby the buffer liquid is moving (Figure 5.8) inside a microchannel or microchamber. We will see later in this chapter the difference in the processes of capture.

Figure 5.9 schematizes one of the first biochips. At the beginning, there were only a few recognition sites, but nowadays some biochips have more than 10,000 recognition sites and thus can recognize many DNA sequences or antigens in one run.

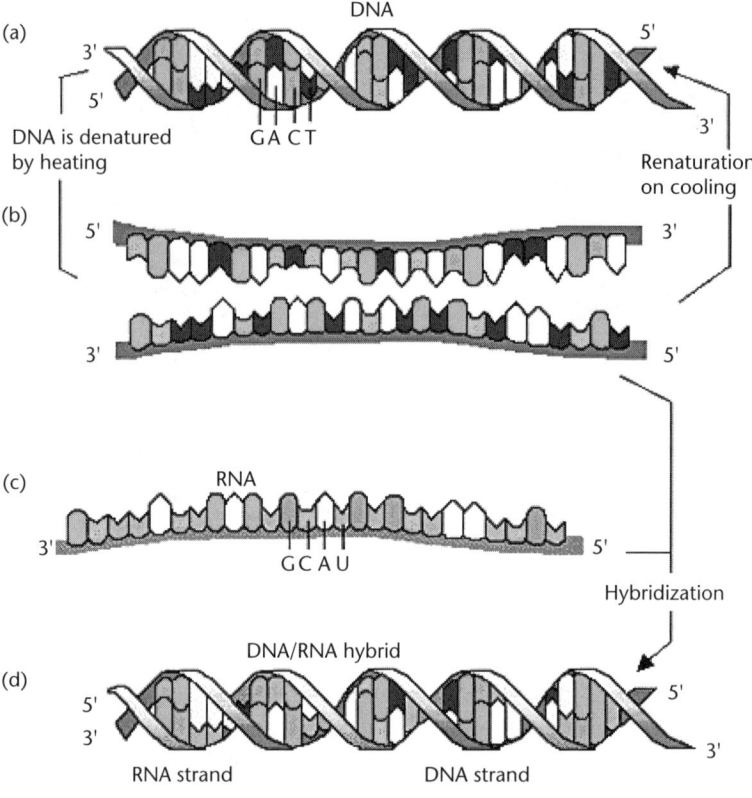

Figure 5.3   (a–d) DNA denaturing and hybridization with a complementary sequence.

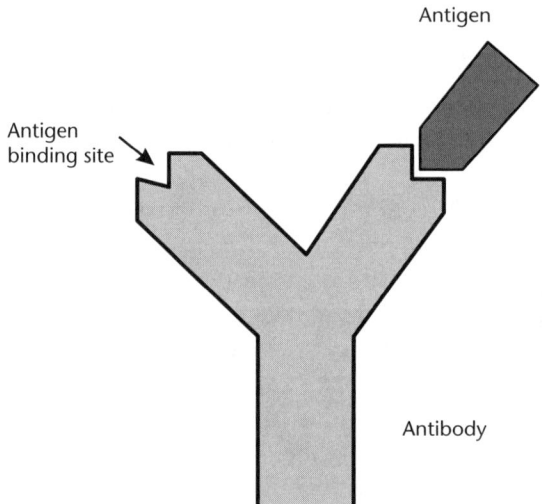

Figure 5.4   Schematic view of antibody-antigen binding.

The presentation of biochip technology would not be complete if the detection problem was left aside. Both the requirements of miniaturization and those of

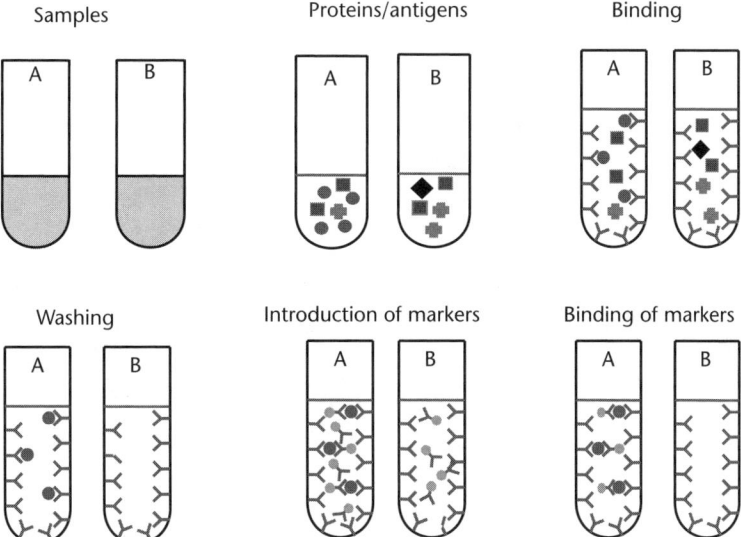

Figure 5.5   Principle of biorecognition (in an Eppendorf tube).

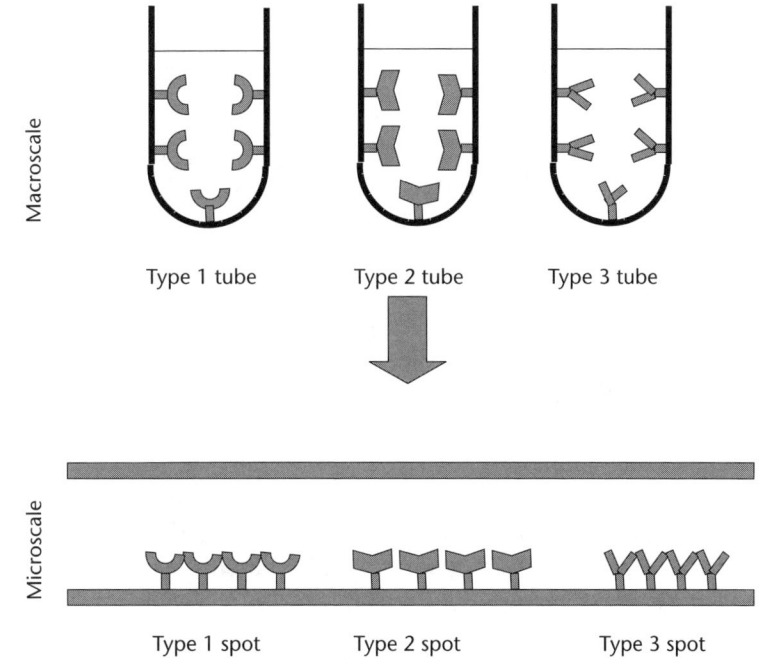

Figure 5.6   Macroscale immunoreactions in Eppendorf tubes to microscale immunoreactions in microfluidic channel.

ultraprecise sensitivity have led to improvements in detection methods. Ultrasensitive detection is a subject of many studies and is not the subject of this book. But one has to keep in mind that any development of a biochip must take into account the definition of a detection device. Optic methods by fluorescence are very widely used. The principle is to attach a fluorescent bead to the immobilized target molecule and then to implement a sensitive reception of the emitted light (Figure 5.10).

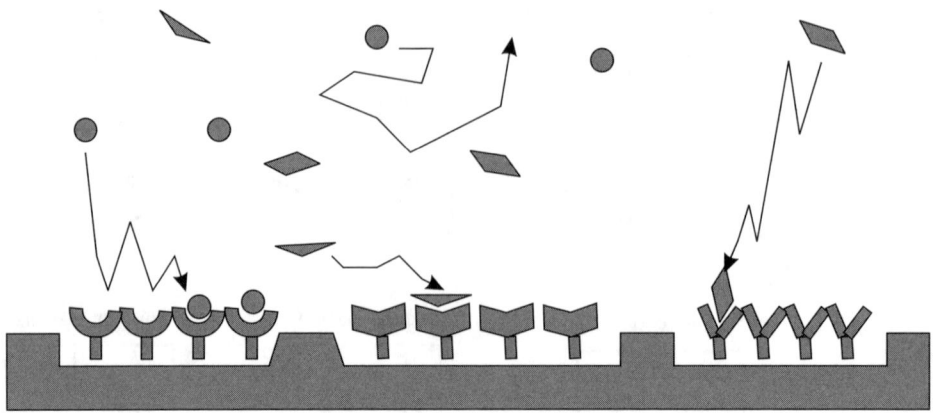

Figure 5.7 Principle of biorecognition of macromolecules submitted to molecular diffusion on a microplate.

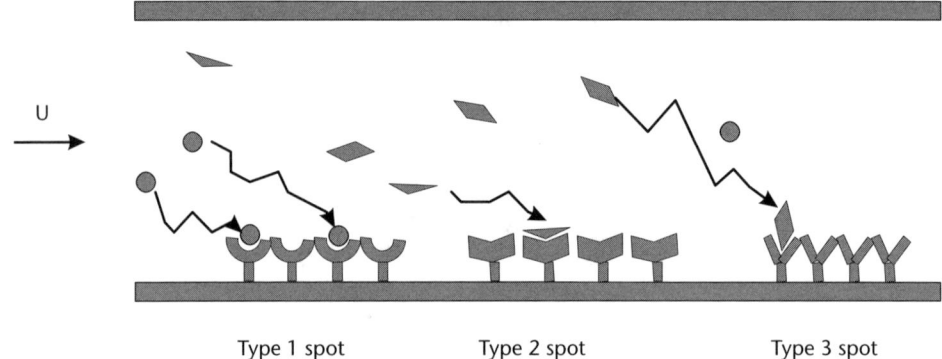

Type 1 spot   Type 2 spot   Type 3 spot

Figure 5.8 Principle of biorecognition of macromolecules submitted to diffusion-advection in a circulating microchannel.

5.3 Biochemical Reactions

Biochemical reactions can be extremely complex. We will not go into the details of these reactions, but only treat the reactions kinetics. The precise chemical interactions leading to reaction is the domain of the biologists and chemists. In biotechnology, we are mostly interested in the kinetics of the reaction to know and improve its efficiency and reduce its duration.

5.3.1 Rate of Reaction

5.3.1.1 Definition

Consider a chemical reaction of the form

$$A + nB \rightarrow mC + D \tag{5.1}$$

and note the molar concentration of a participant J at some instant by the symbol $[J]$. The rate of consumption of a reactant at a given time is defined by $-d[R]/dt$, where R is either A or B; this rate is a positive quantity (Figure 5.11). The rate of

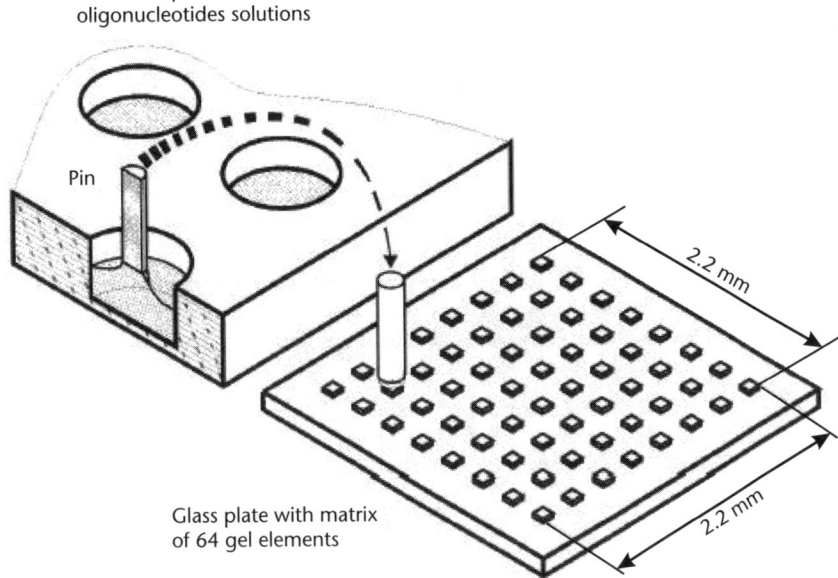

Figure 5.9 Schematic view of one of the first "biochips." Miniaturization is not yet achieved; the dimensions of the plate are millimetric.

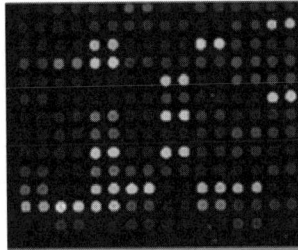

Figure 5.10 Examples of DNA detection by fluorescence in a DNA biochip. Fluorescent spots correspond to a positive reaction and can be treated by image processing.

formation of one of the products C or D—which we denote P—is defined by $d[P]/dt$ and is also a positive quantity (Figure 5.12).

By considering the stoichiometry of (5.1), we deduce the relations

$$\frac{d[D]}{dt} = \frac{1}{m}\frac{d[C]}{dt} = -\frac{d[A]}{dt} = -\frac{1}{n}\frac{d[B]}{dt} \tag{5.2}$$

and the rate of reaction is uniquely defined by

$$v = v_D = \frac{1}{m}v_C = v_A = \frac{1}{n}v_B \tag{5.3}$$

where

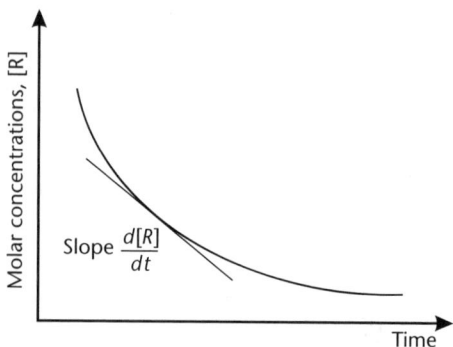

Figure 5.11   Definition of the reaction rate as the slope of the tangent drawn to the curve, showing the variation of concentration of reagent with time.

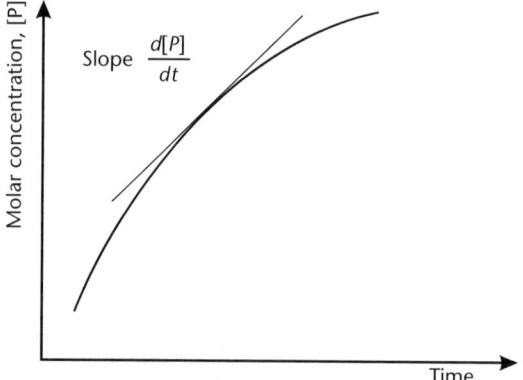

Figure 5.12   Definition of the reaction rate as the slope of the tangent drawn to the curve, showing the variation of concentration of product with time.

$$v_D = \frac{d[D]}{dt}; v_C = \frac{d[C]}{dt}; v_A = -\frac{d[A]}{dt}; v_B = -\frac{d[B]}{dt} \qquad (5.4)$$

5.3.1.2   Rate Laws

Rate of reaction is essential to determine the kinetics of the reaction; it is also a guide to the mechanism of the reaction, for any proposed mechanism should be consistent with the observed rate law. Formally, the rate of reaction is a function of the concentration of the species present in the reaction [3]

$$v = f([A],[B],[C],[D]) \qquad (5.5)$$

For gases, the rate of reaction can be deduced from gas kinetics theory [4] and is expressed by the simple expression

$$v = k[A][B] \qquad (5.6)$$

In a liquid phase, the reaction rate is more empirical and is often—but not always—obtained by an expression of the type

$$v = k[A]^a [B]^b \tag{5.7}$$

where k, a, and b are coefficients independent of time. The parameter k is called the rate constant.

5.3.1.3 Reaction Order

If a reaction rate is described by a formula of the type (5.7)—and this is a frequent case—it is possible to define the order of the reaction with respect to a species by its exponent and an overall reaction order by the sum of the exponents corresponding to each species

$$o = a + b \tag{5.8}$$

Note that the order is not necessarily an integer; for example, the reaction rate may be

$$v = k[A]^{\frac{1}{2}} [B]$$

and the order of the reaction is 3/2. Some reactions may obey a zero order rate law (i.e., the reaction rate does not depend on the concentrations of the species, just on the parameter k). Some comments on the "constant" k are needed at this stage.

First, the unit of k depends on the order of the reaction. For a zero order reaction, k is expressed in mole/m^3/sec; for a first order reaction, k is dimensionally a frequency and expressed in s^{-1}; for a second order reaction, the unit of k is m^3/mol/sec.

Second, the magnitude of k for a given reaction order may be very different; the reason is that for a chemical reaction to proceed, the concerned molecules must have a closely defined state (for example, orientation). This condition is related to a probabilistic behavior, which in turn depends on the activation energy—sometimes called the *Arrhenius* factor. Because the range of activation energy is large, the rate constants take very different values.

5.3.1.4 Temperature Dependence of the Reaction Constant

A closer look at the Arrhenius factor is obtained by considering the dependency of k on the temperature T. This property is often used to modify the equilibrium state of biochemical reactions like DNA hybridization, as we will later see. The dependency of the rate constant k on the temperature T is given by the Arrhenius law

$$\ln k = \ln A - \frac{E_a}{RT} \tag{5.9}$$

The two parameters A and E_a/R are called the Arrhenius parameters; more specifically, A is the "frequency" factor and E_a/R is the activation energy. These parameters are usually determined graphically from experimental results, as shown in Figure 5.13. The intercept is $\ln A$ and the value of the slope $-E_a/R$.

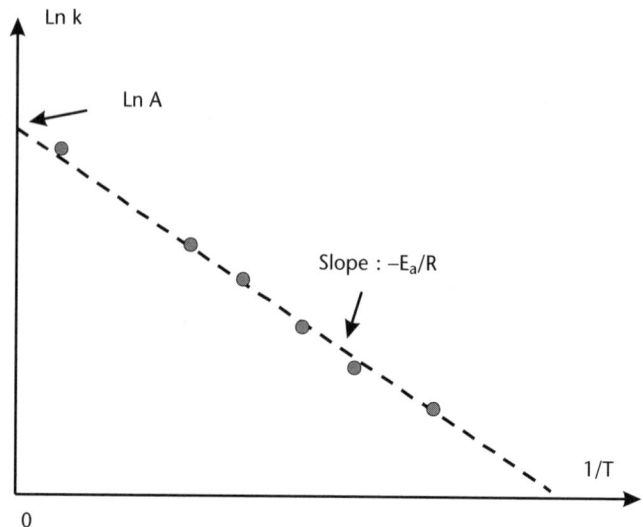

Figure 5.13   Schematic plot of (5.9) showing the Arrhenius parameters A and E_a/R.

High activation energy corresponds to a very steep slope and a very important dependency of k on the temperature. For the rate constant k to really be a constant, the activation energy must be zero. Another form of (5.9) is

$$k = Ae^{-\frac{E_a}{RT}} \tag{5.10}$$

Under this form, an interpretation of the rate constant is the rate of *successful collisions* between reacting molecules [3]. The activation energy represents the minimum kinetic energy that reactants must have in order to form products, and the frequency term A corresponds to the rate at which collisions occur.

5.3.1.5   Rate Laws and Reaction Kinetics

Rate laws are differential equations, and their integration is the concentration as a function of time (i.e., the reaction kinetics). However, their integration is seldom possible analytically. Take the example of the first-order unimolecular reaction

$$A \rightarrow P$$

We have the following reaction rate

$$\frac{d[A]}{dt} = -k[A]$$

and the reaction kinetics is

$$[A] = [A]_0\, e^{-kt} \tag{5.11}$$

5.3.1.6 Near Equilibrium Reactions

Very often a reaction is partly reversible—the product is formed and at the same time dissociate according to

$$A \rightarrow P$$
$$P \rightarrow A$$

The rate laws for the two reactions are

$$v = \frac{-d[A]}{dt} = k[A]$$

$$v' = \frac{-d[P]}{dt} = k'[P]$$

The concentration $[A]$ is reduced by the first reaction and increases by the reverse reaction, and the net rate of change is

$$\frac{d[A]}{dt} = -k[A] + k'[P] \tag{5.12}$$

If the initial concentration is $[A]_0$, and if there is no initial concentration of $[P]$, then

$$[A] + [P] = [A]_0$$

and the reaction kinetics is given by

$$\frac{d[A]}{dt} = -k[A] + k'([A]_0 - [A]) = (-k + k')[A] + k'[A]_0 \tag{5.13}$$

This type of kinetics is classic, and its mathematical structure will be investigated more closely in Section 5.3.3.1.

5.3.1.7 Consecutive Reactions

Some reactions proceed through the formation of an intermediate. Consider, for example, the consecutive reactions

$$A \rightarrow I \rightarrow P$$

with

$$v = \frac{-d[A]}{dt} = k_A[A]$$

$$v' = \frac{-d[I]}{dt} = k_I[I]$$

Let us consider the concentration in the intermediate product $[I]$

$$\frac{d[I]}{dt} = k_A[A] - k_I[I] \tag{5.14}$$

and the concentration $[P]$ is given by the differential equation

$$\frac{dP}{dt} = k_I[I] \tag{5.15}$$

The first of the rate laws is an ordinary first-order decay and, from (5.11), we can write

$$[A] = [A]_0\, e^{-k_A t} \tag{5.16}$$

Substitution of (5.16) in (5.14) yields

$$\frac{d[I]}{dt} = k_A[A]_0\, e^{-k_A t} - k_I[I]$$

and upon integration, assuming that $[I]_0 = 0$

$$[I] = \frac{k_A}{k_I - k_A}\left(e^{-k_A t} - e^{-k_I t}\right)[A]_0 \tag{5.17}$$

If we notice that, at all times,

$$[A_0] = [A] + [I] + [P]$$

We obtain the kinetics of production of $[P]$

$$[P] = \left\{1 + \frac{k_A e^{-k_I t} - k_I e^{-kA t}}{k_I - k_A}\right\}[A]_0 \tag{5.18}$$

Concentrations in $[A]$, $[I]$, and $[P]$ are sketched in Figure 5.14.

This example corresponds to the decay of radioactive elements, such as the reaction

$$^{239}U \rightarrow\, ^{239}Np \rightarrow\, ^{239}Pu$$

5.3.2 Michaelis-Menten Model

5.3.2.1   Presentation of the Model

Now, we can tackle a very important group of reactions in biotechnology—the enzymatic reactions. Enzymatic reactions are of utmost importance in biotechnology for two reasons: first, they are used to break proteins into smaller pieces called peptides that can be analyzed by a mass spectrometer; second, they are a powerful method for amplifying detection in biorecognition processes. In fact,

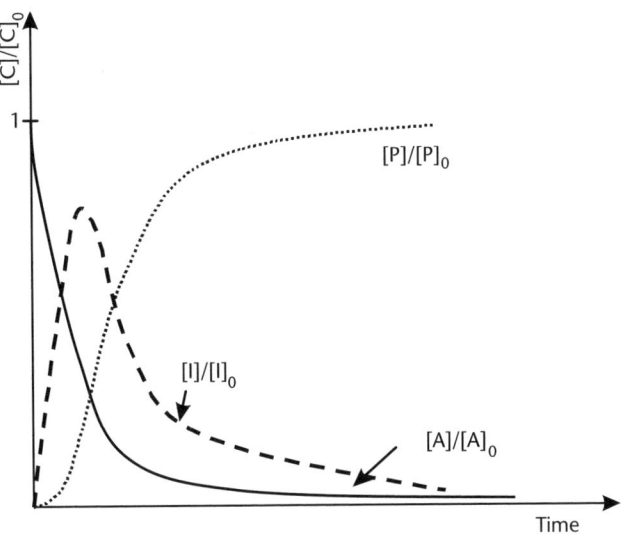

Figure 5.14   Concentrations of *A, I,* and *P* as a function of time.

enzymes are only catalysts for the reaction. In an enzyme-catalyst reaction, a substrate is converted into products, and the reaction rate depends on enzyme concentration.

The catalyst-driven reaction can be written by the symbolic expression

$$E + S \leftrightarrow ES \rightarrow P + E \tag{5.19}$$

where *E, S,* and *P* refer respectively to enzyme, substrate, and product concentration. Note that the name *substrate* corresponds to the species that is undergoing the chemical reaction. The notation *ES* refers to an intermediate state where the substrate *S* is bonded to the enzyme *E* (Figure 5.15).

If we note k_1, k_{-1}, and k_2, the rate constant of the two reactions of (5.19), the kinetics of *ES* binding is given by

$$\frac{d[ES]}{dt} = k_1 [E][S] - k_2 [ES] - k_{-1} [ES] \tag{5.20}$$

and the product concentration kinetics is

$$V = \frac{d[P]}{dt} = k_2 [ES] \tag{5.21}$$

The Michaelis-Menten approach is based on the simplification that assumes that the rate of production of *ES* concentration is constant—that is,

$$\frac{d[ES]}{dt} = k_1 [E][S] - k_2 [ES] - k_{-1} [ES] = 0 \tag{5.22}$$

Then, we have the relation

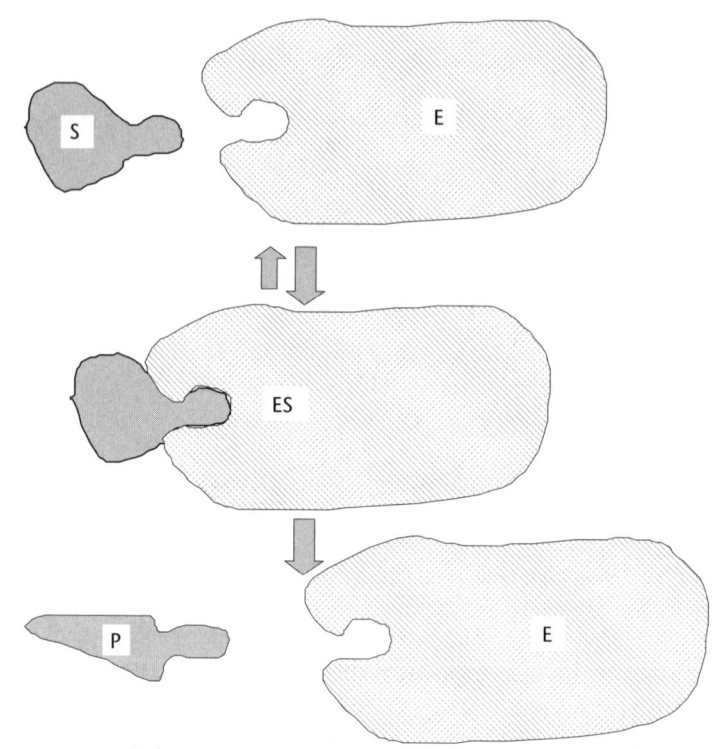

Figure 5.15 Schematic view of the enzymatic reaction.

$$[E][S] = \frac{k_2 + k_{-1}}{k_1}[ES] \qquad (5.23)$$

If we note that the total (initial) concentration of enzyme is

$$[E]_0 = [E] + [ES] \qquad (5.24)$$

We can eliminate $[E]$ and $[ES]$ from (5.22), (5.23), and (5.24), and we deduce the rate of the reaction

$$V = \frac{k_2[E]_0}{1 + \left(\dfrac{k_2 + k_{-1}}{k_1}\right)\dfrac{1}{[S]}} \qquad (5.25)$$

Introducing the notations

$$V_{\max} = k_2[E]_0 \qquad (5.26)$$

and

$$K_m = \frac{k_2 + k_{-1}}{k_1} \qquad (5.27)$$

We obtain the Michaelis-Menten law

$$V = \frac{V_{max}}{1 + \dfrac{K_m}{[S]}} \tag{5.28}$$

By noting that the concentration of substrate S decreases with the concentration of product P according to

$$[S] = [S]_0 - [P] \tag{5.29}$$

and if we recall from (5.21) that V is the rate of production of P, integration of (5.28) gives the relation between the concentration of product $[P]$ and substrate $[S]$

$$V_{max}t = [P] + K_m Ln \frac{[S]_0}{[S]_0 - [P]} \tag{5.30}$$

The constant has been adjusted to have $[P] = 0$ at $t = 0$. Relation (5.30) is implicit. The kinetics of P derived from (5.30) is schematically represented in Figure 5.16.

It is easy to see that the Michaelis-Menten law can be cast under the form

$$\frac{1}{V} = \frac{1}{V_{max}} \left(1 + \frac{K_m}{[S]} \right) \tag{5.31}$$

This form is called the Lineweaver-Burk expression of the Michaelis-Menten relation. It is convenient to determine the kinetic constants K_m and V_{max}. If we rewrite (5.31) under the form

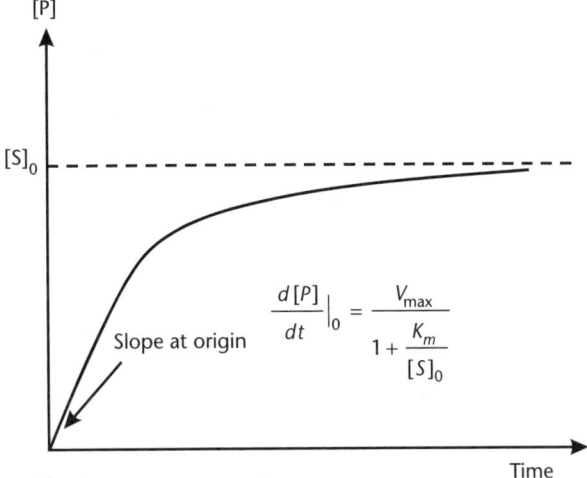

Figure 5.16 Michaelis-Menten kinetics for product concentration.

$$\frac{1}{V} = \frac{1}{V_{max}} + \frac{K_m}{V_{max}} \frac{1}{[S]}$$

we see that the plot of the reciprocal velocity $1/V$ against the reciprocal substrate concentration $1/[S]$ is linear; the intercept is $1/V_{max}$, and the slope is K_m/V_{max} (Figure 5.17).

5.3.2.2 More Insight on the Michaelis-Menten Law

The Michaelis-Menten model is only approximate because of the hypothesis of a constant rate of production of the complex ES. In reality, the kinetics differ somewhat from that of Figure 5.16. By looking closely at a plot of the kinetics of production of concentration P, we find three different parts corresponding to three different regimes (Figure 5.18).

In the following equations, we investigate the physics behind the three different parts of the reaction kinetics [5, 6]. The first part corresponds to the early stage of the reaction $t < t_0$. During this stage, the concentration $[ES]$ may be considered small, and (5.20) collapses to

$$\frac{d[ES]}{dt} = k_1 [E][S] = k_1 [E]_0 [S]_0 = const$$

Integrating this relation and substituting the result in (5.21) yields the parabolic kinetics

$$[P] = k_1 k_2 [E]_0 [S]_0 \frac{t^2}{2} \tag{5.32}$$

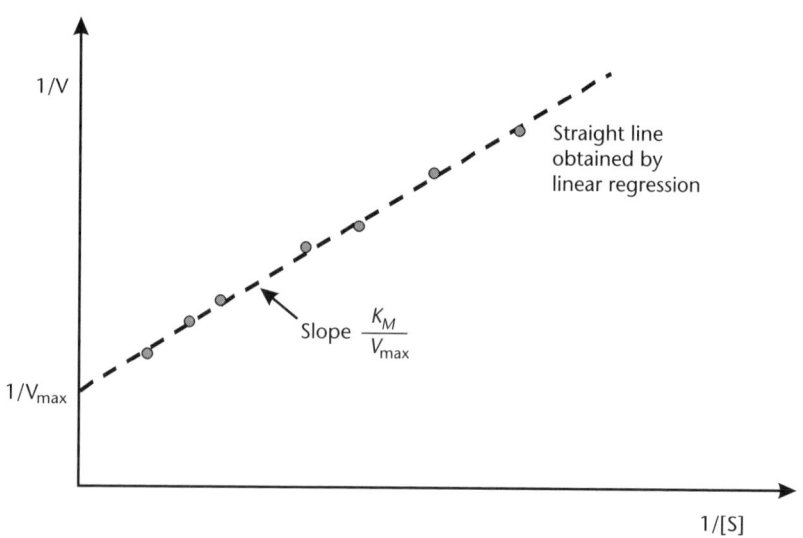

Figure 5.17 Lineweaver-Burk linear relation for an enzymatic reaction. A simple linear regression produces the two Michaelis-Menten parameters K_m and V_{max}.

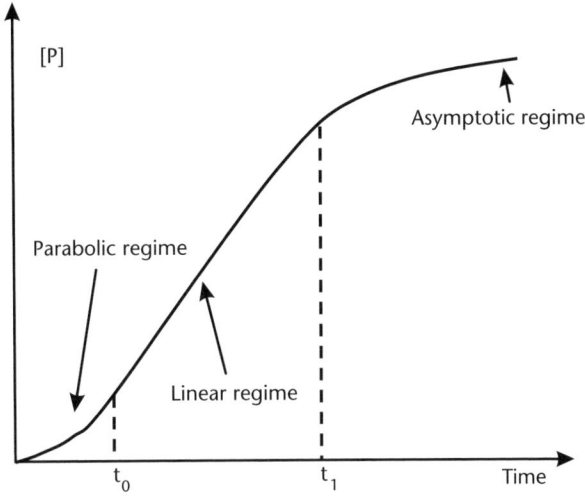

Figure 5.18 The three regimes of the enzymatic reaction.

After this early stage, the reaction acquires a steady-state rate, according to the Michaelis-Menten approach

$$\frac{d[ES]}{dt} = 0$$

We integrate (5.21) to obtain the linear form

$$[P] = at + b$$

Using continuity considerations, the preceding equation can be rewritten as

$$[P] = k_1 V_{max}[S]_0 t_0 \left(t - \frac{t_0}{2} \right) \qquad (5.33)$$

This linear form corresponds to the second regime in Figure 5.18. At the end of the reaction, the substrate S becomes depleted, the concentration $[S]$ may be neglected in (5.20), and the equation collapses to

$$\frac{d[ES]}{dt} = -(k_2 + k_{-1})[ES]$$

This equation can be integrated as an exponential law, and using (5.21), we find

$$\frac{d[P]}{dt} = \frac{k_2 c_1}{(k_2 + k_{-1})} \exp(-(k_2 + k_{-1})t)$$

where c_1 is a constant. Integrating once more, we obtain the form

$$[P] = -\frac{k_2 c_1}{(k_2 + k_{-1})} \exp\left(-(k_2 + k_{-1})t\right) + c_2$$

Using the value of the asymptote $[P]_\infty$ and continuity at time t_1, the *asymptotic* regime is defined by

$$[P] = [P]_\infty - \frac{V_{max}[S]_0 t_0}{K_m} \exp\left[-k_1 K_m (t - t_1)\right] \tag{5.34}$$

So the three different regimes and their assumptions have been identified: parabolic at first when $[ES]$ is small, then linear when $d[ES]/dt = 0$, and finally asymptotic when $[S]$ becomes small. The system of (5.32), (5.33), and (5.34) is more accurate that the Michaelis-Menten law but requires the knowledge of four parameters instead of two: V_{max}, K_m, k_2, and $[P]_\infty$ (it can be shown that the times t_0 and t_1 may be deduced from considerations on the derivability of the kinetic curve). In the following section, we show an example of the difference between the two models.

5.3.2.3   Example of an Enzymatic Reaction

In this example, we set up a catalyst reaction between a synthetic protein and an enzyme. Consider a substrate composed of molecules of benzoyl-arginyl-ethyl-esther (BAEE), which is a synthetic protein, reacting in the presence of trypsine, which is an enzyme. The experiment consists of mixing the substrate S (synthetic protein) with the enzyme E (trypsine) in a small beaker (Figure 5.19).

Reaction kinetics is measured by an optical method based on the absorbance of light by the reaction product B. Figure 5.20 shows the absorbance curves at different times for three different initial concentrations of BAEE. Kinetics plots are then deduced from light absorbance.

Michaelis-Menten model and the piecewise analytical model from preceding section have been used to interpret the experimental data (Figure 5.21). The piecewise analytical model fits better with the experimental data. However, as we mentioned earlier, it requires more physical parameters than the Michaelis-Menten model, which remains a good tradeoff between simplicity and precision.

5.3.3   Adsorption and the Langmuir Model

5.3.3.1   Langmuir Model

Another very important class of reactions in biotechnology is the adsorption of molecules on a solid functionalized surface. In particular, it is the case of DNA

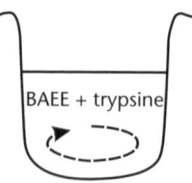

Figure 5.19   Mixing BAEE and trypsine.

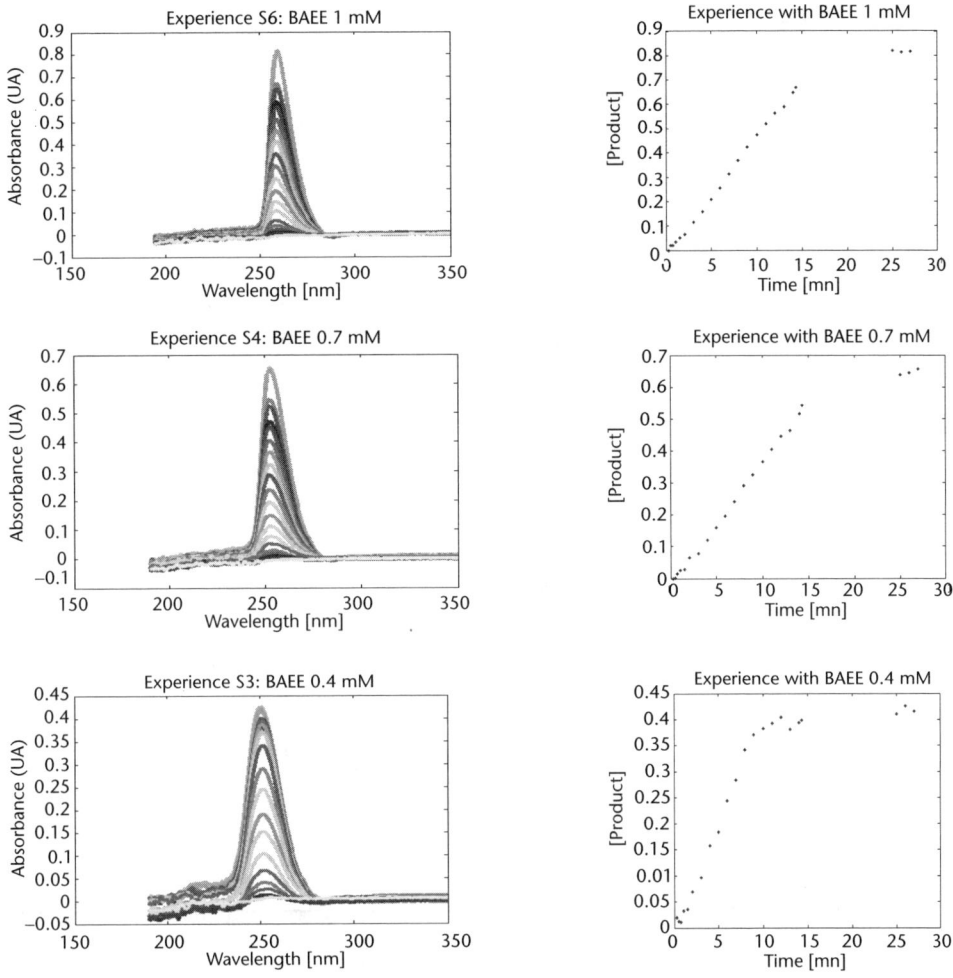

Figure 5.20 BAEE reacting with trypsine in a beaker. Left, measurements of absorbance of light at 500 nm; right, experimental kinetics curves (concentration of Benzoyl versus time). From top to bottom: initial concentrations of substrate 1, 0.7, and 0.4 mM. (Courtesy of CEA/LETI.)

hybridization. In such a reaction, there are three components: first, a "free" substrate in a buffer fluid sometimes called *target* or *analyte*, in concentration $[S]$; second, a surface concentration $[\Gamma]_0$ of ligands—or capture sites—immobilized on a functionalized surface; and third, a product that is the surface concentration of adsorbed targets that we denote $[\Gamma]$ (Figure 5.22). Note that $[S]$ is a volume concentration (unit mole/m^3) whereas $[\Gamma]$ and $[\Gamma]_0$ are surface concentration (unit mole/m^2). Such a kinetic is called a Langmuir-Hinshelwood mechanism.

The reaction is weakly reversible, because targets are constantly captured by ligands, and they constantly dissociate (at a smaller rate). The reaction may be symbolized by

$$S \rightarrow \Gamma$$
$$\Gamma \rightarrow S$$

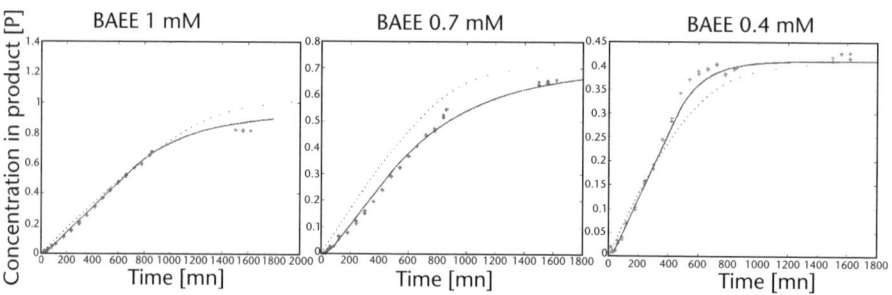

Figure 5.21 Comparison of reaction kinetics between experiments (dots), Michaelis-Menten model (dotted line), and piecewise analytical model (continuous line).

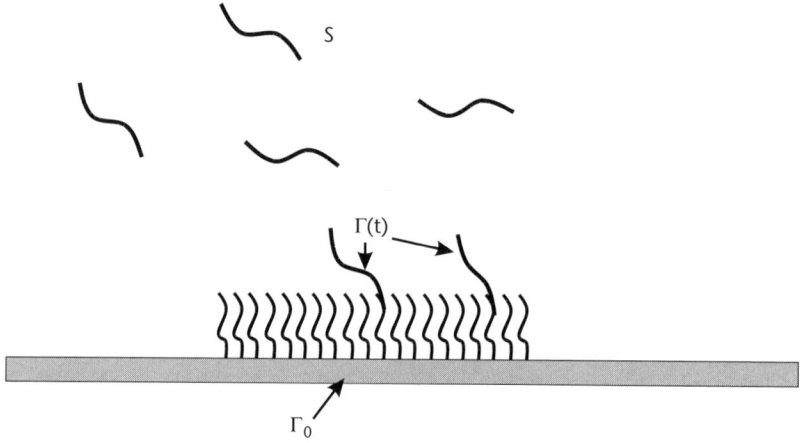

Figure 5.22 Adsorption of targets on a surface functionalized with immobilized ligands.

In the case of adsorption, the definition of the reaction rates are somewhat different from the definition of the usual chemical rates, mainly because the immobilization of the substrate S depends not only on the volume concentration at the wall, but also on the available sites for adsorption. Thus, we can write

$$v = \frac{-d[S]}{dt} = k_{on}([\Gamma]_0 - [\Gamma])[S]_w$$
$$v' = \frac{-d[\Gamma]}{dt} = k_{off}[\Gamma]$$

(5.35)

where k_{on} and k_{off} are called respectively the adsorption and dissociation rates, and $[S]_w$ is the concentration at the wall. For simplicity we will note $\Gamma = [\Gamma]$, $c = [S]$, and $c_0 = [S]_w$. The net rate of adsorption is then

$$\frac{d\Gamma}{dt} = k_{on}c_0(\Gamma_0 - \Gamma) - k_{off}\Gamma$$

(5.36)

Equation (5.36) can be rewritten under the form

$$\frac{d\Gamma}{dt} = k_{on}c_0\Gamma_0 - \left(k_{on}c_0 + k_{off}\right)\Gamma \tag{5.37}$$

Equation (5.37) can be integrated and we obtain

$$\frac{\Gamma}{\Gamma_0} = \frac{k_{on}c_0}{k_{on}c_0 + k_{off}}\left[1 - e^{-\left(k_{on}c_0 + k_{off}\right)t}\right] \tag{5.38}$$

Using (5.38), we obtain the surface concentration kinetics shown in Figure 5.23.

At small times, the exponential term in (5.38) can be developed in a Taylor expansion, and the surface concentration kinetics is the linear function of the time defined by

$$\Gamma = k_{on}c_0\Gamma_0 t \tag{5.39}$$

Equation (5.39) indicates that the kinetics described by the Langmuir equation (5.36) is rapid if the term $k_{on}c_0$ is large (i.e., when the adsorption constant on the surface and the concentration in molecules are large). For longer times, the surface concentration approaches an asymptotic value defined by

$$\frac{\Gamma_\infty}{\Gamma_0} = \frac{k_{on}c_0}{k_{on}c_0 + k_{off}} \tag{5.40}$$

It can be verified in (5.38) that in the case where k_{off} is zero, the asymptotic value is then Γ_0, and the surface is becomes totally saturated. The larger the coefficient k_{off}, the smaller the value of Γ_∞/Γ_0.

5.3.3.2   Adsorption and Desorption

Suppose that after the hybridization has reached its asymptotic value, the remaining targets or analytes in the solution are suddenly washed out. Desorption is then the driving mechanism, and the corresponding kinetics is schematized by Figure 5.24.

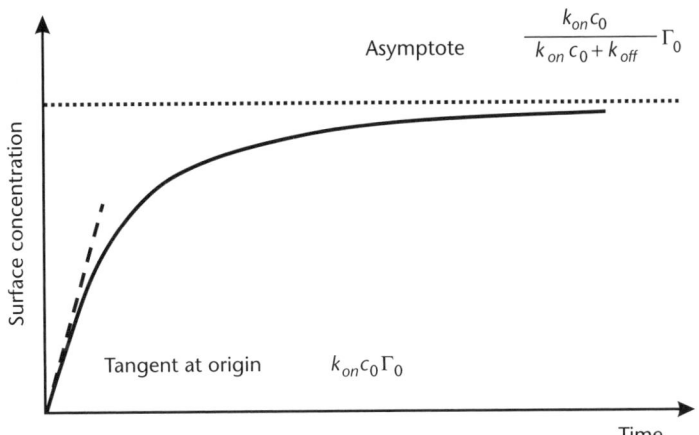

Figure 5.23   Kinetics of surface concentration from (5.38).

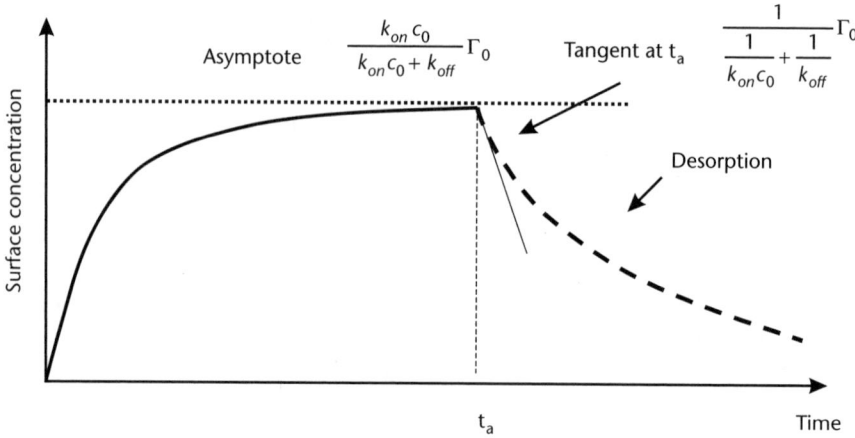

Figure 5.24 Kinetics of adsorption and desorption.

The starting time for desorption is the time t_a, and the surface concentration at this instant is Γ_a

$$\Gamma_a = \frac{k_{on}c_0}{k_{on}c_0 + k_{off}}\Gamma_0 \qquad (5.41)$$

The Langmuir equation for desorption is

$$\frac{d\Gamma}{dt} = -k_{off}\Gamma \qquad (5.42)$$

and the kinetics of desorption is

$$\frac{\Gamma}{\Gamma_a} = e^{-k_{off}(t-t_a)} \qquad (5.43)$$

Desorption kinetics follows an inverse exponential law (Figure 5.24). The tangent to the desorption kinetic curve at $t = t_a$ is given by

$$\frac{\Gamma}{\Gamma_a} = 1 - k_{off}(t - t_a) \qquad (5.44)$$

and the derivative at $t = t_a$ is

$$\frac{d\Gamma}{dt}\bigg|_{t=t_a} = -k_{off}\Gamma_a = -\frac{k_{off}k_{on}c_0}{k_{on}c_0 + k_{off}}\Gamma_0 \qquad (5.45)$$

This last formula may be written under the form

$$\frac{d\Gamma}{dt}\Big|_{t=t_a} = -\frac{\Gamma_0}{\dfrac{1}{k_{on}c_0} + \dfrac{1}{k_{off}}} \tag{5.46}$$

When desorption follows adsorption, the kinetics of desorption depends not only on the desorption coefficient k_{off}, but also on the values of Γ_0 and k_{on}. This property in shown in Figure 5.25, where different desorption kinetics are sketched, depending on the value of the saturation level.

5.3.4 Biological Reactions

5.3.4.1 Introduction

In the preceding sections, we have dealt with chemical and biochemical reactions, in the sense that the reactants were chemical or biochemical. In biology, there are slightly different types of reactions, mainly because one has to take into account the rate of birth or death of living organisms by introducing a source or sink term in the reaction equations. However, these reactions have basically a mathematical formulation similar to chemical and biochemical reactions.

5.3.4.2 Predator-Prey Systems and the Lotka-Volterra Equations

Volterra developed this model in 1925 to predict the evolution of populations of animals in biology (fish population in the Adriatic Sea); nearly at the same time, Lotka derived the same model for some chemical reactions [7, 8]. In the frame of this book, we are mostly interested in the biochemical aspect of the model, and we present it briefly to introduce the special form of the competition terms in the system of Lotka-Volterra equations. We will show later that competition-displacement reactions for immunoassays present similarity with the predator-prey model, and we will use the competition terms extracted from the Lotka-Volterra model.

Biologists have developed models to predict the evolution of two interconnected populations, especially if one population is the prey and the other is the predator. It

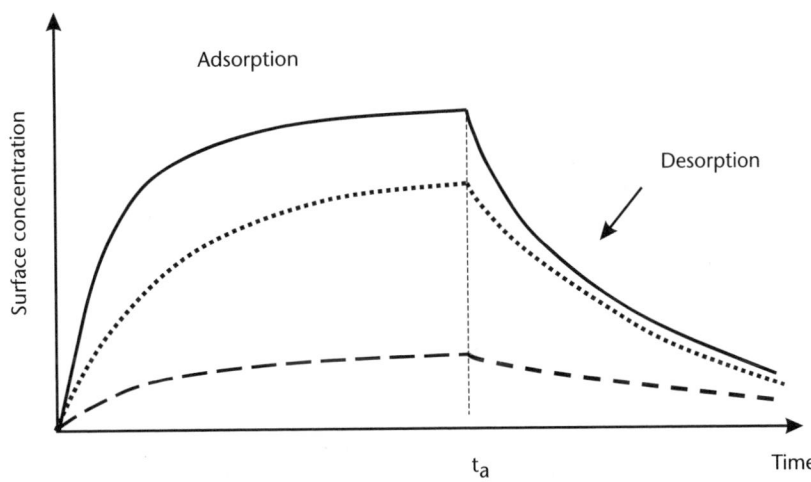

Figure 5.25 Different adsorption and desorption kinetics, depending on the kinetic constants.

has been observed that the fluctuations of the two populations are closely linked (Figure 5.26).

The simplest—yet very interesting—model is that of Lotka-Volterra. If A and B represent the populations of prey and predators, their time evolution is given by [9, 10]

$$\frac{\partial A}{\partial t} = aA - bAB$$
$$\frac{\partial B}{\partial t} = -cB + dAB$$

(5.47)

In (5.47) the term aA represents the growth of population A if predators are absent—a being the rate of birth—and the term $-bAB$ represents the decrease in the number of prey due to the action of the predators (for these reasons, it is proportional to A and to B). On the other hand, the term $-cB$ represents the mortality rate of predators—b being the rate of deaths—and the term dAB represents the prey contribution—as a source term—to the predator growth rate (proportional to A and B).

Mathematically speaking, the system (5.47) is strongly coupled and nonlinear. It is also structurally not stable. However, it bears lots of the physics of the evolution of the prey-predator system. A first step in analyzing the Lotka-Volterra model is to render the system nondimensional by introducing the new parameters

$$\tau = at; \alpha = \frac{c}{a}$$
$$u = d\frac{A}{c}; v = b\frac{B}{a}$$

(5.48)

Substituting (5.48) in (5.47) yields

$$\frac{\partial u}{\partial \tau} = u(1 - v)$$
$$\frac{\partial v}{\partial \tau} = av(u - 1)$$

(5.49)

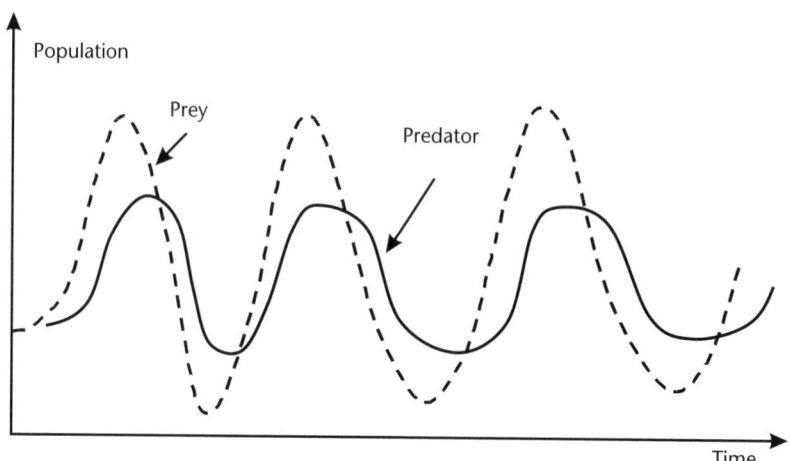

Figure 5.26   Time evolution of populations of prey and predators.

In the (u, v) plane, we obtain

$$\frac{dv}{du} = \alpha \frac{v(u-1)}{u(1-v)}$$

The variables u and v can be separated

$$\frac{(1-v)}{v} dv = \alpha \frac{(u-1)}{u} du \qquad (5.50)$$

Integration of (5.50) produces the phase trajectories

$$\alpha u + v - \ln(u^{a} v) = H \qquad (5.51)$$

For a given H, the trajectories in the phase plane are closed, as illustrated in Figure 5.27.

The diagram of Figure 5.27 shows that the two populations are linked and form the shifted cycles of Figure 5.26. For our concerns here, we will keep in mind that the nonlinear terms bAB and dAB represent the interactions between species A and B, especially the first term bAB, which represents the competition between the species.

5.4   Biochemical Reactions in Microsystems

In the preceding section, we have investigated the kinetics of biochemical reactions. However, in reality they can seldom be considered alone without taking into account other physical phenomena like diffusion or transport. Indeed, the reactants are usually injected into the microchamber in which they later diffuse and react. Thus, it is important to consider the global problem of advection-diffusion, coupled

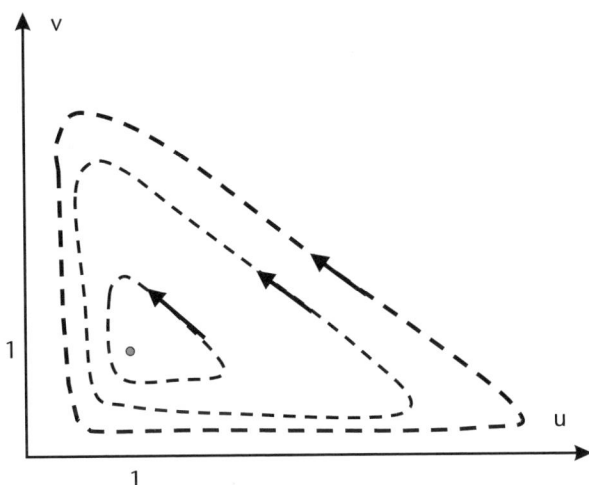

Figure 5.27 Closed phase plane trajectories from (5.45) with various H, corresponding to the Lotka-Volterra system. The arrows denote the direction of change with increasing time τ.

with the biochemical reaction itself. We will consider next the kinetics of these coupled problems.

The reaction itself may be performed in two different ways: first, in the whole volume of the reaction chamber; second, on a functionalized surface located on the wall of the reaction chamber. The first type is called a *homogeneous reaction*; the second is called a *heterogeneous reaction*.

Thus, we will consider successively the reactions kinetics coupled with advection-diffusion phenomena for homogeneous or heterogeneous situations.

5.4.1 Homogeneous Reactions

5.4.1.1 Governing Equations

Let us consider a second-order reaction of the type

$$A + nB \rightarrow mC$$

occurring in a fluid volume where the reactants A and B are transported by a flow of velocity u. If we recall the advection-diffusion equation (Chapter 4), and notice that there is now a sink-source term for concentration, the governing equations are [11]

$$\frac{\partial c_A}{\partial t} + u \nabla c_A = D_A \Delta c_A - k c_A c_B$$

$$\frac{\partial c_B}{\partial t} + u \nabla c_B = D_B \Delta c_B - n k c_A c_B \qquad (5.52)$$

$$\frac{\partial c_C}{\partial t} + u \nabla c_C = D_C \Delta c_C + m k c_A c_B$$

where D_A, D_B, and D_C are the diffusion coefficients of species A, B, and C, and k is the reaction rate. In (5.52), we have adopted the concentration notations $c_A = [A]$, $c_B = [B]$, and $c_C = [C]$. Note that the sink-source term has the characteristic form of a second-order reaction $k[A][B]$. The advection-diffusion equations are in this case nonlinear due to the nature of the sink-source term. Moreover, the two first equations in $[A]$ and $[B]$ are strongly coupled via their sink term. The third equation for $[C]$ is only weakly coupled to the two others. The solution of the system is not easy and requires a numerical approach.

Typically, there are two main cases of problems. Note that τ_C is the characteristic time of the reaction and τ_M is the mixing time, which depends on dynamic fluid motion or only diffusion, and define a nondimensional number by

$$Da = \frac{\tau_C}{\tau_M} \qquad (5.53)$$

Da is called the Dammköhler number.

For a purely diffusive situation, the diffusion mixing time τ_M is of the order of

$$\tau_M \approx \frac{L^2}{D}$$

where L is the characteristic dimension of the microsystem, and D is the order of magnitude of the diffusion coefficients of the reactants. After substitution, one obtains

$$Da = \frac{D\tau_c}{L^2}$$

If Da is large, the reaction time is much larger than the mixing time. The concentrations in [A] and [B] can then be considered uniform in the reacting volume, and (5.52) collapses to

$$\frac{\partial c_A}{\partial t} \approx -kc_A c_B$$

$$\frac{\partial c_B}{\partial t} \approx -nkc_A c_B \qquad (5.54)$$

$$\frac{\partial c_C}{\partial t} \approx mkc_A c_B$$

This system is considerably easier to solve, since it does not requires the knowledge of the velocity field and of the diffusion process.

If Da is small, the picture is much more complicated. There are reaction fronts that form and diffuse progressively before obtaining a homogeneous final state [12]. Numerical treatment is usually required for such systems.

5.4.1.2   Reaction-Diffusion at a Front Separating Two Reactants

Start from the same second-order reaction

$$A + B \rightarrow C$$

and suppose that it takes place in a volume at rest (no convective transport), as sketched in Figures 5.28 and 5.29.

In the case of a one-dimensional space (Figure 5.29), we have indicated the solution for the concentration alone in Chapter 3. This solution was an error (*erf*)

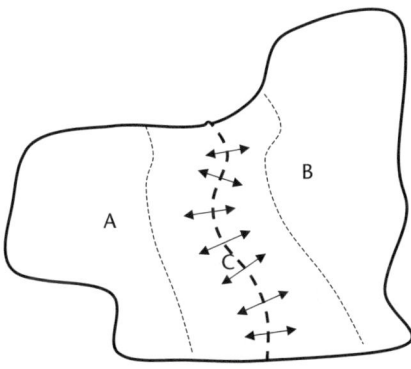

Figure 5.28   Reaction $A + B \rightarrow C$ on a diffusion front.

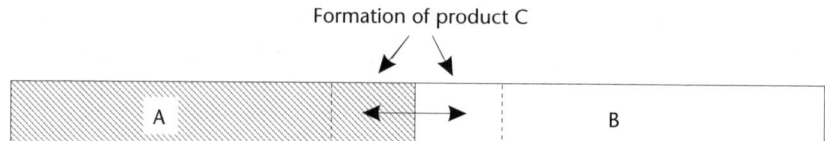

Figure 5.29   Reaction $A + B \rightarrow C$ in one-dimensional geometry.

function, and the concentration spreads proportionally to the square root of time. Now we add a second-order reaction. The equations governing this type of reaction diffusion in one-dimensional geometry are

$$\frac{\partial c_A}{\partial t} = D_A \frac{\partial^2 c_A}{\partial x^2} - kc_A c_B$$

$$\frac{\partial c_B}{\partial t} = D_B \frac{\partial^2 c_B}{\partial x^2} - kc_A c_B \tag{5.55}$$

together with the initial conditions

$$t = 0 \quad c_A(x,0) = c_{A,0} \quad c_B(x,0) = 0 \quad \text{for} \quad x < 0$$

$$t = 0 \quad c_A(x,0) = 0 \quad c_B(x,0) = c_{B,0} \quad \text{for} \quad x > 0 \tag{5.56}$$

The system of (5.55) can be transformed under a more symmetrical form by using the following transformations

$$x^* = \frac{x}{L}, t^* = \left(\frac{D}{L^2}\right)t, D = \sqrt{D_A D_B}, L = \left(\frac{D}{k\sqrt{c_{A,0} c_{B,0}}}\right)^{\frac{1}{2}} \tag{5.57}$$

and we obtain the equivalent system of equations

$$\frac{\partial c_A}{\partial t^*} = \chi \frac{\partial^2 c_A}{\partial x^{*2}} - \beta^{-1} c_A c_B$$

$$\frac{\partial c_B}{\partial t^*} = \chi^{-1} \frac{\partial^2 c_B}{\partial x^{*2}} - \beta c_A c_B \tag{5.58}$$

and we see that the system depends only on the two nondimensional parameters χ and β defined by

$$\chi = \frac{D_A}{D_B}, \beta = \sqrt{\frac{c_{A,0}}{c_{B,0}}} \tag{5.59}$$

Contrary to the diffusion problem, there is no known analytical solution for (5.58). We just mention here that, for very long times, there exists an approximate solution [13] using the variable $R = c_A c_B$.

5.4.1.3 Reaction and Advection-Diffusion in a Microchannel

In biotechnology, chemical reactions are very often performed in microchannels with parallel flowing buffer fluids (Figure 5.30).

If y is the cross direction and x is the direction of the flow, the two reactants A and B produce the component C by reaction-diffusion in the transverse direction y (Figure 5.31).

Suppose a reaction symbolized by

$$A + B \rightarrow C$$

For a steady state flow of reactants, the system of equations for the reaction is [14, 15]

$$U\frac{\partial c_A}{\partial x} = D_A \frac{\partial^2 c_A}{\partial y^2} - kc_A c_B$$

$$U\frac{\partial c_B}{\partial x} = D_B \frac{\partial^2 c_B}{\partial y^2} - kc_A c_B \qquad (5.60)$$

$$U\frac{\partial c_C}{\partial x} = D_C \frac{\partial^2 c_C}{\partial y^2} + kc_A c_B$$

where, for the diffusion term, the second derivative in x has been neglected in front of that in y. We use a change of variables resembling to that of (5.57)

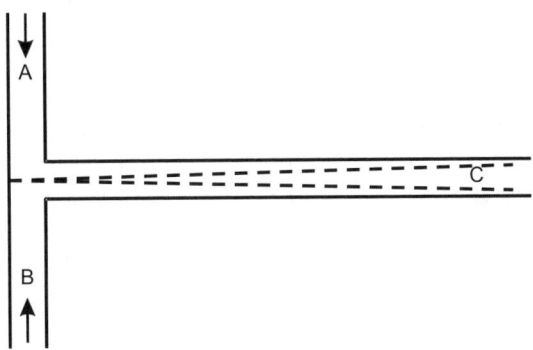

Figure 5.30 Reaction of two components flowing in parallel in a microchannel.

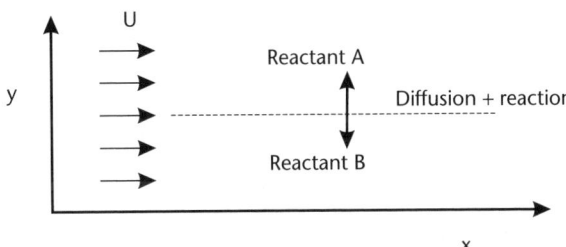

Figure 5.31 Sketch of the homogeneous reaction in a flowing solution.

$$x^* = \frac{x}{\frac{UL^2}{D}}, \ y^* = \frac{y}{L}, \ D = \sqrt{D_A D_B}, \ L = \left(\frac{D}{k\sqrt{c_{A,0} c_{B,0}}} \right)^{\frac{1}{2}} \qquad (5.61)$$

and, if we use the notations

$$c^*_A = \frac{c_A}{c_{A,0}}, \ c^*_B = \frac{c_B}{c_{B,0}}, \ \chi = \frac{D_A}{D_B}, \ \beta = \sqrt{\frac{c_{A,0}}{c_{B,0}}}$$

the system (5.60) becomes

$$\frac{\partial c^*_A}{\partial x^*} = \chi \frac{\partial^2 c^*_A}{\partial y^{*2}} - \frac{1}{\beta} c^*_A c^*_B$$

$$\frac{\partial c^*_B}{\partial x^*} = \frac{1}{\chi} \frac{\partial^2 c^*_B}{\partial y^{*2}} - \beta c^*_A c^*_B \qquad (5.62)$$

This system is nondimensional and depends only on the two parameters χ and β. Again, this system is nonlinear and strongly coupled. The use of numerical methods is required to solve such systems. In Figure 5.32, we show the computed solution obtained with the numerical software FEMLAB.

5.4.2 Heterogeneous Reactions

Biochemical reactions are said heterogeneous when a ligand is immobilized on the solid walls and the targets (also called analytes or reactants) are in solution. Heterogeneous reactions are widely used in microsystems [16]. They appear to be often more convenient than homogeneous reactions, because, for one thing, the ligand at the wall can be reused after washing the reaction chamber. Also, it is easier to proceed in two steps: immobilization of the ligands on a "reaction" surface at the wall and introduction of the analytes by a carrier fluid, rather than designing a complex micofluidic system where the targets and ligands merge and mix at the same time in the reaction chamber. Finally, it is also often more convenient to detect the binded couples (targets-ligands) when they are immobilized on a wall surface.

In this section, we give some examples of how kinetics of heterogeneous reactions are calculated. The first example is that of a diffusion-reaction problem of the Langmuir type with concentration depletion.

5.4.2.1 Example of Concentration Depletion

There are two general trends in the treatment of biochemical reactions in biotechnology: First, the volumes are getting smaller, and the ratio between the reaction surface (functionalized or labeled surface) and the volume is increasing. Second, the number of target molecules or particles is getting smaller in order to increase the specificity and efficiency of the biochip. It follows from these two considerations that the concentration in targets may be affected by depletion during

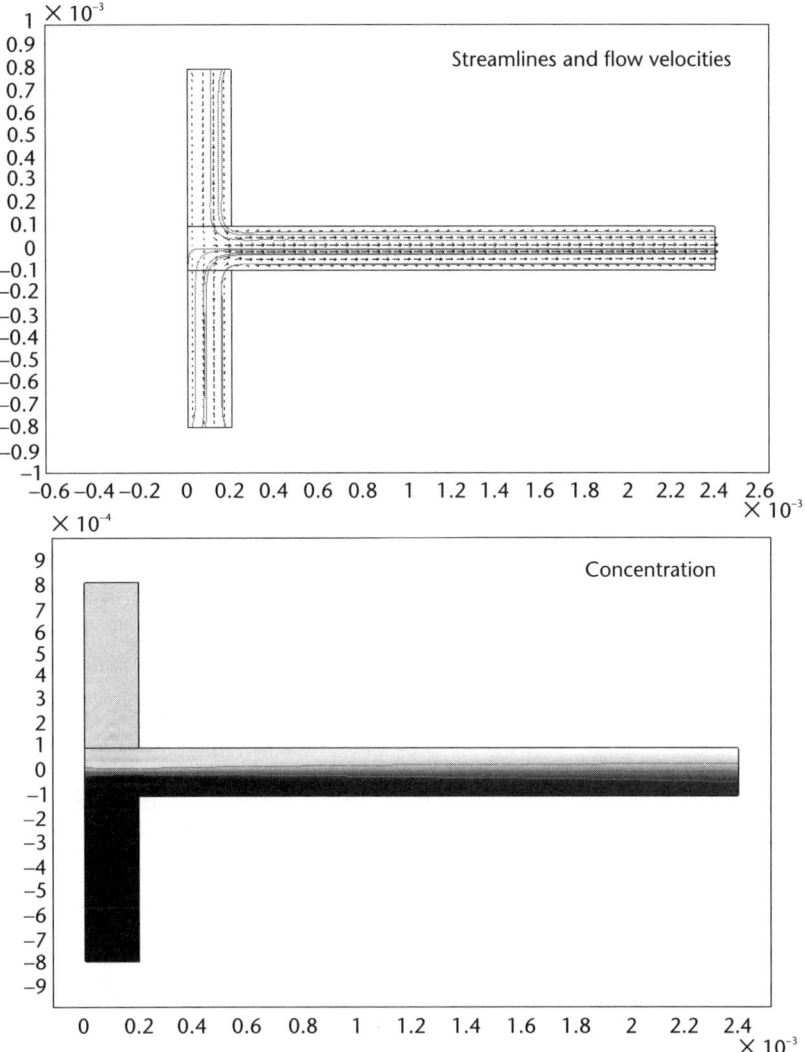

Figure 5.32 View of the flow velocities (top) and the reaction zones (bottom) in the T-shape channel. Calculation performed with the FEMLAB numerical software.

the reaction [17]. It is no more a constant as was supposed when we produced the solution of the Langmuir equation.

We investigate here the solution to the Langmuir equation in the case of a uniform concentration in the liquid volume that is decreasing with time, and we show that there exists a closed-form solution to the Langmuir equation in the case of depletion, assuming the concentration is spatially homogeneous. Consider the schematic case of Figure 5.33.

The mathematical formulation of the problem is obtained by replacing the constant concentration c_0 by a variable concentration c in the Langmuir equation.

$$\frac{d\Gamma}{dt} = k_{on}c(\Gamma_0 - \Gamma) - k_{off}\Gamma \qquad (5.63)$$

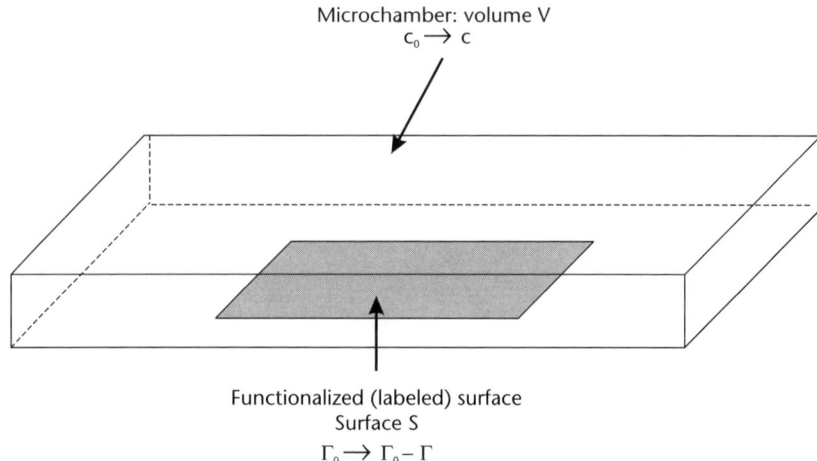

Microchamber: volume V
$c_0 \rightarrow c$

Functionalized (labeled) surface
Surface S
$\Gamma_0 \rightarrow \Gamma_0 - \Gamma$

Figure 5.33 Schematic view of the microchamber. Because of depletion, the concentration c in the chamber decreases.

The concentration c is obtained by the mass balance taking into account the depletion due to immobilization of the targets [18]:

$$S \frac{d\Gamma}{dt} = -V \frac{dc}{dt} \tag{5.64}$$

where S is the functionalized surface, and V is the volume of the microchamber. Taking into account the initial conditions, integration of (5.64) leads to

$$c = c_0 - \frac{S}{V}\Gamma \tag{5.65}$$

where $c_0 = c(t = 0)$ is the initial (uniform) concentration. Upon substituting (5.65) in (5.63), one obtains the differential equation for the surface concentration Γ

$$\frac{d\Gamma}{dt} = k_{on}c_0\Gamma_0 - \left(k_{on}c_0 + k_{off} + k_{on}\frac{S}{V}\Gamma_0\right)\Gamma + k_{on}\frac{S}{V}\Gamma^2 \tag{5.66}$$

and for the concentration

$$\frac{dc}{dt} = k_{off}c_0 + \left(k_{on}c_0 - k_{off} - k_{on}\frac{S}{V}\Gamma_0\right)c - k_{on}c^2 \tag{5.67}$$

By considering an infinite volume $V = \infty$, (5.66) collapses to the usual Langmuir equation. The two differential equations (5.66) and (5.67) are of the mathematically well-known Ricatti type [19]. Ricatti equations can be solved if a particular solution is known. In such a case, a change of variable using the particular solution transforms the Ricatti equation in a Bernoulli equation, which has a closed-form solution.

Usually, in order to promote hybridization, the temperature of a reaction is set to a value well beneath the *fusion* temperature (i.e., the temperature where dissociation dominates), so that k_{off} is usually kept small. However, k_{off} is not necessarily vanishing in front of $k_{on} c$, especially in our case where the initial concentration c_0 is small. Thus, all of the terms in (5.67) have their importance. Let us mention that values of k_{on} and k_{off} for immunoassays have been investigated thoroughly in the literature [20, 21].

The algebraic manipulations to obtain to the solution are somewhat long. We just indicate briefly the approach. First, we search the solution of the second-order characteristic polynomial in c:

$$k_{on} c^2 - \left(k_{on} c_0 - k_{off} - k_{on} \frac{S}{V} \Gamma_0 \right) c - k_{off} c_0 = 0 \tag{5.68}$$

The two roots of the polynomial are given by

$$c^- = \frac{1}{2} \left(c_0 - \frac{k_{off}}{k_{on}} - \frac{S}{V} \Gamma_0 \right) - \frac{1}{2} \sqrt{\left(c_0 - \frac{k_{off}}{k_{on}} - \frac{S}{V} \Gamma_0 \right)^2 + 4 \frac{k_{off}}{k_{on}} c_0}$$

$$c^+ = \frac{1}{2} \left(c_0 - \frac{k_{off}}{k_{on}} - \frac{S}{V} \Gamma_0 \right) + \frac{1}{2} \sqrt{\left(c_0 - \frac{k_{off}}{k_{on}} - \frac{S}{V} \Gamma_0 \right)^2 + 4 \frac{k_{off}}{k_{on}} c_0} \tag{5.69}$$

Note that these two roots have the dimension of a concentration and that $\dfrac{k_{off}}{k_{on}}$ is the equilibrium constant. In order to simplify the notations, we note

$$\hat{c} = \sqrt{\left(c_0 - \frac{k_{off}}{k_{on}} - \frac{S}{V} \Gamma_0 \right)^2 + 4 \frac{k_{off}}{k_{on}} c_0}$$

$\hat{c}$ also has the dimension of a concentration. Using the notations $\hat{c}$, c^+, and c^-, it is possible to show that the solution of (6.67) is

$$c = \frac{1}{\dfrac{1}{c_0 - c^-} e^{-k_{on} \hat{c} t} + \dfrac{1}{\hat{c}} \left(1 - e^{-k_{on} \hat{c} t} \right)} + c^- \tag{5.70}$$

The kinetics of surface concentration is obtained by substituting the concentration kinetics from (5.70) in (5.65)

$$\Gamma = \frac{V}{S} (c_0 - c) \tag{5.71}$$

At first glance, (5.70) seems somewhat complicated, but this is not really the case. Let us examine three different cases:

First, it is easy to see that if $V \to \infty$, then $S/V \to 0$, and from (5.71) we obtain $c = c_0$, which is the expected result for a semi-infinite case.

Second, suppose that $k_{off} = 0$ and that the number of targets is larger than the number of ligands (initial hybridization sites).

$$N = \frac{n_{targets}}{n_{ligands}} = \frac{c_0 V}{\Gamma_0 S} > 1$$

The functionalized surface will end being saturated by the immobilized targets. One finds first that $c^- = 0$ and $\hat{c} = c_0 - \frac{S}{V}\Gamma_0$. Then, (5.62) collapses to

$$c = \cfrac{1}{\cfrac{1}{c_0} e^{-k_{on}\left(c_0 - \frac{S}{V}\Gamma_0\right)t} + \cfrac{1}{\left(c_0 - \frac{S}{V}\Gamma_0\right)}\left(1 - e^{-k_{on}\left(c_0 - \frac{S}{V}\Gamma_0\right)t}\right)}$$

By letting $t \to \infty$ in the preceding equation, $c \to c_0 - \frac{S}{V}\Gamma_0$ and $\Gamma \to \Gamma_0$.

Third, $k_{off} = 0$ and the number of targets is smaller than the number of ligands (initial hybridization sites)

$$N = \frac{n_{targets}}{n_{ligands}} = \frac{c_0 V}{\Gamma_0 S} < 1$$

The functionalized surface cannot be saturated. It is easy to see that $c^- = c_0 - \frac{S}{V}\Gamma_0$ and $\hat{c} = -\left(c_0 - \frac{S}{V}\Gamma_0\right)$

Then the solution is

$$c = \cfrac{1}{\cfrac{1}{\frac{S}{V}\Gamma_0} e^{-k_{on}\left(\frac{S}{V}\Gamma_0 - c_0\right)t} - \cfrac{1}{\left(c_0 - \frac{S}{V}\Gamma_0\right)}\left(1 - e^{-k_{on}\left(\frac{S}{V}\Gamma_0 - c_0\right)t}\right)} + c_0 - \frac{S}{V}\Gamma_0$$

By letting $t \to \infty$ in the preceding equation, one finds $c \to 0$ and $\Gamma \to \frac{V}{S}c_0 < \Gamma_0$.

As for the classic Langmuir kinetics, the kinetic curve approaches an asymptote when time is sufficiently important. The value of the asymptote is obtained by taking $t = \infty$ in (5.70)

$$c_\infty = \hat{c} + c^-$$

After substitution of the value of $\hat{c}$, the asymptotic value of the concentration is

$$c_\infty = c^+ \qquad (5.72)$$

This result shows that the asymptote is simply given by the second root of the polynomial. This result can be derived directly from (5.67): When a permanent regime is attained, the time derivative of c vanishes, and the asymptotic value for the concentration then verifies the zero right-hand side of (5.67). Thus, the positive root

(c^+) of the characteristic polynomial is the value of the asymptote. The asymptotic value of the surface concentration—if there is no saturation—is then given by

$$\Gamma_\infty = \frac{V}{S}\left(c_0 - c^+\right) \tag{5.73}$$

The kinetics at $t = 0$ is the same as that of the Langmuir model

$$\frac{d\Gamma}{dt}\Big|_{t=0} = k_{on}c_0\Gamma_0$$

As we have seen, the ratio between the surface of capture and the volume of the chamber is an important parameter. Targeted concentration levels in biochips are usually in the range $c_0 = 10^{-9}$ to 10^{-12} mole/l. Depending on whether the targets are DNA strands or antibodies, the surface density of ligands lies in the range $\Gamma_0 = 500$–$1,000$ molecules/μm^2. Typical range of values of the ratio S/V is obtained by considering two types of microsystems: the first one is that of Figure 5.34 with a round functionalized spot—the surface of the functionalized spot is approximately $S = \pi R^2$, $R = 2$ mm, and V is of the order of $10 \times 10 \times 1$ mm³, so that $S/V = 0.13$ m⁻¹. The second case is that of a capillary tube of radius $R = 100$ μm, functionalized along one-tenth of its length. The ratio S/V is then equal to $S/V = 2/R = 2,000$ m⁻¹.

The depletion model shows that precautions should be taken when the detection device is operated under a closed-volume condition (no circulation of fluid) or under a closed-loop condition. In such cases, depletion in targets/analytes has to be taken into account in the kinetics. Langmuir kinetics and (5.70) do not agree if the S/V ratio is sufficiently large. A comparison of the kinetics of binding is shown in Figure 5.35 using $S/V = 10$ m⁻¹ and four different values of k_{off}. The other parameters

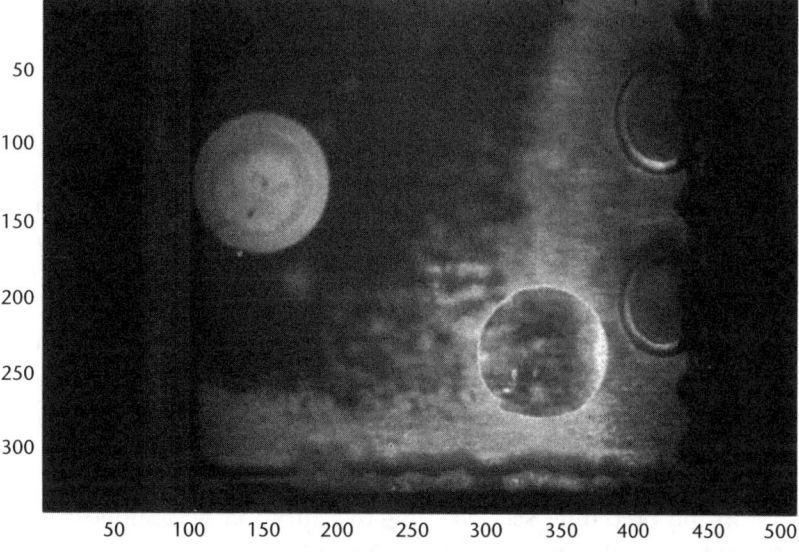

Figure 5.34 Capture of marked DNA strands on a round functionalized spot. (Courtesy LETI/BioMérieux.)

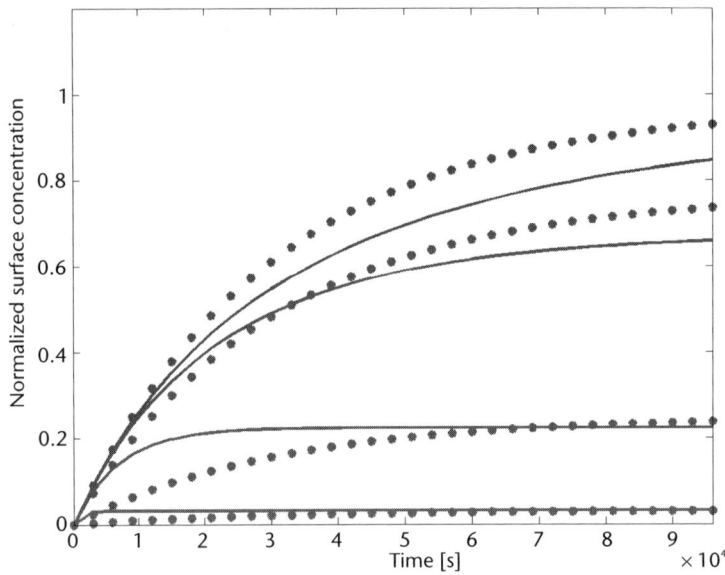

Figure 5.35 Comparison of adsorption kinetics between semi-infinite Langmuiran model (dotted line) and present model (continuous line) for four different values of k_{off}: 10^{-6}, 10^{-5}, 10^{-4}, and 10^{-3} s^{-1} and a ratio $S/V = 10$ m^{-1}.

correspond to DNA hybridization and typical values of hybridization of 32-bp DNA segments have been used: $k_{on} = 110$ $m^3/mole/sec$, $\Gamma_0 = 1.668 \cdot 10^{-8}$ $mole/m^2$, and $c_0 = 0.3 \cdot 10^{-6}$ $mole/m^3$.

Kinetics obtained by the Langmuir model and the depletion model are similar only in the case of a large k_{off}. For such values of k_{off}, depletion is not very important because of the large desorption (dissociation) process.

On the other hand, if we plan to find the kinetic constants by a fit of the kinetic curves, care should be taken if depletion occurs. The fitted values will differ if a Langmuir model is used instead of a depletion-modified Langmuir model.

5.4.2.2 Diffusion Limited Reaction

Introduction

In a biochip, heterogeneous reaction kinetics depend not only on the reaction rate—as we have seen in the first section of this chapter—but also on the diffusion of species toward the surface where the reaction occurs. If diffusion is fast (e.g., in the case of very small diffusing molecules), the flux of these reactant molecules at the wall is such that there is no delay in the reaction. On the other hand, if the diffusion process is slow, there will be depletion near the reacting surface and the biochemical reaction will be slowed down. In the first case, the process is limited by the reaction rate; in the second case, it is limited by the diffusional flux. This problem is the subject of an abundant literature [22–26].

In the preceding section, the concentration was supposed uniform (but time dependant) in the microchamber volume. In this section, we treat the case where a depletion layer is observed near the functionalized surface (Figure 5.36).

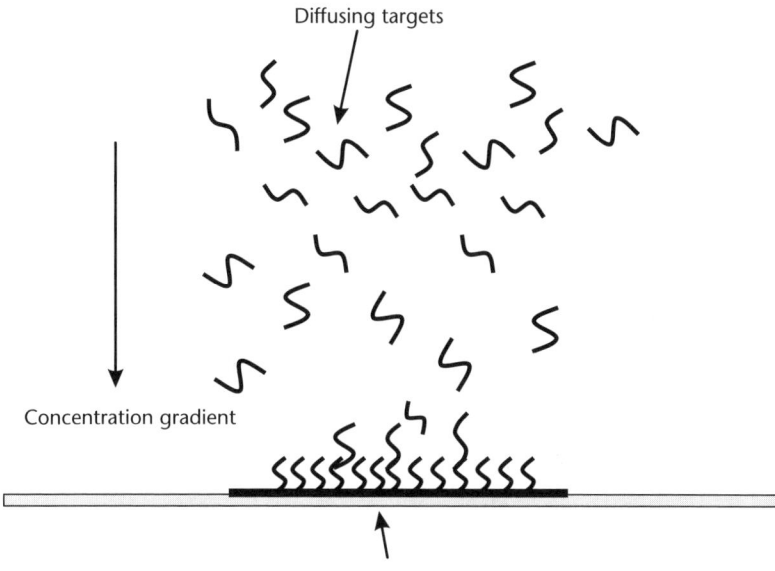

Figure 5.36   Schematic view of the diffusing targets and hybridization on the functionalized surface.

Governing Equations

Concentration in the fluid volume is obtained through the usual diffusion equation

$$\frac{\partial c}{\partial t} = D\Delta c \tag{5.74}$$

where c is the concentration of substrate, and D is its diffusion coefficient. In a two-dimensional Cartesian system, this equation can be rewritten as

$$\frac{\partial c}{\partial t} = D\left(\frac{\partial^2 c}{\partial x^2} + \frac{\partial^2 c}{\partial y^2}\right) \tag{5.75}$$

On the other hand, at the functionalized wall, the Langmuir model for binding yields

$$\frac{d\Gamma}{dt} = k_{on}c_w\left(\Gamma_0 - \Gamma\right) - k_{off}\Gamma \tag{5.76}$$

where Γ is the concentration in immobilized analytes, Γ_0 is the initial concentration in available hybridization sites, k_{on} is the adsorption coefficient at the wall, k_{off} is the desorption coefficient at the wall—also called elution—and c_w is the concentration at the wall.

The two equations (5.76) and (5.75) are not independent. They are coupled by Fick's law

$$\frac{d\Gamma}{dt} = -D\nabla c\big|_w \tag{5.77}$$

Equations (5.76) can be substituted in (5.77), and we obtain the value of the wall concentration as a function of the wall concentration (and its derivative)

$$\Gamma = \frac{D\nabla c|_w + k_{on}c_w\Gamma_0}{\left(k_{on}c_w + k_{off}\right)} \tag{5.78}$$

Equation (5.78) shows that there is some kind of equilibrium between the value of the concentration near the wall and the surface concentration in hybridized targets: if the concentration near the functionalized surface decreases—for any reason, like an interruption in the arrival of targets—there will be a temporary depletion of targets near the wall, and the immobilized DNA strands will start to dissociate. On the other hand, if there is a large supply of targets near the wall, the rate of hybridization will increase.

Numerical Approach

Numerical methods must be set up to solve such problems. If one has access to a finite element software, the numerical approach is straightforward. If not, and if the geometry of the microchamber is simple, a numerical formulation based on a finite difference approach can be set up using the following discretization, based on the grid of Figure 5.37.

First, using a Crank-Nicholson semi-implicit scheme [19], the advection-diffusion equation (5.75) can be discretized under the form

$$\frac{c_{i,j}^{n+1} - c_{i,j}^{n}}{\Delta t} = \frac{D}{2}\left[\frac{c_{i+1,j}^{n+1} - 2c_{i,j}^{n+1} + c_{i-1,j}^{n+1}}{(\Delta x)^2} + \frac{c_{i,j+1}^{n+1} - 2c_{i,j}^{n+1} + c_{i,j-1}^{n+1}}{(\Delta y)^2}\right]$$
$$+ \frac{D}{2}\left[\frac{c_{i+1,j}^{n} - 2c_{i,j}^{n} + c_{i-1,j}^{n}}{(\Delta x)^2} + \frac{c_{i,j+1}^{n} - 2c_{i,j}^{n} + c_{i,j-1}^{n}}{(\Delta y)^2}\right] \tag{5.79}$$

In (5.79), the subscripts n or $n + 1$ refer to the time step. Next, using an implicit scheme, (5.76) becomes

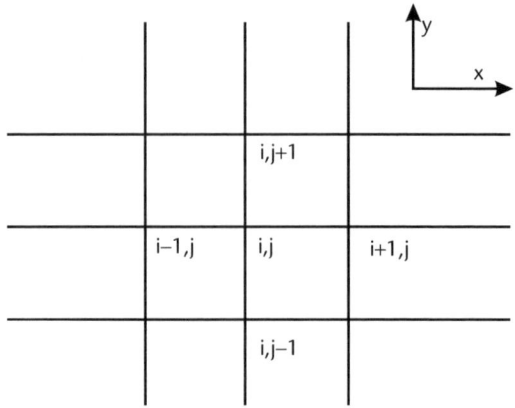

Figure 5.37 Schematic view of the discretization grid.

$$\Gamma_i^{n+1} = \frac{\Gamma_i^n + k_{on}c_{i,0}^{n+1}\Gamma_0 \Delta t}{1 + \left(k_{on}c_{i,0}^{n+1} + k_{off}\right)\Delta t} \tag{5.80}$$

where the notation $c_{i,0}$ refers to the concentration at the wall. Fick's law can be discretized by

$$\frac{\Gamma_i^{n+1} - \Gamma_i^n}{\Delta t} = -\frac{D}{\Delta y}\left[\frac{c_{i,0}^{n+1} - c_{i,1}^{n+1}}{2} + \frac{c_{i,0}^n - c_{i,1}^n}{2}\right] \tag{5.81}$$

After substitution of (5.81) in (5.80), we obtain a linear relation between $c_{i,0}^{n+1}$ and $c_{i,1}^{n+1}$, and the whole system can be cast under the matrix form

$$[A]\{c^{n+1}\} = \{S^n\} \tag{5.82}$$

where the vector $\{s^n\}$ depends on the concentrations at the preceding time step. By using the relevant boundary conditions, and by inversing the system [27], one obtains the concentration distribution at the new time step $n + 1$.

Example of Diffusion Limited Reaction

Suppose a microchamber with a round functionalized spot, as shown in Figure 5.38. Hybridization kinetics is monitored by fluorescence (Figure 5.39).

If the dimensions of the chamber are sufficiently large and the diffusion coefficient is sufficiently small, the reaction is slowed down by the depletion of targets in the vicinity of the reactive surface. This case is called diffusion limited reaction. It can be shown [24] that the nondimensional Dammkohler number characterizes the type of reaction

$$Da = \frac{D/\delta}{k_{on}\Gamma_0} \tag{5.83}$$

where δ is the vertical dimension of the microchamber. If Da is large, the reaction time is larger than the diffusion time, and diffusion does not delay the reaction kinetics. On the other hand, if Da is small, the reaction kinetics is limited by diffusion.

A very interesting note can be made by analyzing (5.83). If the vertical dimension of the microchamber is sufficiently small, the Dammkohler number becomes large, and diffusion does not limit the reaction kinetics. So, it is best to design a microchamber with a vertical size smaller than

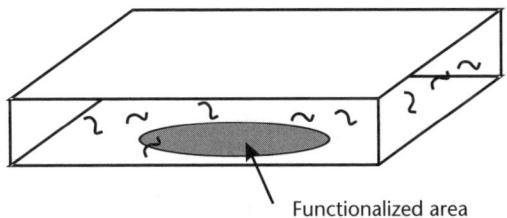

Functionalized area

Figure 5.38   Sketch of a functionalized surface in a microchamber where the buffer fluid is at rest.

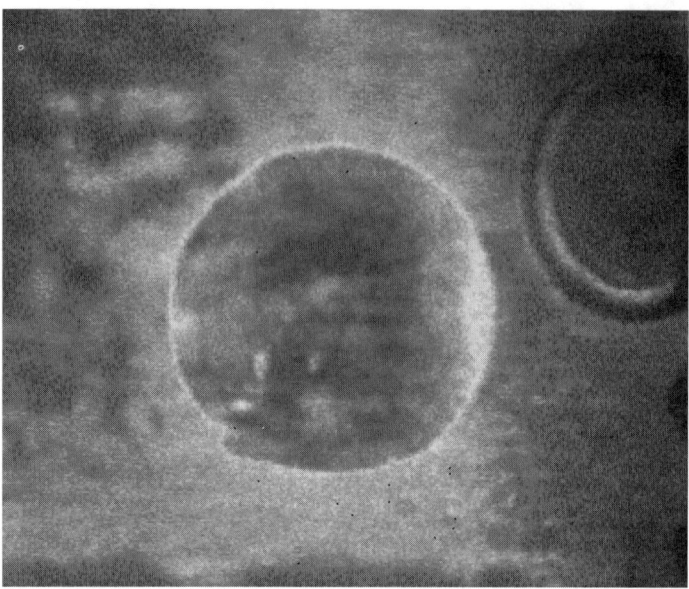

Figure 5.39 View under the microscope of a functionalized surface during the process of hybridization: the intensity of the laser light diffracted relates to the surface concentration of hybridized DNA. (Courtesy of CEA-LETI/BioMérieux.)

$$\delta << \frac{D}{k_{on}\Gamma_0} \tag{5.84}$$

Usually, for oligonucleotides, the dimension δ should be smaller than 50–70 μm.

When the reaction is diffusion limited, a region of concentration depletion forms at the vicinity of the reactive surface. Figure 5.40 shows a contour plot of the concentration in such a case.

If the reaction were not diffusion limited, the Langmuir kinetics would be the right one. If the reaction is diffusion limited, the kinetics curves depart from that of the Langmuir model (Figure 5.41).

The numeric model may be used to predict the reaction parameters. They can be adjusted to fit the experimental results (Figure 5.42).

5.4.2.3 Advection-Diffusion and Biochemical Reactions

Introduction

Consider the case of a buffer fluid flowing through a microchamber comprising one or more reactive surfaces (Figures 5.43 and 5.44). Usually such experimental devices are used to find the kinetic constants k_{on} and k_{off} for a given sequence of DNA.

Governing Equations

The advection-diffusion equation for concentration in the fluid volume is

$$\frac{\partial c}{\partial t} = D\Delta c - v\nabla c \tag{5.85}$$

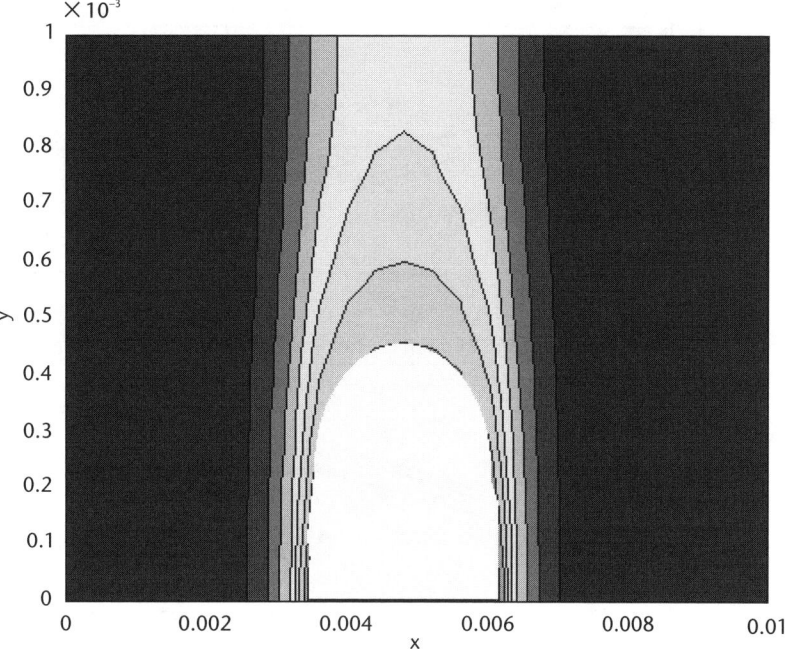

Figure 5.40 Concentration depletion in the case of a diffusion limited reaction. Caution: The representation is not to scale, as the vertical axis has been extended for visualization.

where c is the concentration of targets or analytes, D is its diffusion coefficient, and v is the flow velocity. Very often the channel is such that we can use a Hagen-Poiseuille velocity field (see Chapter 1). In the case of a flat channel limited by two parallel plates separated by a distance d, the flow velocity is

$$v(y) = \frac{3}{2}\bar{v}\left[1 - \left(\frac{y}{d/2}\right)^2\right]$$ (5.86)

where $\bar{v}$ is the average velocity in the channel. Because the fluid velocity is directed along the x-axis, the advection-diffusion equation can be cast into the form

$$\frac{\partial c}{\partial t} = D\left(\frac{\partial^2 c}{\partial x^2} + \frac{\partial^2 c}{\partial y^2}\right) - v\frac{\partial c}{\partial x}$$ (5.87)

At the functionalized wall, the Langmuir model for binding yields

$$\frac{d\Gamma}{dt} = k_{on}c_w\left(\Gamma_0 - \Gamma\right) - k_{off}\Gamma$$ (5.88)

where Γ is the concentration in immobilized analytes, Γ_0 is the initial concentration in available hybridization sites, k_{on} is the adsorption coefficient at the wall, k_{off} is the desorption coefficient at the wall—also called elution—and c_w is the concentration at the wall.

The coupling between the two equations is realized by Fick's law

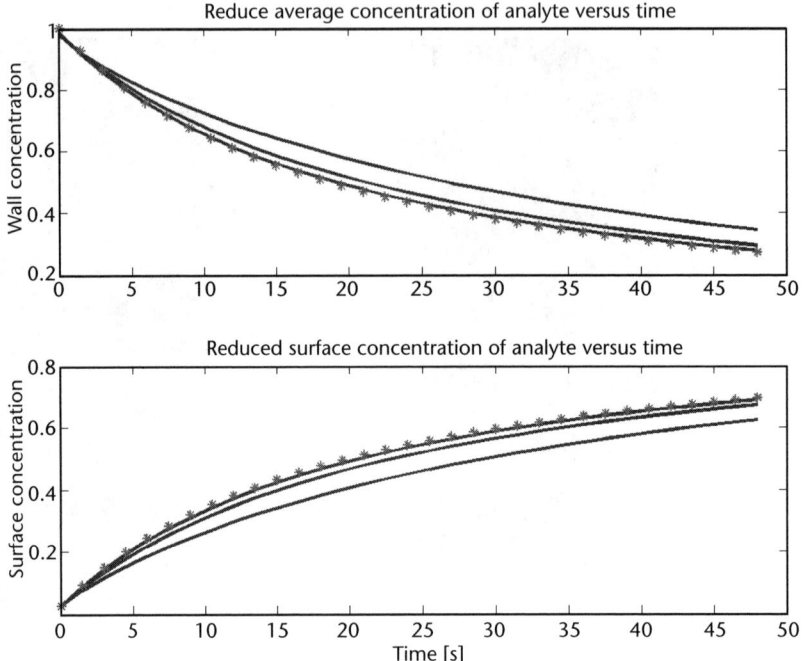

Figure 5.41 Top: Average concentration of analyte at the wall (c_p/c_{bulk}) versus time. The dotted line is the Langmuir solution (kinetics limited only by the biochemical reaction). The three lines correspond respectively to $D = 10^{-11}$, $3 \cdot 10^{-11}$, and 10^{-10} m^2/s, and to the three Dammkoehler numbers 1/2, 1/0.66, and 1/0.2. Bottom: Same as the top but for the surface concentration of hybridized sites (Γ/Γ_0) versus time. Note that the Langmuir solution does not depend on D.

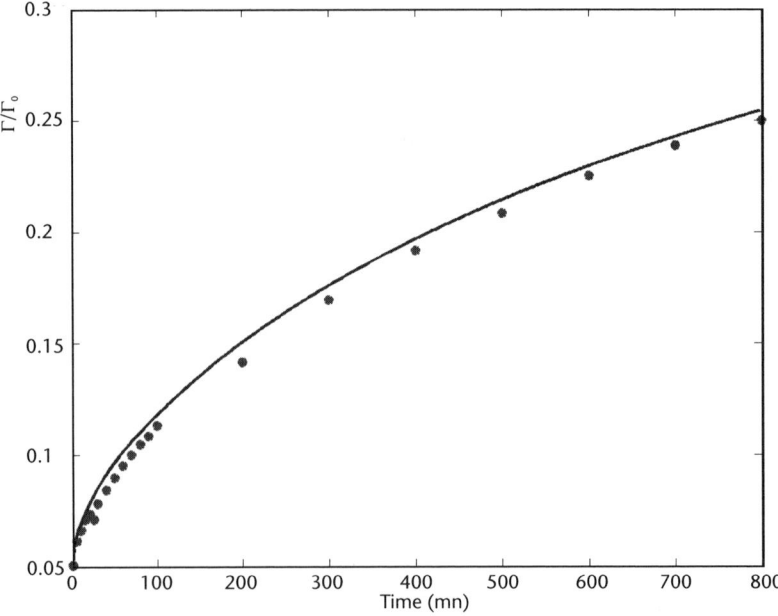

Figure 5.42 Kinetics of hybridization of DNA segments on a solid flat surface functionalized with complementary DNA strands. The dots correspond to the experimental results and the continuous line to the calculation.

$$\frac{d\Gamma}{dt} = -D\nabla c\big|_w \tag{5.89}$$

Equations (5.89) can be substituted in (5.88), and we obtain the value of the wall concentration as a function of the wall concentration (and its derivative)

$$\Gamma = \frac{D\nabla c\big|_w + k_{on}c_w\Gamma_0}{\left(k_{on}c_w + k_{off}\right)}$$

The same remarks as those in the preceding section can be made concerning the fluctuating equilibrium near the wall.

Numerical Approach

Numerical methods must be set up to solve such problems. If one has access to a finite element software, the numerical approach is straightforward. If not, and if the geometry of the microchamber is simple, a numerical formulation based on a finite difference approach can be set up using the following discretization based on the grid defined in Figure 5.45. The method is very similar to that of the preceding section.

First, using a Crank-Nicholson semi-implicit scheme [15], the advection-diffusion equation (5.87) can be discretized under the form

$$\begin{aligned}
\frac{c_{i,j}^{n+1} - c_{i,j}^n}{\Delta t} &= \frac{D}{2}\left[\frac{c_{i+1,j}^{n+1} - 2c_{i,j}^{n+1} + c_{i-1,j}^{n+1}}{(\Delta x)^2} + \frac{c_{i,j+1}^{n+1} - 2c_{i,j}^{n+1} + c_{i,j-1}^{n+1}}{(\Delta y)^2}\right] \\
&+ \frac{D}{2}\left[\frac{c_{i+1,j}^n - 2c_{i,j}^n + c_{i-1,j}^n}{(\Delta x)^2} + \frac{c_{i,j+1}^n - 2c_{i,j}^n + c_{i,j-1}^n}{(\Delta y)^2}\right] \\
&- v_{i,j}\frac{\left(c_{i,j}^{n+1} - c_{i-1,j}^{n+1}\right)}{\Delta x}
\end{aligned} \tag{5.90}$$

In (5.90), the subscripts n or $n + 1$ refer to the time step. Note that the velocity term must be discretized following the flow direction. Next, using an implicit scheme, (5.89) becomes

$$\Gamma_i^{n+1} = \frac{\Gamma_i^n + k_{on}c_{i,0}^{n+1}\Gamma_0\Delta t}{1 + \left(k_{on}c_{i,0}^{n+1} + k_{off}\right)\Delta t} \tag{5.91}$$

where $c_{i,0}$ refers to the concentration at the wall. Fick's law can be discretized by

$$\frac{\Gamma_i^{n+1} - \Gamma_i^n}{\Delta t} = \frac{D}{\Delta y}\left[\frac{c_{i,0}^{n+1} - c_{i,1}^{n+1}}{2} + \frac{c_{i,0}^n - c_{i,1}^n}{2}\right] \tag{5.92}$$

After substitution of (5.92) in (5.91), we obtain the linear relation between $c_{i,0}^{n+1}$ and $c_{i,1}^{n+1}$, and the whole system can be cast under the matrix form

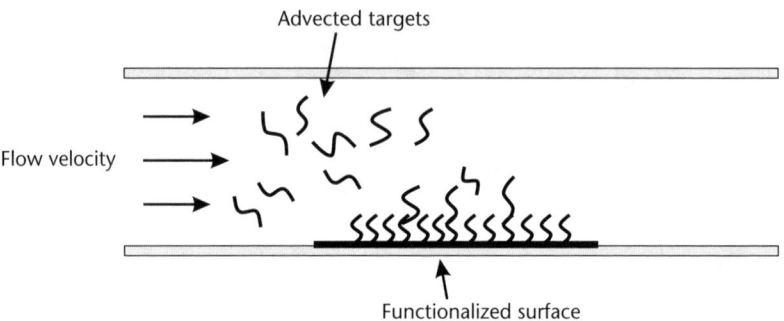

Figure 5.43   Schematic view of DNA hybridization in a microchannel (side view).

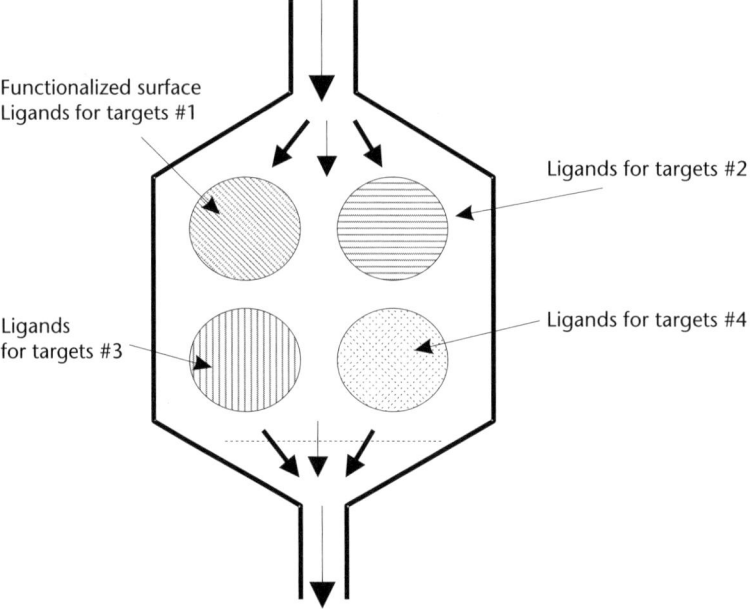

Figure 5.44   Schematic view of a typical hybridization microchamber, with different types of "spots" for hybridization.

$$[A]\{c^{n+1}\} = \{S^n\} \tag{5.93}$$

where the vector $\{s^n\}$ depends on the concentrations at the preceding time step. By using the relevant boundary conditions, and by inversing the system [22], one obtains the concentration distribution at the new time step $n + 1$.

Example of Advection-Diffusion-Reaction Kinetics

In this example, we show how experimental records of hybridization kinetics combined with the numerical model of the preceding section can be used to find the kinetics constants of different DNA strands. The experiment set up corresponds to that of Figure 5.46. In this experiment, a constant buffer fluid flow carries different types of DNA strands, with different sequences and length. The average flow velocity is 1 mm/sec (10 μl/mn), and the dimensions of the microchamber are

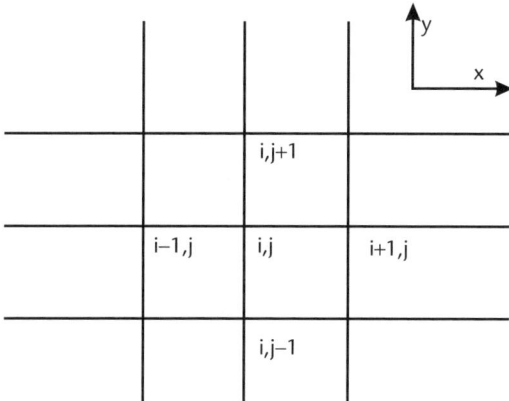

Figure 5.45   Schematic view of the discretization grid.

$10 \text{ mm} \times 10 \text{ mm} \times 1 \text{ mm}$. The flow is turned on for 50 minutes, then it is stopped for 310 minutes, and it is again turned on for the rest of the experimental time. Hybridization kinetics is monitored by fluorescence. Four different kinetics are obtained for the four different types of oligonucleotides (Figure 5.46).

The approach is to use the numerical model of the preceding section and to fit the kinetics curves by varying the parameters c_0 (bulk concentration), D (diffusion coefficient), k_{on} (constant of hybridization), and k_{off} (constant of desorption). At the experiment temperature, k_{off} can be considered negligible, so that the fit is performed by varying three parameters only (Figure 5.47). A few trials are enough to find the values of c_0, D, and k_{on}.

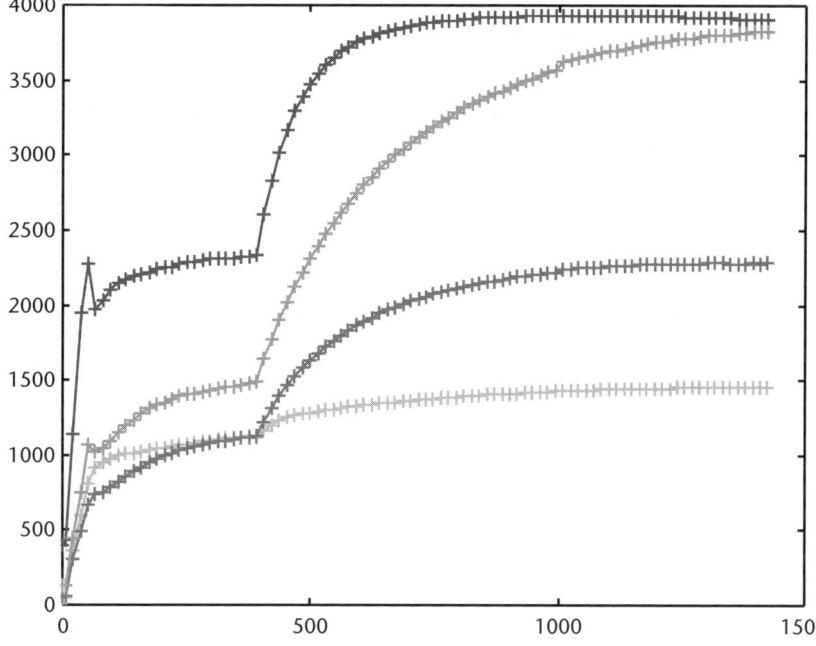

Figure 5.46   Experimental hybridization kinetics for different DNA strands. Fluorescence versus time (in minutes).

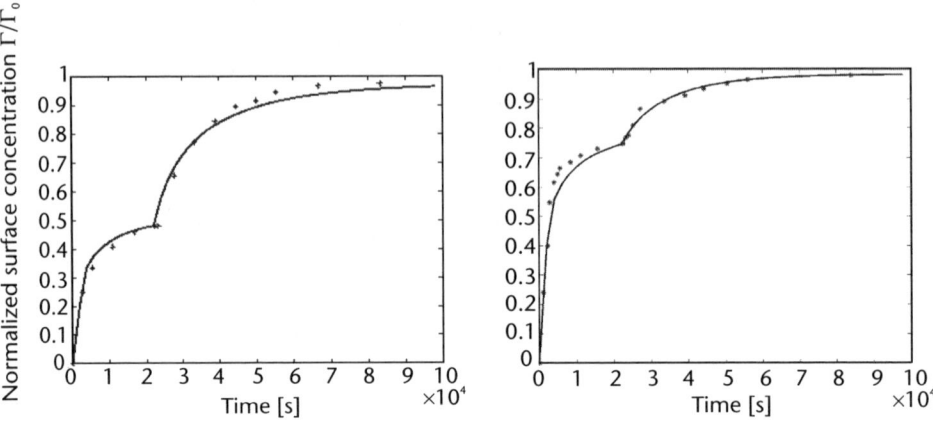

Figure 5.47 Fit of the experimental kinetics for 29 and 15 basis pairs of DNA strands.

Table 5.1 shows the typical values for DNA strands of different length obtained by this approach. One of the conclusions of this analysis is that the adsorption constant depends not only on the length of the DNA strand, but also on the nature of the basis pairs (A, C, G, T).

5.4.2.3 Displacement (Competition) Reactions

Introduction and Principle

In the preceding sections, we have dealt with a class of heterogeneous reactions that may be called *sandwich reactions*. Up to date, sandwich reactions are the most common. This class of reactions is schematized in Figures 5.48, 5.49, and 5.50. It is done in three steps: (1) functionalization of a zone on the solid wall, (2) hybridization or capture, and (3) detection by a fluorescent tag (or another method). The first step is schematized in Figure 5.48: A surface of the solid wall is functionalized with ligands having an affinity with the target macromolecule.

In a second step, the buffer fluid sample is injected in the microchamber, and the targets are adsorbed and immobilized (temporarily) by an equilibrium reaction on the available sites (Figure 5.49).

Finally, detection of the concentration in immobilized targets is performed (Figure 5.50).

Sandwich reactions were historically first. However, in the 1980s, a new, more sophisticated concept was invented to avoid tagging the biological targets with a marker, which is often a complicated process and which is not always possible (e.g.,

Table 5.1 Typical Values of Advection-Diffusion and Reaction Parameters for Short DNA Strands

Number of Basis Pairs	$D\ (m^2/sec)$	k_{on} $(m^3/mole/sec)$	C_0 $(mole/m^3)$
32	$0.85 \cdot 10^{-10}$	110	$0.34 \cdot 10^{-5}$
14	$0.80 \cdot 10^{-10}$	60	$0.24 \cdot 10^{-5}$
15	$0.75 \cdot 10^{-10}$	125	$0.46 \cdot 10^{-5}$
29	$0.70 \cdot 10^{-10}$	75	$0.29 \cdot 10^{-5}$

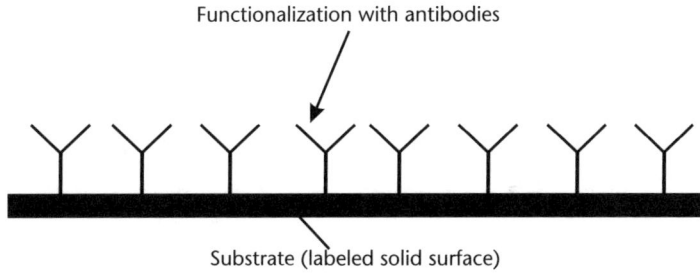

Figure 5.48   Step 1: Functionalization of the solid surface with ligand antibodies.

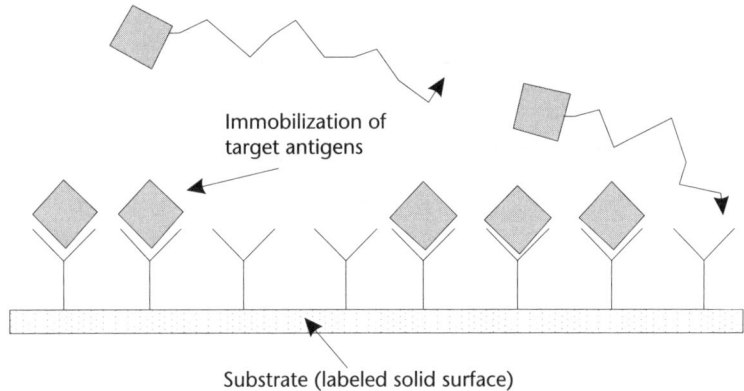

Figure 5.49   Step 2: Capture of the targets transported by a buffer fluid.

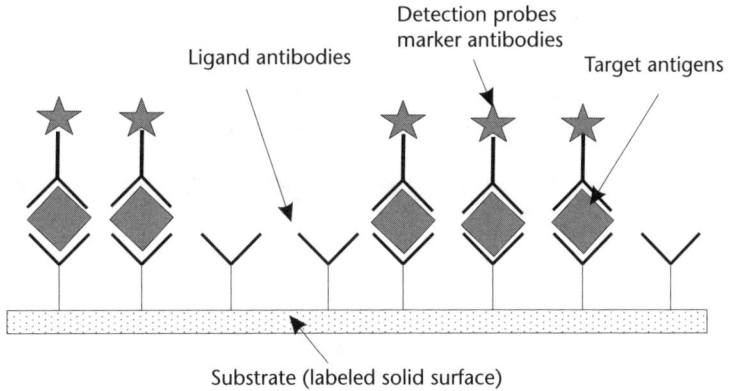

Figure 5.50   Step 3: Detection step—marking the targets with fluorophores.

biochips aimed at bioterrorism targets detection have no a priori knowledge of the target, thus it cannot be tagged). This new type of reaction process is called *displacement reaction*, or sometimes *competition reaction* [28–31]. The principle is shown in Figures 5.51 to 5.54. The first step is the same as for sandwich reactions: functionalization of a surface of capture with ligands. The second step is most of the time the most difficult to set up, because it requires finding molecules that are "analog" to the targets, with an affinity to the ligands smaller than the targets. This is the tricky part that biologists and chemists have to resolve. These analogs can be

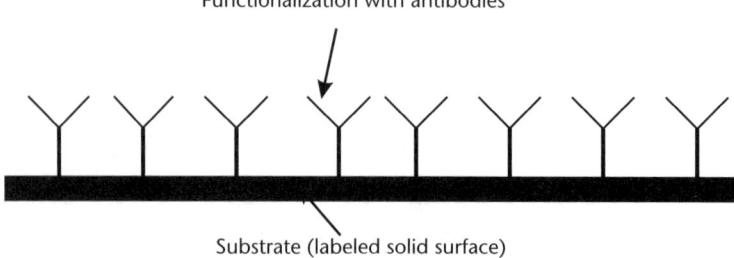

Figure 5.51   Functionalization of the surface of capture with ligands.

Figure 5.52   Saturation of the functionalized sites with analogs tagged with a marker.

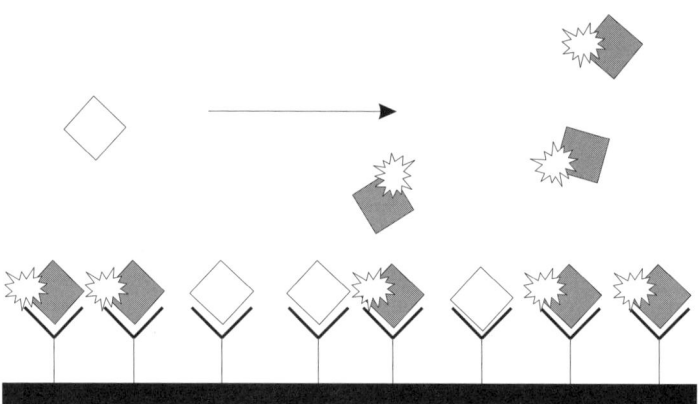

Figure 5.53   Targets are transported in the reaction chamber by a buffer flow. They start displacing the analogs because they have a higher affinity with the ligands.

marked, usually by a fluorescent marker. They are then immobilized to the ligands by a typical sandwich-type reaction. It is searched to have the highest possible number of analogs immobilized: the higher the saturation level, the better for the sensitivity of the detection. When the targets are brought in the vicinity of the surface of capture, they displace the analogs, because they have a higher affinity with the ligands. One can say they "compete" with the analogs to bind to the ligands, hence the name *competition*.

The result of the displacement is a change in the fluorescence level. At the location of the surface of capture, the level of fluorescence decreases due to a decrease in the marked analog surface concentration. Further down the microchannel, the displaced analogs carry their fluorescent markers, and there is a fluorescence increase at the microchannel outlet.

Detection of fluorescence emission

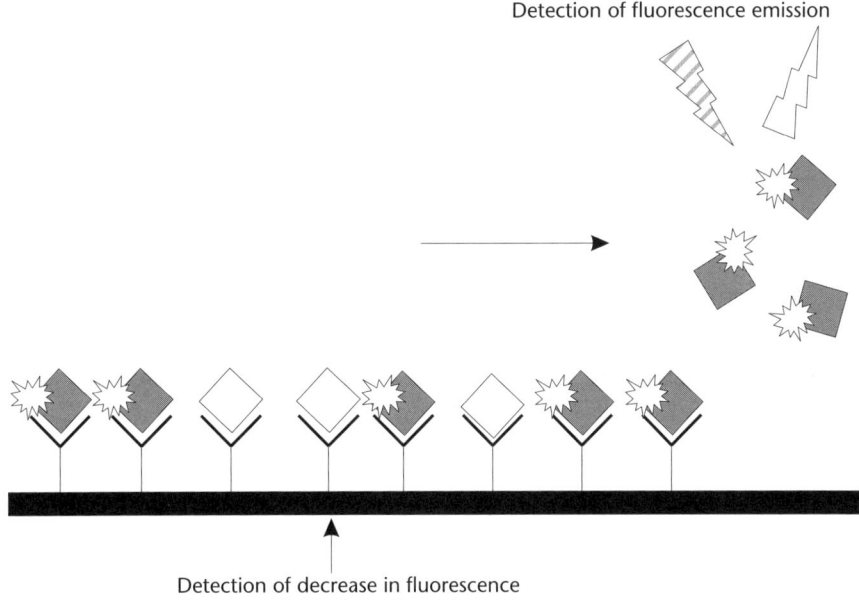

Detection of decrease in fluorescence
directly at the surface of capture

Figure 5.54 Detection can be performed by two methods: by measurement of the decrease in fluorescence of the capture surface or by measurement of the increase in fluorescence at the channel outlet.

In conclusion, *the targets are not marked*; only the analogs are marked, which is much more convenient. However, the process requires finding the best analogs possible and takes longer to set up. In a sense, displacement reactions are the opposite of sandwich reactions.

Displacement Reaction Using Fluorescence Resonance Energy Transfer

A very interesting type of displacement reaction uses the fluorescence resonance energy transfer (FRET) principle [32, 33].

When a fluorophore is excited by a light at the right wavelength, it emits light at a slightly shifted wavelength, as sketched in Figure 5.55.

However, if another fluorophore is placed next to this fluorophore, it quenches the light emission by energy transfer (Figure 5.56). This is the principle of FRET.

Using the FRET principle, it is possible to set up the protocol for a displacement reaction [33]. The first step consists of the functionalization of the surface of capture with tripods formed by an antibody and a fluorophore (Figure 5.57). In a second step analogs—labeled with *quencher* fluorophores—are introduced in the microchamber (or the microchannel). The analogs immobilize on tripods by chemical affinity (Figure 5.58), resulting in a quenching of the fluorescence of the tripods.

In a third step, targets are in turn introduced inside the microchamber. They displace the analogs—with their fluorophores—which are progressively removed from the microchamber. The fluorescence level of the surface of capture increases again (Figure 5.59).

Finally, depending on the particular application of the reaction, a fourth step may be done: this step consists of automatically regenerating the surface to be able

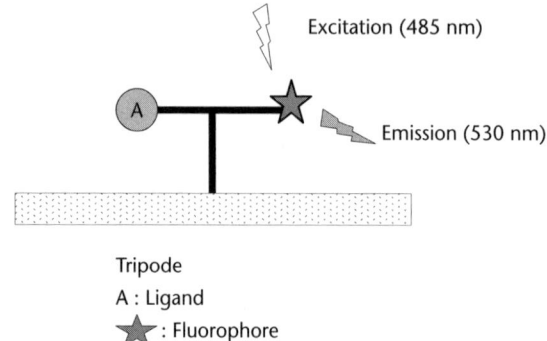

Figure 5.55 Schematic view of a tripod for functionalization of the solid surface. Fluorophore excitation and emission wavelengths are slightly shifted in frequency.

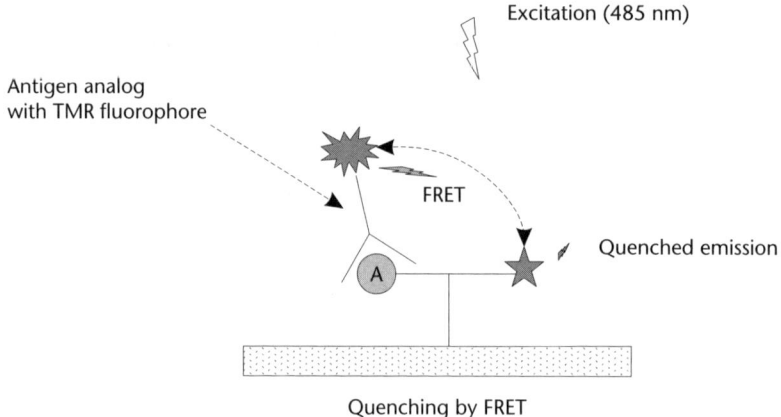

Figure 5.56 Principle of fluorescence quenching by FRET. Most of the fluorescent activity is quenched by the proximity of tetramethyrhodamine (TMR).

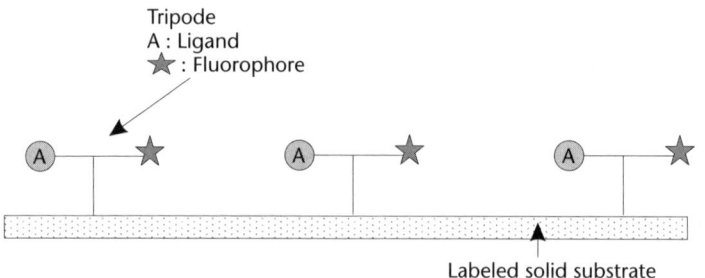

Figure 5.57 Step 1: Functionalization of the surface with tripods.

to perform successive reactions (Figure 5.60). Analogs are reintroduced in the microchamber and the same process is reinitiated.

The advantage of this type of reaction is that it is a typical displacement reaction, plus the functionalized surface of capture can be regenerated just by reintroducing the analogs in the microchamber. The microsystem can thus be reused.

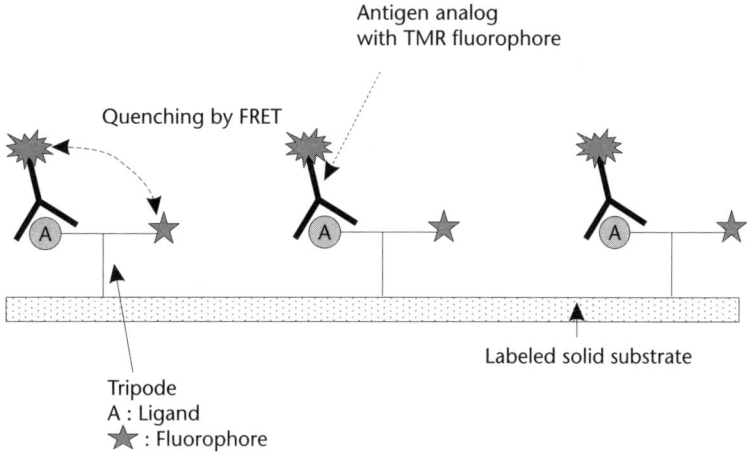

Figure 5.58 Step 2: Analogs are introduced in the microchamber provoking the quenching of the fluorescence by FRET.

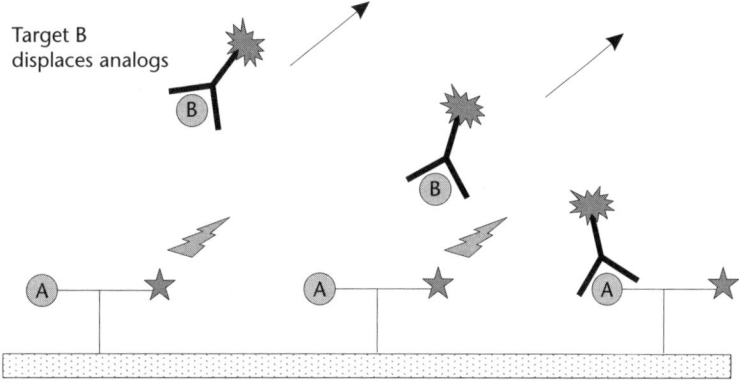

Figure 5.59 Step 3: Targets B displace analogs; detection is associated with the restoration of fluorescence.

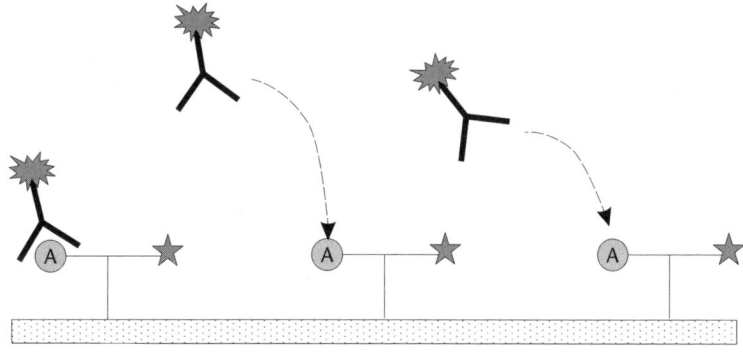

Figure 5.60 Step 4: Regeneration of the functionalized surface. Fluorescence activity is quenched once more time.

In conclusion, this type of displacement reaction has two major advantages: no labeling of the targets and automatic regeneration. It has the drawback that it is sometimes difficult to find the right analogs. Up to now, displacement reactions are less sensitive than sandwich reactions, but their performances are steadily increasing.

Modeling Displacement Reactions

If the technology of displacement reactions is now well known, and biochips using this type of reactions are currently used [30, 31], the modeling of such reactions is still under development.

The model presented here is based on the analogy with the competition term in the Lotka-Volterra equations [34]. First, suppose that the analogs and the targets are together in a solution and consider the Langmuir equations for each antigen A^* and A (A^* refers to the target, and A refers to the analog).

$$\frac{d\Gamma_1}{dt} = k_1 c_1 \left(\Gamma_0 - \Gamma_1 - \Gamma_2 \right) - k_{-1}\Gamma_1$$

$$\frac{d\Gamma_2}{dt} = k_2 c_2 \left(\Gamma_0 - \Gamma_1 - \Gamma_2 \right) - k_{-2}\Gamma_2 \tag{5.94}$$

The first equation in (5.94) considers that the analog A has the kinetics coefficients k_1 and k_{-1} and can bind to ligands in a concentration of $(\Gamma_0 - \Gamma_1 - \Gamma_2)$. The second equation has the same meaning for the targets this time. In such an approach, there is no displacement because each type of molecule binds to the ligands independently.

In reality, because the analogs are immobilized first, and the targets arrive after the washing of the channel, the Langmuir system of equations should be

$$\frac{d\Gamma_1}{dt} = -k_{-1}\Gamma_1$$

$$\frac{d\Gamma_2}{dt} = k_2 c_2 \left(\Gamma_0 - \Gamma_1 - \Gamma_2 \right) - k_{-2}\Gamma_2 \tag{5.95}$$

Still there is no competition between the two species in (5.95). Now let us introduce a competition term by similarity with the Lotka-Volterra equations. By analogy, it can be assumed that the target A^* is the predator and the analog A the prey. A competition term is then introduced in the equations to indicate that there is a competition between A^* and A. The system (5.94) becomes

$$\frac{d\Gamma_1}{dt} = k_1 c_1 \left(\Gamma_0 - \Gamma_1 - \Gamma_2 \right) - k_{-1}\Gamma_1 - \beta c_2 \Gamma_1$$

$$\frac{d\Gamma_2}{dt} = k_2 c_2 \left(\Gamma_0 - \Gamma_1 - \Gamma_2 \right) - k_{-2}\Gamma_2 \tag{5.96}$$

where the coefficient β has the same dimension as the adsorption constant k_1 (m^3/mole/sec). Displacement of antigens A by the targets A^* is then equal to the displacement rate β multiplied by the concentration of free targets A^* and immobilized analogs A.

In the particular operation sequence that we have defined here, the displacement equations are

$$\frac{d\Gamma_1}{dt} = -k_{-1}\Gamma_1 - \beta c_2 \Gamma_1$$

$$\frac{d\Gamma_2}{dt} = k_2 c_2 \left(\Gamma_0 - \Gamma_1 - \Gamma_2\right) - k_{-2}\Gamma_2$$

(5.97)

The difference with the Lotka-Volterra model is that we have not introduced cross terms in the second equation for A^* because the targets are assumed not to be affected by the analogs A.

The first equation can be integrated (c_2 is assumed constant)

$$\frac{\Gamma_1}{\Gamma_a} = e^{-(k_{-1} + \beta c_2)t}$$

(5.98)

where Γ_a is the initial surface concentration of immobilized analogs. Under the form (5.98), the model shows that desorption of analogs A under the action of competition of the targets A^* is faster than if the analogs were alone.

After substitution of (5.98) in (5.97), we are left with a single differential equation for the kinetics of adsorption of the targets

$$\frac{d\Gamma_2}{dt} = k_2 c_2 \left(\Gamma_0 - \Gamma_a e^{-(k_{-1} + \beta c_2)}t - \Gamma_2\right) - k_{-2}\Gamma_2$$

(5.99)

Equation (5.99) can be integrated under a closed form, using some algebraic developments. The method consists of writing

$$\Gamma_2 = \Gamma_2^* + \varepsilon$$

where Γ_2^* is the solution of the equation

$$\frac{d\Gamma_2^*}{dt} = k_2 c_2 \left(\Gamma_0 - \Gamma_2^*\right) - k_{-2}\Gamma_2^*$$

which means that Γ_2^* is the solution of the Langmuir equation for the targets A^* alone. It can be shown that the solution of the differential equation (5.99) is

$$\frac{\Gamma_2}{\Gamma_0} = \frac{k_2 c_2}{k_2 c_2 + k_{-2}}\left(1 - e^{-(k_2 c_2 + k_{-2})t}\right) +$$

$$\frac{k_2 c_2}{k_{-1} - \left(k_2 c_2 + k_{-2}\right)}\frac{\Gamma_a}{\Gamma_0}\left(e^{-k_{-1}t} - e^{-(k_2 c_2 + k_{-2})t}\right)$$

(5.100)

The first term is the term $\dfrac{\Gamma_2^*}{\Gamma_0}$ and the second term (always negative) is a correction taking into account the occupancy of the surface sites (ligands) by the analogs A. Note that the asymptotic value of Γ_2 is

$$\frac{\Gamma_2}{\Gamma_0}\Big|t \to \infty = \frac{k_2 c_2}{k_2 c_2 + k_{-2}}$$

which is the same asymptote as for Γ_2^*: At a very long time, all of the analogs have dissociated, and the targets' hybridization kinetics follow a simple Langmuir law.

Example of Displacement Reaction

Efficiency of displacement reactions for different targets, mostly biochemical, have been tested in microdevices using a continuous flow—hence the name *continuous flow immunoassays* (CFI).

In order to estimate the displacement rate, the analytes (targets) are introduced in the channel by concentration "bursts" of different levels (Figure 5.61). Each burst of analyte concentration displaces a certain number of analogs. So, at the microchannel outlet, bursts of concentration in marked analogs are detected. The displacement rate is determined by comparison between the concentration level in targets and the corresponding concentration level in analogs.

The concentration in analogs is given by a series of differential equations of the type, corresponding respectively to periods of simple desorption and periods of displacement.

$$\frac{d\Gamma_1}{dt} = -k_{-1}\Gamma_1 \quad \text{for} \quad t < t_i^-$$

$$\frac{d\Gamma_1}{dt} = -k_{-1}\Gamma_1 - \beta c_{2,i}\Gamma_1 \quad \text{for} \quad t_i^- \le t \le t_i^* \qquad (5.101)$$

$$\frac{d\Gamma_1}{dt} = -k_{-1}\Gamma_1 \quad \text{for} \quad t < t_i^+$$

where $c_{2,i}$ is the burst concentration in analytes between times t_i^- and t_i^+. The solution of the system (5.101) for the first burst is

$$\Gamma_1 = \Gamma_a e^{-k_{-1}t} \quad \text{for} \quad t < t_1^-$$

$$\Gamma_1 = \Gamma_a e^{\beta c_{21}t_1^-} e^{-(k_{-1}+\beta c_{21})t} \quad \text{for} \quad t_1^- \le t \le t_1^+ \qquad (5.102)$$

$$\Gamma_1 = \Gamma_a e^{-\beta c_{21}(t_1^+ - t_1^-)} e^{-k_{-1}t} \quad \text{for} \quad t < t_1^-$$

The general solution for an experiment with i successive bursts is

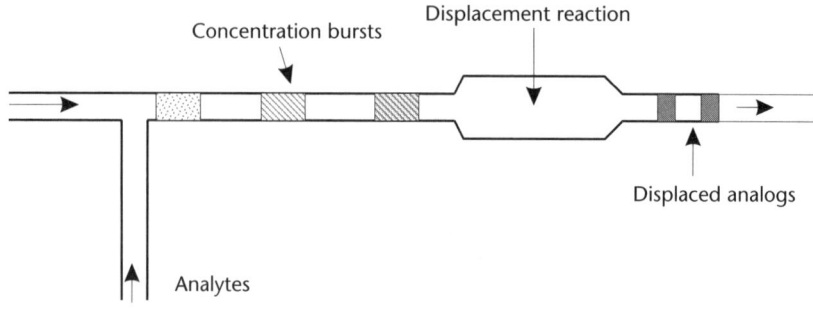

Figure 5.61 Schematic view of an experimental displacement reaction microdevice.

$$\Gamma_1 = \Gamma_a e^{-k_{-1}t} \left[\prod_{j=1}^{i-1} e^{-\beta c_{2j}\left(t_j^+ - t_j^-\right)} \right] e^{-\beta c_{2i}\left(t-t_i^-\right)} \quad \text{for} \quad t_i^- \leq t \leq t_1^+$$

(5.103)

$$\Gamma_1 = \Gamma_a e^{-k_{-1}t} \left[\prod_{j=1}^{i-1} e^{-\beta c_{2j}\left(t_j^+ - t_j^-\right)} \right] \quad \text{for} \quad t > t_1^+$$

In Figure 5.62, we show calculated kinetics of displaced analogs. In such a case, the bursts of targets had the same concentration. The base line is the kinetics of desorption of analogs alone (without the displacement due to the targets). The successive bursts of displaced analogs are decreasing in importance because the immobilized concentration in analogs is decreasing.

In Figure 5.63, results of the model have been compared to experimental results [35, 36]. In this case, the different concentrations of the bursts of targets [cyclotrimethylenetrinitramine (RDX)] changed. The peaks in concentration of displaced analogs are well predicted by the model. These peaks are clearly linked to the successive peaks of RDX entering the microchannel.

5.5 Conclusion

The most important application of biochips is biorecognition, and biorecognition is based on key-lock type of reactions. To this regard, we have investigated the physics of biochemical reactions and determined the kinetics of the most important reactions, like DNA hybridization and enzymatic reactions for proteins. These kinetics, however, can be modified by the concentration of reacting species (analytes or targets), and the coupling between biochemical reactions and advection-diffusion of reagents in the biochip is essential. Finally, we have distinguished between two types of reactions, the sandwich reaction, which derives directly from the key-lock approach, and the displacement reactions that are more complex and require the use of an analog to the target.

At this point, it is essential to point out that detection is an important part of the conception of any biochip. It is not sufficient to have a very efficient capture—by hybridization or immunorecognition—if there is no sensitive detection associated

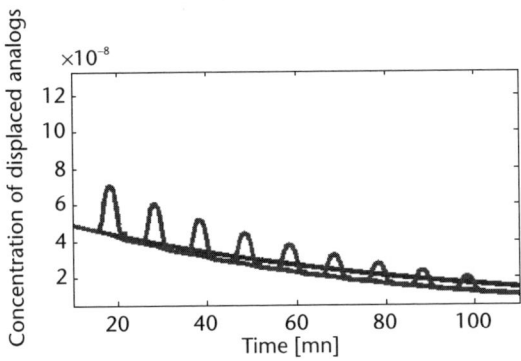

Figure 5.62 Calculated bursts of displaced analogs. Compare with the case of desorption alone. Peaks correspond to burst of targets, displacing the analogs.

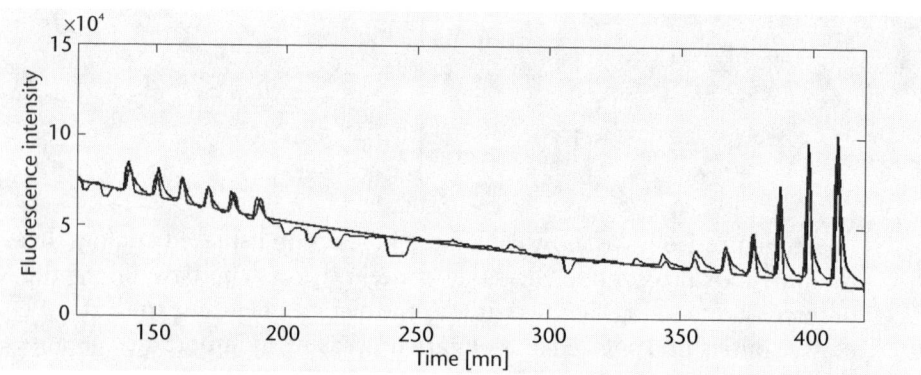

Figure 5.63   Kinetic curves of displacement of RDX analogs [36]. Compare between experimental results and calculated kinetics. The base line corresponds to the desorption of RDX analogs alone.

with it. The reading of the biochip reactive surface should be at least as sensitive as the reaction itself. Detection is not the subject of this book; let us just mention that the research on detection for biochip recognition is the topic of abundant literature, and is a field that is constantly improving. Developments are aimed in two directions: first, improvements in the detection method itself, like improving the fluorescence by using new fluorophores (e.g., quantum dots) or developing enzymatic amplification for detection; second, improvements of the design and materials, like improved waveguides in the case of detection by fluorescence or the use of complementary metal oxide semiconductor (CMOS) detectors for photons emitted by the fluorophores or by the enzymatic revelation.

References

[1]   http://www.cheng.cam.ac.uk/research/groups/laser/Teaching/metrology/immuno_label.pdf.

[2]   Harlow, E., and D. Lane, "Using Antibodies: A Laboratory Manual," Cold Spring Harbor Laboratory, 1999.

[3]   Atkins, P. W., *Physical Chemistry*, New York: Freeman, 1998.

[4]   Laidler, K. J., *Chemical Kinetics*, New York: Harper and Row, 1987.

[5]   Berthier, J., P. Combette, and L. Blum, "Numerical Calculation of a Microfluidic Protein Reactor: Is the Classical Michaelis-Menten Integral Relation Sufficiently Accurate?" *Lab-on-a-Chip and Microarrays Conference*, Zurich, Switzerland, January 14–16, 2002.

[6]   Berthier, J., P. Combette, and L. Blum, "A Model for the Kinetics of Heterogeneous Enzymatic Reaction in a Protein Microreactor," *Fourth LETI Annual Review*, CEA, Grenoble, France, June 2002.

[7]   Sharov, A. A., Virginia Tech, http://www.gypsymoth.ento.vt.edu/~sharov/PopEcol/lec10/lotka.html.

[8]   Lotka, A. J., *Elements of Physical Biology*, Baltimore, MD: Williams and Wilkins, 1925.

[9]   Edelstein-Keshet, L., *Mathematical Models in Biology*, Birkhäuser Mathematical Series, New York: McGraw-Hill, 1987.

[10] Murray, J. D., *Mathematical Biology*, New York: Springer Verlag, 1989.

[11] Tabeling, P., *Introduction à la Microfluidique*, Paris, France: Belin, 2003.

[12] Park, S. H., et al., "Gel-Free Experiments of Reaction-Diffusion Front Kinetics," *Physical Review E.*, Vol. 64, 2001, pp. 055102-1–055102-4.

[13] Galfi, L., and Z. Racz, "Properties of the Reaction Front in an A + B → C Type Reaction-Diffusion Process," *Physical Review A*, Vol. 38, No. 6, 1988.

[14] Baroud, C. N., et al., "Reaction-Diffusion Dynamics: Confrontation Between Theory and Experiment in a Microfluidic Reactor," *Physical Review E.*, Vol. 67, 2003, pp. 060104-1–060104-4.

[15] Kamholz, A. E., et al., "Quantitative Analysis of Molecular Interaction in a Microfluidic Channel: The T-Sensor," *Anal. Chem.*, Vol. 71, 1999, pp. 5340–5347.

[16] Butler, J. E., "Solid Supports in Enzyme-Linked Immunosorbent Assay and Other Solid-Phase Immunoassays," *Methods*, Vol. 22, 2000, pp. 4–23.

[17] Ruckstuhl, T., M. Rankl, and S. Seeger, "Highly Sensitive Biosensing Using a Supercritical Angle Fluorescence (SAF) Instrument," *Biosensors and Bioelectronic*, Vol. 18, 2003, pp 1193–1199.

[18] Berthier, J., et al., " A Modified Langmuir Equation for Microfluidics Systems," *Proc. 8th World Congress on Biosensors*, Granada, Spain, May 24–26, 2004.

[19] Press, W. H., et al., *Numerical Recipes*, Cambridge: Cambridge University Press, 1987.

[20] Sapsford K. E., et al., "Kinetics of Antigen Binding to Arrays of Antibodies in Different Sized Spots," *Anal. Chem.*, Vol. 73, 2001, pp. 5518–5524.

[21] Winzor, D. J., "Determination of Binding Constants by Affinity Chromatography," *Journal of Chromatography A.*, Vol. 1037, 2004, pp. 351–367.

[22] Mason, T., et al., "Effective Rate Models for the Analysis of Transport Dependent Biosensor Data," *Mathematical Biosciences*, Vol. 159, 1999, pp. 123–144.

[23] Glaser, R., "Antigen-Antibody Binding and Mass Transport by Convection and Diffusion to a Surface: A Two-Dimensional Computer Model of Binding and Dissociation Kinetics," *Analytical Biochemestry*, Vol. 213, 1993, pp. 152–161.

[24] Hibbert, D. B., J. J. Gooding, and P. Erokhin, "Kinetics of Irreversible Adsorption with Diffusion: Application to Biomolecule Immobilization," *Langmuir*, Vol. 18, 2002, pp. 1770–1776.

[25] Stenberg, M., L. Stiblert, and H. Nyguen, "External Diffusion in Solid-Phase Immunoassays," *J. Theor. Biol.*, Vol. 120, 1986, pp. 129–140.

[26] Lionello, A., et al., "Adsorption of Proteins in a Microchannel," *Lab-on-a-Chip*, Vol. 5, 2005, p. 254.

[27] MATLAB, The MathWorks, version 6, September 2000.

[28] Narang, U., et al., "Multianalyte Detection Using a Capillary-Based Flow Immunosensor," *Analytical Biochemestry*, Vol. 255, 1998, pp. 13–19.

[29] Selinger, J. V., and S. Y. Rabbany, "Theory of Heterogeneity in Displacement Reactions," *Anal. Chem.*, Vol. 69, 1997, pp. 170–174.

[30] Kusterbeck, W., et al., "A Continuous Flow Immunoassay for Rapid and Sensitive Detection of Small Molecules," *J. of Immunological Methods*, Vol. 135, 1990, pp. 191–197.

[31] Ligler, F. S., et al., "Integrating Waveguide Biosensor," *Anal. Chem.*, Vol. 74, 2002, pp. 713–719.

[32] Volland, H., et al., "Solid-Phase Immobilized Tripod for Fluorescent Renewable Immunoassay: A Concept for Continuous Monitoring of an Immunoassay Including a Regeneration of the Solid Phase," *Anal. Chem.*, Vol. 77, 2005, pp. 1896–1904.

[33] Neuburger, L.-M., et al., "A New Concept for Continuous Flow Immunosensors," *Proc. 8th World Congress on Biosensors*, Granada, Spain, May 24–26, 2004.

[34] Berthier, J., et al., "An Analytical Model for CFIs," *Proc. 8th World Congress on Biosensors*, Granada, Spain, May 24–26, 2004.

[35] Narang, U., "Fiber Optic-Based Biosensor for Ricin," *Biosensor and Bioelectronics*, Vol. 12, 1997, pp. 937–945.

[36] Wemhoff, G. A., et al., "Kinetics of Antibody Binding at Solid-Liquid Interfaces in Flow," *J. of Immunological Methods*, Vol. 156, 1992, pp. 223–223.

CHAPTER 6

Experimental Approaches to Microparticle-Based Assays

The microparticles we deal with in this book can be distinguished into two categories: In the first category are the particles of biological interest that are naturally present in the biological systems and on which it is necessary to obtain some information. The second category deals with artificial particles that are manufactured by chemical synthesis or by genetic modification, as tools to perform a function in the process (observation, characterization, or manipulation). Dealing with micro- and nanoparticles means that the objects are not only smaller, they have intrinsic properties witnessing these length scales (mechanical, optical, and so forth) that are emphasized here.

In this chapter, which is more oriented toward experimental situations, we first present the biological objects limiting ourselves to major biopolymers and to some aspects of cells. Then, we introduce a few basic physical notions and definitions, and we review some synthetic particles and their use. The next part is devoted to techniques used to characterize these objects, and we end with a few words on micromanipulation techniques. This chapter should be read as an introduction to these experimental techniques. It can give an idea of what is possible along the main lines described in the present book but is in no way an exhaustive picture. The interested reader is encouraged to go into the matter more closely with the classical books or review papers listed in the references at the end of the chapter or throughout the text. Furthermore, the electric field based techniques are described in Chapter 8, while Chapter 7 is devoted to magnetic particles and related techniques.

6.1 A Few Biological Targets

In this part, we focus on a few examples that are a major concern in many studies of this area. We only review here some aspects of three families of the most important biological macromolecules: DNA, RNA, and proteins. In the second part, we deal with some aspects of live cells.

With the sequencing of the genome of many organisms in the last 30 years, there has been a need to better understand not only the function of the different genes but also, more ambitiously, the way the different constituents of cells or organisms interact and organize themselves into complex networks. On this "functional genomics" point of view, physics and engineering are everywhere, from the concepts of DNA or protein arrays and their interpretation to the modeling of the various functions and interactions in the biochemical networks.

6.1.1 Biopolymers

Classically, protein expression is described by the following sequence: the genetic information carried on the DNA sequence is "read" by a protein assembly called *RNA polymerase*; this transcription gives rise to the RNA molecules. The messenger RNA finds its way out of the nucleus in the cytoplasm and is translated into functional proteins by the ribosome. We will give a few elements on the structure of these macromolecules here.

6.1.1.1 DNA Strands

Desoxy ribo nucleic acid (DNA) molecules are made of two strands twisted around each other (see Figure 6.1). Each of these strands is constituted of four bases—adenine ([A]), thymine ([T]), guanine ([G]), and cytosine ([C])—on a phosphate backbone. They are arranged in a double helix, where the bases are located inside and paired exclusively ([A]-[T] and [G]-[C]); they are called Watson-Crick base pairs (see Figure 6.1). The two strands are oriented and arranged in antiparallel directions. The arrangement of the base pairs along the strand bears the genome of an individual. It contains all of the person's genetic information.

Above a certain denaturation temperature, the two strands separate. This property is used in the PCR technique to amplify the number of copies of DNA molecules in a given sample. In this technique, after the denaturation step, the sample is cooled down and *primers* (short complementary sequences) bind to the beginning and the end of the region of the DNA to be amplified. An enzyme (the polymerase) then reads the single strand and matches it with its complementary sequence using the free nucleotides in solution. So, starting from one double strand, we end up with two. The same process is cycled 30 to 40 times, leading to an exponential amplification of the number of copies of the initial sample.

At a much larger scale, DNA is a polymer. When sufficiently diluted, DNA chains in solution adopt a coil configuration (see Figure 6.2) whose radius, called the radius of gyration R_g, is directly related to the size of the monomers b and their number N through the relation [1]:

$$R_g = b \cdot N^\nu \tag{6.1}$$

For polymers in *good solvent* (meaning that the interactions between a monomer and a solvent molecule are favored compared with interactions between two

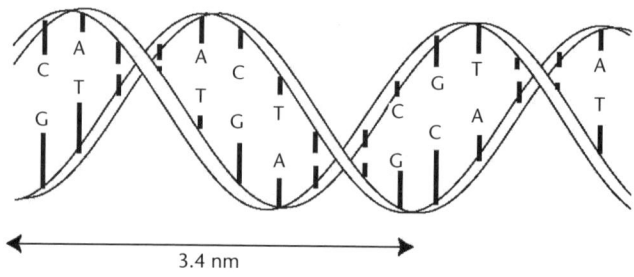

Figure 6.1 Double-helix structure of a DNA molecule.

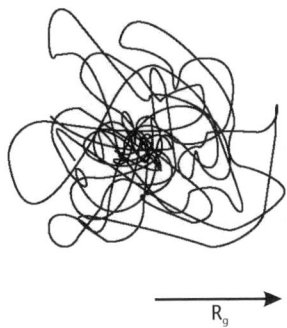

Figure 6.2   Coil configuration of a single polymer chain in solution.

monomers), this exponent ν is 3/5. In some cases, however, these interactions are effectively comparable, and the chain is said to be ideal; the exponent ν is then 1/2. Importantly, although for different physical reasons, the case of DNA double-strand molecules falls into this last category. This arises because of the semiflexible nature of this chain, and it can then be shown that only unrealistically long chains would be in good solvent [2] (see Figure 6.3). The persistence length of these polymers is 50–100 nm, depending on the characteristics of the solution (pH, salinity, and so forth).

This picture of DNA molecules is meaningful only in a buffer solution. In a cell's nucleus, DNA in the form of chromosomes is packed extremely tightly by histones, a particular class of compaction proteins.

DNA microarrays that have been used for several decades exist in several versions. Generically, it consists of depositing spots of single-strand DNA sequences on

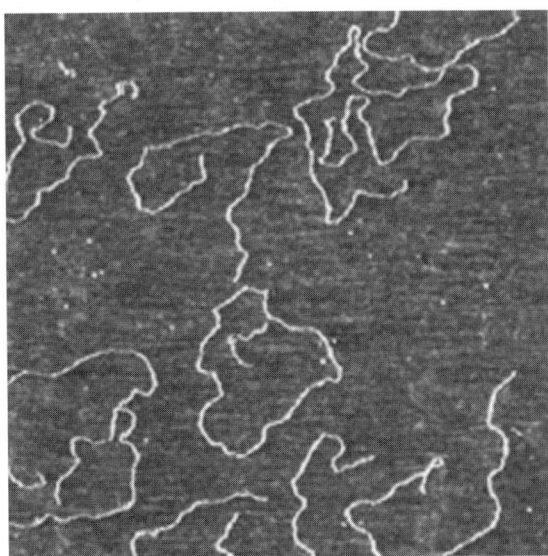

Figure 6.3   Atomic force microscopy of DNA molecules adsorbed on a solid surface (see a description of the technique in Section 6.3.1.3). The image is color coded in height so that the height difference between the white color and the dark color is ~1 nm. The size of the image is 1.5 μm. From such images, the contour length of these molecules can be accurately measured, along with their interactions with proteins if need be. (*From:* [3]. © 2002 Wiley-VCH. Reprinted with permission.)

a solid substrate, such as a glass slide, and in measuring the hybridization efficiency with DNA or RNA single strands, by fluorescence or radioactive labeling. When the target is itself DNA, these arrays can be used to identify genes such as those implicated in some diseases. The quantification of the level of these genes can be a diagnostic tool, for instance, with suspected precancerous cells [4, 5].

6.1.1.2 RNA

Ribo nucleic acid (RNA) molecules are similar to single-strand DNA from the point of view of their sequence. One of the four bases is different, with thymine being replaced by uracyle. The phosphate backbone also presents some slight differences. The messenger RNAs (mRNAs) are the molecules resulting from the transcription process. They contain the same genetic information as DNA and come out of the nucleus to be translated into proteins in the cytoplasm. Other smaller RNA molecules, such as transfer RNA (tRNA), are used in this translation process. The bases of RNA, like the ones of DNA, associate themselves but because of the single-strand structure of RNA molecules, these interactions lead to partially folded structures (see Figure 6.4). The shape imposed on these molecules by this folding plays an important role in the translation into proteins.

The folding of these molecules is fixed by their sequence, hence by the DNA sequence. Their structures correspond to a minimum of energy and can now be accurately computed for reasonably long molecules (up to a few thousand nucleotides) [6, 7].

Because it is an indicator of gene expression, mRNA has a major position in functional genomics, and it is one of the major targets of the DNA arrays described earlier. By hybridization with the small DNA sequences spotted on the array, the levels of particular RNAs are measured, and, from there, one gets some information on the amount of the translated functional proteins. Although an indirect measurement, it is much easier to work with RNA than with proteins, and, since their levels are strongly correlated, it is a quite useful information [4].

6.1.1.3 Proteins

Proteins are the product of the translation of RNA by ribosomes in the cytoplasm. Along the RNA strand, three consecutive nucleotides, a codon, are translated into an amino acid in a very robust way using the so-called genetic code.

The primary sequence of proteins thus mirrors that of the initial DNA molecule. This sequence determines the three-dimensional structure of the proteins: Similarly to the case of RNA that we saw earlier, interactions between amino acids make the protein fold. Here, these interactions are more diverse: Van der Waals interactions, electrostatic interactions, and hydrogen bonds combine to shape the proteins into their functional form. There are several levels of organization; first, the interactions between neighboring parts of the chain can lead to regular structures such as the α-helices or the β-sheets. These structures (the secondary structures) can arrange themselves into more complex domains that are common to some extent to many different proteins. For a given protein, these domains are arranged in a particular way to define the full three-dimensional tertiary structure. Very often, these

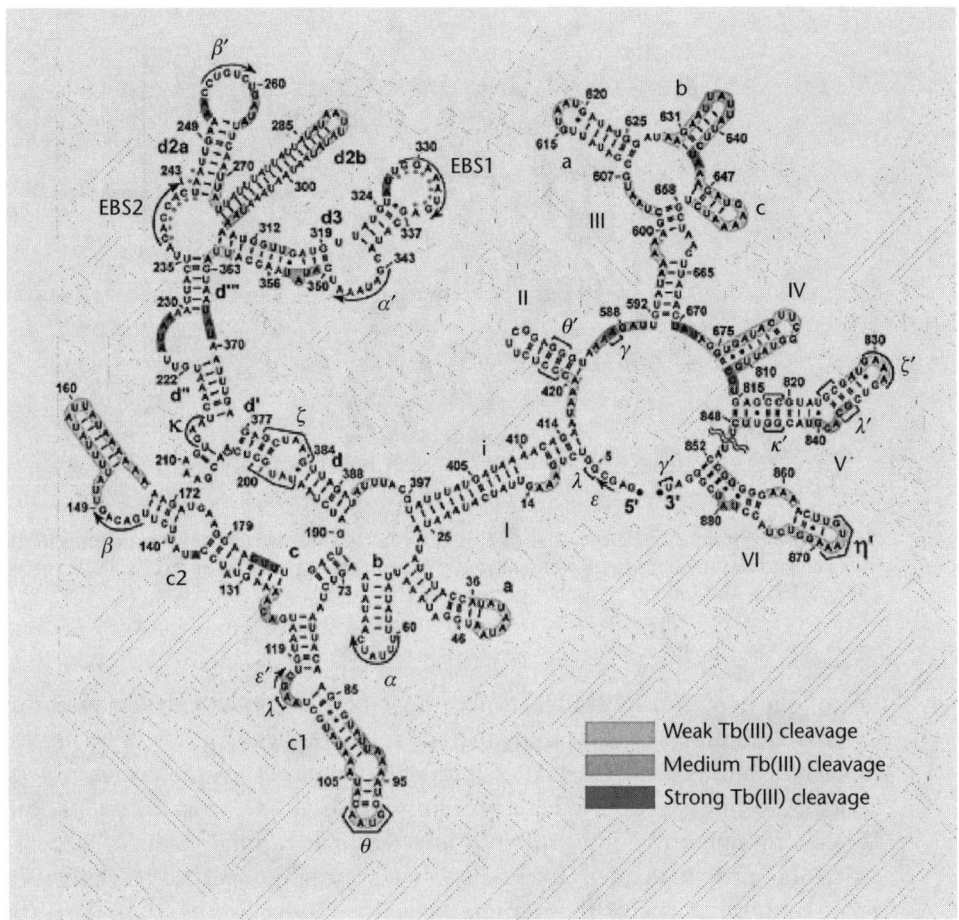

Figure 6.4   Pairing of RNA bases leads to a complex molecular structure (*From:* [8]. © 2000 Nature Publishing Group. Reprinted with permission.)

structures are not functional by themselves. The functional protein is a multimer (sometimes called the quaternary structure) of several of these units that are then called monomers (not to be mistaken with the monomer as a single unit of a polymer chain). Some proteic assemblies result from only a few monomers (sometimes only one); some others result in the self-assembly of many of them: this is the case, for instance, of the actin filaments or of microtubules that form the cell cytoskeleton and have a very dynamic assembly-disassembly behavior inside the cells. This is also the case of viral capsids.

As the three-dimensional structure imposes the protein function, it is of great interest to experimentally access it. Practically, the techniques used are X-rays diffraction, electron microscopy, and, for smaller proteins, nuclear magnetic resonance (NMR). We have seen in the preceding part that it was already difficult to compute the shape of RNA molecules where there are only a limited number of possible interactions. This is, of course, even more the case , and it is actually quite difficult to predict the three-dimensional structure of a protein from its sequence [9] (see Figure 6.5).

Some proteins are not only targets that deserve characterization, but they also can become tools in the hands of biologists. Enzymes, for instance, are catalysts that

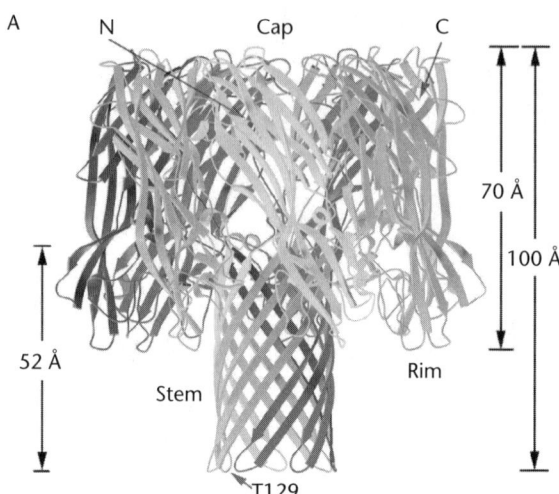

Figure 6.5 Structure of α-hemolysin. This complex is a heptamer that forms pores in membranes: the *stem* crosses the lipid membrane, and the *cap* is in contact with the extracellular medium. (*From:* [10]. © 1996 AAAS. Reprinted with permission.)

are of particular importance, since most of the biological processes are highly dynamic. Members of this family include restriction enzymes that are heavily used in molecular biology as molecular scissors able to cut DNA molecules at very specific spots. Another example where proteins are used as "workhorses" is antibodies: When a foreign body or a protein is injected in an animal, some of its cells produce particular proteins called antibodies that are highly specific to this *antigen*. This property is the basis of the immunological response and is often used to produce antibodies that can be tagged and then specifically targeted to the protein of interest in cells.

We mentioned earlier the use of DNA microarrays to quantify the level of gene expression. Measuring directly the level and the activity of proteins with protein microarrays is the next step in this process. Here, the sequences of DNA spotted on the surface are replaced by proteins or by molecules with which these proteins can interact. One can then access protein-protein interactions, proteins interactions with small molecules, and the like.

The information one gets is richer than that obtained by measuring the levels of RNA. Indeed, after their production, proteins mature, are modified, and interact—all processes that can slightly affect their function or activity [11].

6.1.2 Some Aspects of Cells

Cells are extremely complex arrangements and, of course, living entities (the smallest there are). Among species, prokaryotic cells do not have nucleus. They include bacteria and archae and are much simpler than eukaryotic cells, in which DNA is packed in a nucleus.

Both cell types have a barrier protecting them from the exterior: a phospholipid soft membrane for eukaryotes and a more rigid wall for bacteria.

6.1.2.1   Eukaryotic Cells

It is often required to sort and characterize particular cells. An interesting example of this process is the one of micrometastases in the field of early cancer diagnosis. In some cancers, such as breast cancer, isolated tumor cells disseminate in the body of the patient by being conveyed by the circulating blood. They can also be found in their bone marrow. The challenge becomes the detection of these cells, estimated to be 1–10 cells per 1 million nucleated cells. Their detection relies on specific markers in their cytoplasm or on their membrane and uses fluorescence-based techniques such as the ones reviewed later in this chapter. However, the number of pathologic cells is so low that not only does their detection require a particularly sensitive and specific technique, but it is also of prime importance to first enrich the medium with these cells. This results in a two-phase process whereby the enrichment does not have to be highly specific but should take all of the suspect cells and a second step where the true detection needs to be highly specific. An efficient strategy in this line is to use immunomagnetic enrichment: suspect cells are attached to magnetic beads by antigen-antibody coupling and then sorted with a magnet. This example illustrates the current efforts to move and control populations of single cells, isolate them, and quantify their amount and their characteristics [12].

Live cells can also be used as sensors: In cell-based biosensors, the cell itself is the detector. For instance, some cells are extraordinarily sensitive to particular chemicals and can be used as a sensor for traces of these molecules. In others cases, if a drug has to be tested on a particular organ, the individual response of cultured cells of this organ is monitored by electrical or optical means, and, although the details of the response are not always fully understood, they carry directly the biological information rather than an indirect characterization [13].

Cell arrays take advantage of the high parallelism of the microarrays technology with cells. They consist of arrays of live cells or cell clusters that are each transfected with a different gene. The response of these different cells to external stimuli can then be monitored in parallel [14].

6.1.2.2   Bacteria

Bacteria are smaller than eukaryotic cells (typically a couple of micrometers versus about 10 μm). The bacteria family is very diverse; their mode of locomotion varies from one species to the next. *Escherichia coli*, for instance, swims by a succession of "runs" and "tumbles." Its body is covered with flagellae individually connected to motors. According to the direction of rotation of the motors, the motion of bacterium is quite different: When they rotate in one direction, they do it synchronously, and the cell "runs" in straight lines. However, they lose this synchronicity when they rotate in the other direction, and the bacterium "tumbles." As a whole, the motion of these bacteria is a random walk accurately described by an effective diffusion. At the Reynolds numbers dealt with at the scale of these micron-sized swimmers, there is basically no inertia, and they use viscous forces to their advantage. It can be shown [15] that for a statistics of runs obeying the Poisson distribution of mean duration τ, the effective diffusion coefficient is

$$D = v^2\, \tau/3 \qquad (6.2)$$

where v is the velocity of the cell during a run. With reasonable values $v \sim 10^{-5}$ ms^{-2} and $\tau \sim 1$ second, we obtain $D \sim 10^{-6}$ cm^2/sec, comparable to small molecules' diffusion coefficients.

Not only can these bacteria swim quite efficiently, they also have the ability to swim in the direction of a food or an attractant source. The mechanism of this chemotaxis has long been questioned, as a simple diffusion calculation shows that the gradients across a 1-μm bacterium cannot explain this motion by themselves. It turns out that a bacterium does not compare the concentrations between its two extremities but rather compares them at different times during its runs. If there is more food after a time T in the run, it means that the bacterium runs in the "right" direction, and it therefore makes this run longer before the next tumble. If this is not the case, the tumble is not delayed. This way, an efficient bias is given toward the source of attractants [16].

Although very different from eukaryotic cells, some of the practical situations found with these cells are very similar to concerns mentioned earlier. For instance, the fight against bioterrorism has emphasized the importance of checking for the presence of a few virulent bacteria within relatively large samples.

We have described some properties of bacteria. In a biology laboratory, they count among the biologist's best friend. Since the genome of some of them has been entirely sequenced, it has now become a routine work to modify it using molecular biology tools (e.g., by inserting exogenous genes that can express particular proteins). These bacteria (and *E. coli* in particular) are thus literally transformed into protein factories.

6.2   Biotechnological Tools

Synthetic microparticles find a natural use either as macroscopic "handles" for the manipulation of molecules or cells, or to add or to enhance a signal in the various detection schemes. This last category includes immunofluorescence or immunoelectron microscopy, in which a fluorescent dye or a metal colloid is coupled to an antibody that specifically targets the molecule of interest.

6.2.1   Fluorophores

Fluorescence is one of the major tools used in biochemistry/biotechnology. The high sensitivity of the technique and the numerous available coupling strategies make it a very versatile routine tool. Before describing the probes used in this context, let us first recall a few notions of the physical principle.

6.2.1.1   Fluorescence

Some molecules have the ability to absorb photons at a given wavelength and emit them back at a different (longer) wavelength corresponding to a lower energy (Figure 6.6). Fluorescein, for instance, is a very popular fluorescent dye that shines in yellow-green ($\lambda \sim 520$ nm) when excited by blue light ($\lambda \sim 480$ nm). There is a range

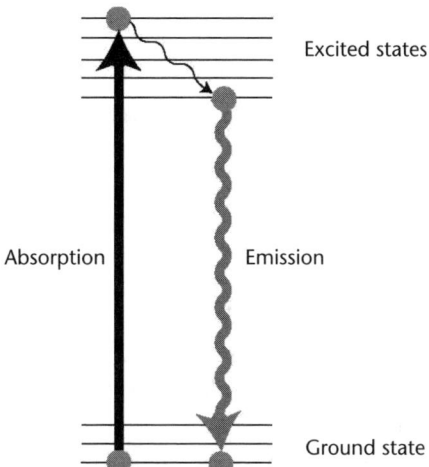

Figure 6.6   Jablonski energy diagram illustrating the principle of fluorescence.

of wavelengths that are absorbed (the excitation spectrum) and a range of emitted wavelengths (emission spectrum) (Figure 6.7).

This property is used extensively in fluorescence microscopy but also with other techniques, such as flow cytometry (see Section 6.3.3.2). The reasons for such extensive use are:

• Extreme sensitivity of fluorescence up to single fluorophore detection;
• The huge panel of possible reactants that chemistry has bought. With the right coupling chemistry, it is possible to couple a dye to virtually any biological object of interest.

Immunofluorescence is a particular coupling strategy that uses biochemistry rather than chemistry. Antibodies are first covalently coupled to a fluorophore and then allowed to interact with the cell or the tissue so that only the protein of interest "becomes" fluorescent.

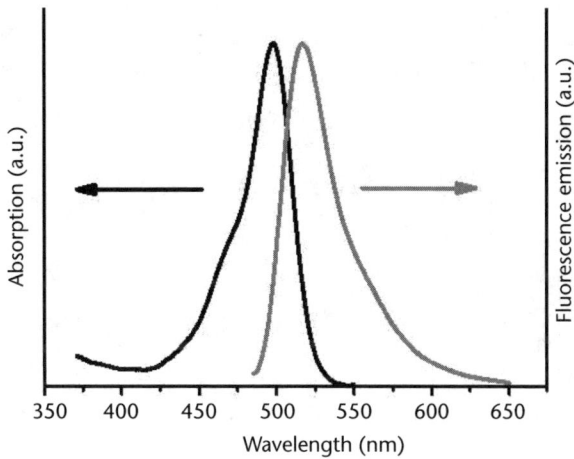

Figure 6.7   Excitation (black) and emission (gray) spectra of fluorescein.

6.2.1.2 Fluorescent Molecules and Particles

Fluorophores are fluorescent molecules characterized by their excitation and emission spectra. The quantum yield, defined as the number of emitted photons per absorbed photon, quantifies their efficiencies. The fluorescence properties of these molecules depend on their immediate environments. They can thus become sensitive and local sensors for pH or viscosity. They can also detect ions in solutions: Calcium or other divalent ions, for instance, are readily and quantitatively detected by the FuraRed molecule even within cells.

Unfortunately, these molecules cannot switch indefinitely between their excited state and their ground state. After a certain number of these transitions (this number is extremely variable from one molecule to the next), they permanently lose their fluorescence properties, a phenomenon called *photobleaching* that is enhanced by the presence of dissolved oxygen in the solution. By carefully degassing solutions and keeping the excitation to a minimum, this effect can be minimized at the cost of much decreased fluorescence intensity.

This effect can, however, be used to one's advantage: by purposely bleaching a spot in the sample by using a high light intensity and observing the dynamics of recovery (with a much lower light intensity), one has access to the local diffusion coefficient of the tracer (Figure 6.8). This technique is called fluorescence recovery after photobleaching (FRAP).

FRET is yet another technique based on fluorescence: It consists of using two dyes in such a way that the emission spectrum of the first dye significantly overlaps the excitation spectrum of the second one. This way, when the two molecules are close enough, exciting the first dye results in a decrease of its emission and,

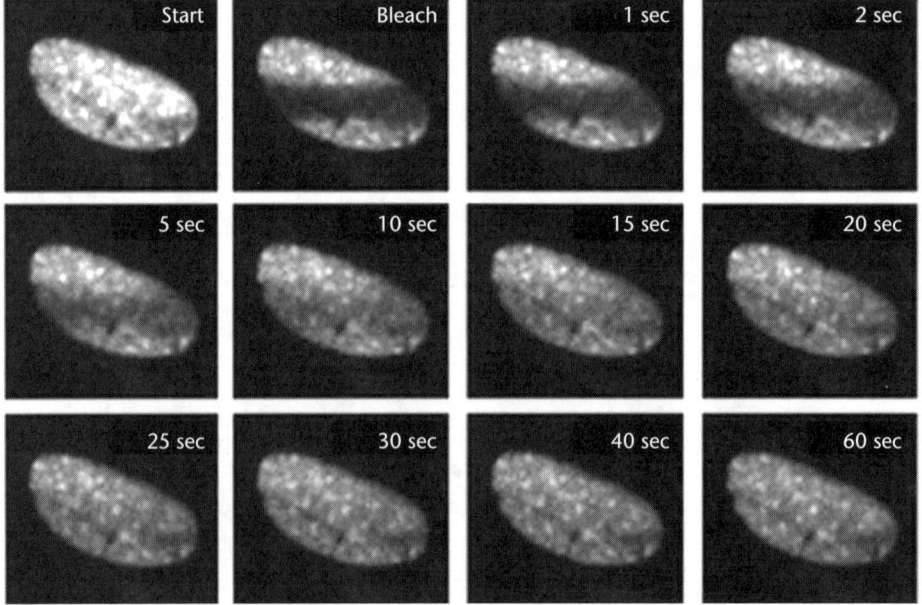

Figure 6.8 FRAP of a protein fused with green fluorescent protein (GFP) in living cells. The 3-μm-wide band bleached across the width of the cell nucleus progressively recovers its fluorescence from the diffusion of proteins initially outside this band. (*From:* [17]. © 2000 Rockefeller University Press. Reprinted with permission.)

conversely, to an increase of emission for the second one. Typically, the molecules can be no further apart than a distance called the Forster radius that is of the order of 5 nm [18]. FRET is thus well adapted to intramolecular distance measurements or to other situations where the interacting molecules are very close.

Quantum dots (QD) are fluorescent nanoparticles (typically 10 nm) whose use in biology-oriented applications is relatively recent. These inorganic particles are made of semiconductors (very often ZnSe crystals surrounded by a thin ZnS shell) and have a few remarkable properties that explain their popularity.

They all share a broad excitation spectrum in the blue and have an extremely bright emission wavelength that depends only on their size. Moreover, they show practically no photobleaching, and they are small enough to be incorporated in many systems, even at the surface of live cells. By using two sizes, two different proteins can thus be labeled and excited with the same wavelength, making colocalization experiments particularly easy. The use of these jewels in biology-related applications has been delayed in particular because of the difficulties in dispersing these hydrophobic particles in water, not to mention their coupling to biomolecules. This is due to a particularly inert surface, but this difficulty has recently been solved by their encapsulation with amphiphilic molecules such as block copolymers [19].

GFP is a naturally fluorescent protein present in the jelly fish *aequorea victoria*. The GFP gene can be fused by genetic engineering to one of the proteins of interest, so that the resulting protein is a fusion of both, coupling the desired function with fluorescence. This way, proteins in living cells can be directly observed by fluorescence. Better efficiency and other colors have been developed by modifying the gene of the original GFP.

6.2.2   Other Micro- and Nanoparticles

We review here some of the particles used in tests or in biotechnology-related applications. Because of their potential, magnetic beads deserve a chapter by themselves and are described in Chapter 7.

6.2.2.1   Latex

Latex particles are plastic micron-sized particles that are commonly used because of their variety: they can be found commercially with different surface chemistries, different sizes, and other distinctive properties. For instance, one can easily get fluorescent or magnetic beads.

These particles are synthesized by emulsion polymerization in which the hydrophobic organic monomer is encapsulated by surfactant molecules in a micelle (see Section 6.2.2.3) and then polymerized in the water phase. This produces nice monodisperse suspensions that unfortunately need the surfactant to stay stable. However, it is possible to add at the polymerization step a monomer that can play the role of the surfactant and that stays at the bead-water interface after copolymerization, stabilizing the particles by electrostatic interactions. Moreover, these chemical functions that are now at the surface of the particles can be used to initiate the coupling of biomolecules on the beads.

For instance, the widely used latex agglutination tests consist of adsorbing anti-bodies to micron-sized latex particles. When the corresponding antigen is present, these beads interact with it and clump together. Because of their size, the particles and the clumps scatter light differently, and a simple visual observation gives the answer on the presence of the antigen.

6.2.2.2 Gold Nanoparticles

The main use of gold nanoparticles is electron microcopy. They can be coupled to antibodies by electrostatic nonspecific adsorption. When the target protein is present, the nanoparticles couple specifically to it and appear as distinctive tiny black spots in transmission electron microscopy.

The particular optical properties of these nanoparticles can also be used to improve agglutination tests classically performed with latex particles (see previous explanation). For instance, they are used in some commercial pregnancy tests. Because of their small size, these particles have a characteristic red color caused by a phenomenon called plasmon resonance that we will review in more detail in Section 6.3.3.1. The urine of pregnant women contains a particular hormone whose corresponding antibody is adsorbed both on the nanoparticles and on micron-sized latex particles. When the hormone is present, the two types of particles coagglutinate and yield the formation of red clumps.

6.2.2.3 Surfactants and Micelles

Surfactants (also called amphiphiles) are molecules composed of two antagonistic parts: a hydrophilic polar head and a hydrophobic nonpolar tail. Phospholipids that are the constituents of the cellular membrane belong to this category. In a water solution, the hydrocarbon tails minimize their interactions with the water, and the molecules self-assemble into structures exposing only the hydrophilic headgroups toward the water. In fact, they tend to aggregate into micelles such as the one depicted in Figure 6.9 as soon as their concentration is high enough.

We can write the equilibrium between a solubilized surfactant molecule S and a micelle consisting of n of these molecules S_n $(n \gg 1)$

$$nS \leftrightarrow S_n$$

With the equilibrium constant

$$K = [S_n]/[S]^n \tag{6.3}$$

If c is the total concentration in surfactants

$$c = [S] + n[S_n] \tag{6.4}$$

let us define c^* as $c^* = (nK)^{1/n}$.

Combining (6.3) and (6.4), we immediately find:

2 nm

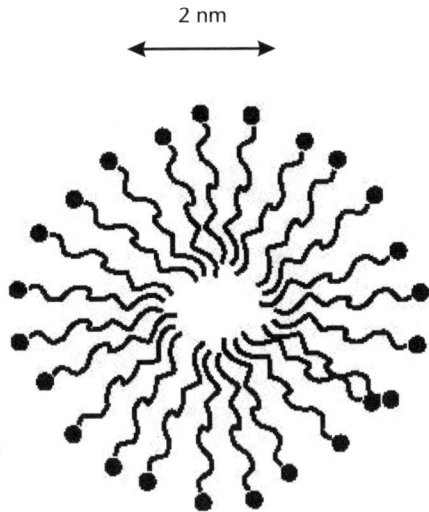

Figure 6.9 Schematic view of a micelle (two-dimensional cross section).

$$\text{If } c \ll c^*, [S] \sim c$$
$$\text{If } c \gg c^*, [S] \sim c^*$$

In other words, below c^*, called the critical micellar concentration, surfactant molecules are individually solubilized; above this concentration, they tend to aggregate in micelles (Figure 6.10). Micelles are very dynamic objects constantly exchanging molecules with the free surfactant molecules in the solution [20].

Micelles are not restricted to small surfactant molecules. In particular some block copolymers can be tailored to make micelles particularly well suited to drug delivery applications. In this last case, chemists have been quite imaginative in designing molecules with the right lengths and the right chemistries that self-assemble in micelles and can contain an organic drug in their core. These micelles can specifically target particular cells in an organ without releasing the drug too early or being destroyed too soon by the defense mechanisms.

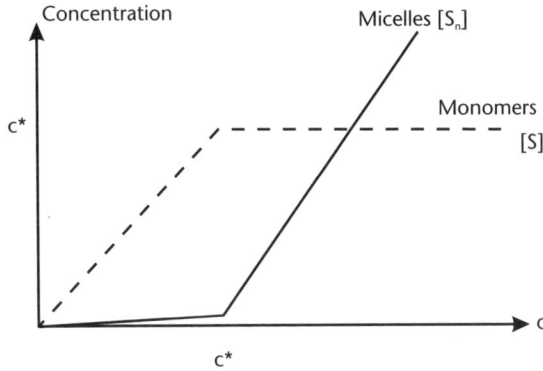

Figure 6.10 Phase diagram of amphiphile molecules. Above the critical micellar concentration, surfactants predominantly assemble in micelles.

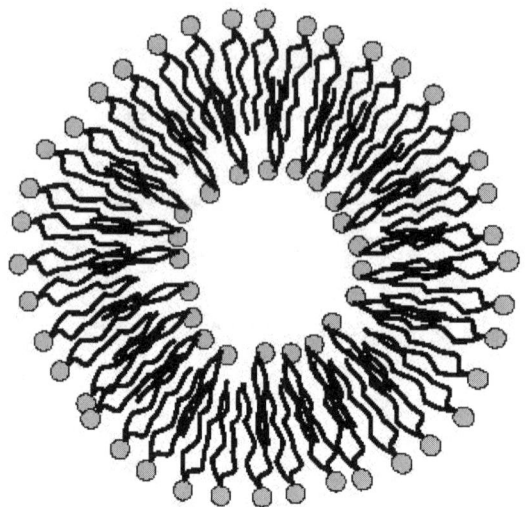

Figure 6.11   Schematic view of a bilayer vesicle.

The surface area per molecule in these aggregates results from a balance between attractive tail-tail interactions and repulsive head-head interactions. From geometric arguments, when the tail group of these molecules is too bulky for optimum packing into micelles, bilayers can self-assemble and lead to the formation of vesicles (Figure 6.11). This is, for instance, the case for most two-chains surfactants, such as phospholipids, that are the major constituents of cell membranes.

Cell membranes based on these structures incorporate membrane proteins that have a hydrophobic domain interacting strongly with the aliphatic chains of the lipids. We saw earlier that the structure determination of proteins is often performed by diffraction techniques, and thus they need to be crystallized. However, the membrane domains of these proteins are so hydrophobic that it makes them irreversibly precipitate as soon as they are removed from this lipidic environment. Their crystallization thus necessitates particular sequences in which the hydrophobic domain remains protected from water by surfactants molecules. These protocols have to be finely tuned for each particular protein, and, for this reason, their structures are still unknown for most of them, whereas soluble proteins whose crystallization conditions can be better rationalized are better known in that respect [21].

6.2.3   Chemical Modification of Surfaces

As in any microsystem, microparticles have a large surface-volume ratio. As a result, surface-related problems and surface chemistry become of paramount importance. The first way that comes to mind to immobilize biomolecules on surfaces is to rely on nonspecific adsorption. Although such a coupling can be efficient enough in some situations, better control is usually required.

Very few surfaces are truly inactive. Very often, they bear chemical groups that can be used for further surface chemistry. In particular, metal surfaces—in particular, gold—and oxide surfaces—in particular, SiO_2—are good templates for these chemical modifications [21]. This last case is of particular interest because these surface treatments are also applied to glass by extension, although the chemistries of these two surfaces are not strictly identical. Polymer surfaces such as the surface of latex beads can also include a sophisticated surface chemistry by the right choice of the monomers used for their synthesis. However, even in this case, direct coupling is generally not possible because very reactive groups would readily hydrolyze in water where the coupling reaction is to take place. Intermediate coupling molecules are thus needed. Generically, these molecules have a reactive group at each extremity. One of them reacts on the solid surface: In the case of gold, it is a thiol group; it is a silane group in the case of silica. The end group at the other extremity of these molecules is exposed toward the exterior world and is used for the coupling to the proteins, for instance.

Chemical grafting on a plane surface and on a microparticle share some common features but also differ in a number of ways. On plane surfaces, the grafting of these molecules results from a collective mechanism where they interrelate together by Van der Waals interactions as they react on the surface [21]. The monolayer can be reinforced by a lateral polymerization illustrated by the case of silanes on silica, where some of the silane groups react on the surface while the others react together forming a *net*. While it is relatively easy to qualitatively modify a surface, achieving good monolayers, which is the first step to a good surface coverage, is a delicate operation, particularly in the silane/silica surface (even more so for the silane /glass system because of the defects and the different chemistry of the glass surface). With this strategy, it becomes possible to change the physical properties of the surfaces, such as transforming hydrophilic surfaces to hydrophobic ones. With particles, there is usually no need for a "perfect" monolayer, and the grafting conditions are less drastic.

A major use of this surface chemistry is to protect surfaces from nonspecific adsorption. In that case, long PEG molecules can be used. To adsorb on the surface, a protein would have to compress this layer, which is entropically very unfavorable. [22].

DNA chips are another example requiring surface modification. DNA oligomers are spotted onto a surface (usually glass) and need to be permanently anchored on it. Glass is negatively charged, and very few molecules would naturally remain stuck on it once it is in contact with a water-based buffer. On the other hand, if the surface is coated with amine groups by silanization, the surface becomes positively charged, and these oligomers then stick irreversibly to it.

In the case of proteins, a covalent reaction is even better. Very often, a group able to react on amines is chosen, as the exterior surface of proteins is rich in these chemical groups. The aldehyde group present in glutaraldehyde and the N-hydroxy-succidimide (NHS) group are common choices for this purpose. Other groups, such as vinyl sulfone, can react on thiols [23]. This strategy, however, has several drawbacks: it needs a high enough density of amine groups on the protein surface; it can interfere with the function of the protein if it reacts precisely on the functional site; and, of course, even if it reacts on some other random place of the protein, the orientation information is lost.

To overcome these difficulties, strategies that involve *molecular glues* by specific and sturdy interactions, such as that of streptavidin with biotin or hexahistidine sequence $(His)_6$ with nitrilo-tri acetic acid (Ni-NTA) can be preferred [24]. Antigen-antibody can also be used to the same end. These strategies are particularly seducing, as groups such as biotin or hexahistidine sequence can be genetically included in the protein during its synthesis by cells and are actually often used to purify them after cells' lysis. The position of these groups on the protein is thus well known; it does not interfere with their function. If the linker of the streptavidin or the NTA to the surface is sufficiently rigid, the orientation of the protein is preserved.

6.3 Experimental Methods of Characterization

6.3.1 Microscopies

Although they are all called microscopies, and all are imaging techniques, there is little in common between optical microscopy, electron microcopy, and atomic force microscopy on the physics point of view.

6.3.1.1 Optical Microscopy

The first microscopy technique that comes to mind to characterize particles is optical microscopy. Optical microscopes, although based on the same basic design, constantly improve, adding new potentialities that the use of lasers as light sources and the computer analysis of images have contributed to enhance. They are an invaluable compromise between ease of use, versatility, and performance.

The basic principle of the formation of images has been basically unchanged since the apparition of the first microscopes in the seventeenth century. This short section is not aimed to replace an optics textbook but merely to remind us of a few key points.

Although microscopes used to be designed in the way presented in Figure 6.12, new models are now corrected to include a region within the microscope where all of the rays are parallel. The reason for using such geometry is fairly easy to understand: The aberrations of the optics limit the performances of a microscope. For good optics, these imperfections are limited, but it is often necessary to include optical components (filters, polarizers, and so forth) in the optical path. In this case, it becomes necessary to position these elements in a region of the microscope where the rays are parallel. The modern research microscopes incorporate this feature and are called infinity corrected systems.

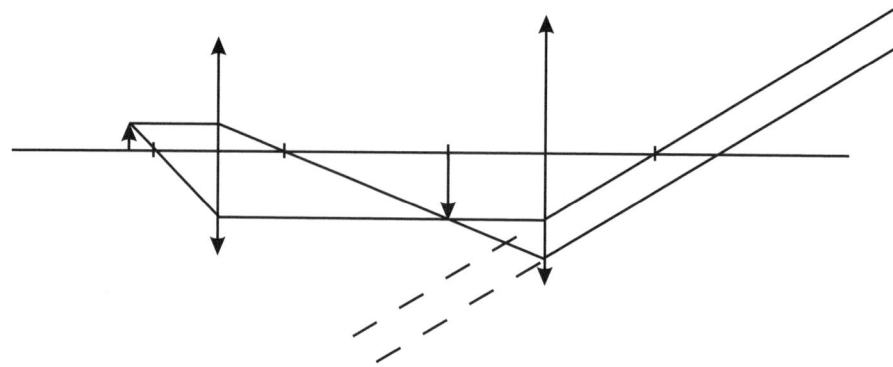

Figure 6.12 Optical path of a microscope. The object to be observed is before the focal point of the objective, close to it. Its image is formed at the focal point of the ocular, sending the final image to infinity.

Because of the diffraction, a single object appears as the convolution of the object shape and a function called the Airy function. This means that, due to the laws of far field optics, a point source will appear to have a finite size in the microscope. To be able to "see" two objects, they thus have to be further apart than the width of this function, whose order of magnitude is the wavelength of light. Thus, there is a separation criterion stating that the ultimate resolution of an optical microscope is given by the classical Rayleigh formula

$$d = 1.22(\lambda/2 \cdot N_a) \qquad (6.5)$$

where d is the smallest possible spacing between the two objects, λ is the wavelength, and N_a is the numerical aperture of the microscope.

To achieve a better resolution, high numerical aperture objectives are needed (in particular oil immersion objectives) as well as the use of blue wavelengths.

Still, any object, regardless of its size, can be observed by optical microscopy, provided that it emits enough photons. Its observed lateral size has nothing to do with the true one, but if one is interested in its dynamic behavior or in a more macroscopic measurement such as concentration, this is not a concern. Fluorescence imaging of single molecules that are now routinely performed in many laboratories is a good illustration of these capabilities. Along the same line, images are now routinely computer analyzed. Optical interferometric techniques, coupled with the right mathematical analysis, can detect tiny displacements down to a few nanometers [25] (Figure 6.13).

It is certainly one of the strengths of optical microscopy to be able to image samples in water-based solutions. This way, fully hydrated, possibly live cells or tissues

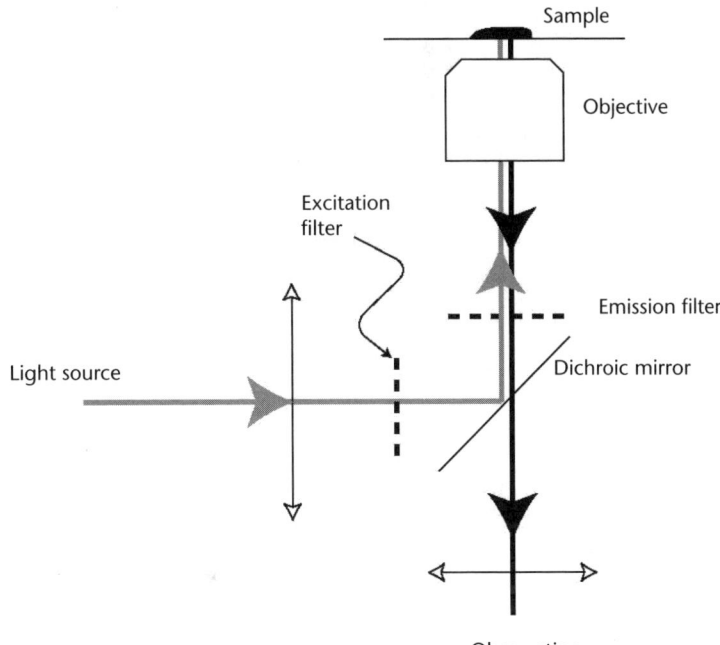

Figure 6.13   Principle of fluorescence microcopy. In this particular configuration, the excitation light and the fluorescence emission go through the objective (epifluorescence).

can be imaged. Fluorescence microscopy enables us to observe only the objects carrying a fluorescent dye. The proteins of interest are then localized by comparing images in fluorescence mode to images in white light mode. By using two dyes and working with two colors, two different objects can be colocalized.

Many solutions exist that can increase the contrasts of a particular image, depending on the characteristics of the sample: The polarization of the light, its angle of incidence, and the interference between two light rays are strategies used alone or in conjunction and are commercially available [26] (Figure 6.14).

The confocal microscope [27] works on a somewhat different principle. The image is formed point after point. Here, the light source is a point source. The point image in the sample emits light that is collected by a detector through a pinhole confocal with the source. This way, only the light that one wants to collect is indeed collected. On the other hand, the light emitted by other parts of the sample whose fluorescence is also excited by the illumination (the parts of the sample in the cone of light produced by the microscope objective) is largely excluded from the detector. The sample is then scanned in order to make the full-scale image. Why is this interesting? In classical microcopy, all of the fluorescence light is collected, and although

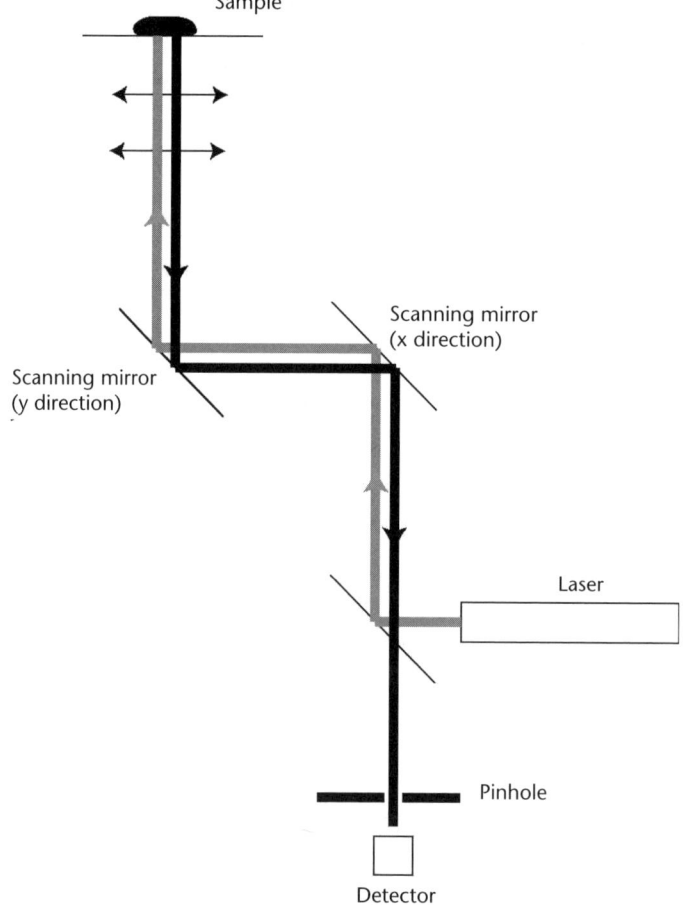

Figure 6.14 Principle of fluorescence confocal microscopy. No image is formed on a screen, but the fluorescence intensity is collected point by point by a sensor such as photomultiplier.

there is a maximum of intensity at the focal point, the image is blurred by this background. The confocal technique is very efficient in rejecting unwanted out-of-focus light. *Slices* at a given height are performed by scanning the sample in the *x-y* plane, and three-dimensional images can then be reconstructed by combining these slices. As it is a scanning technique, it is somewhat slow and thus not very efficient in studying fast dynamics.

A further refinement of confocal microscopy is the two-photon microscopy [27] that uses a nonlinear effect to excite fluorescence only at the focal point, leaving unexposed the rest of the cone of illumination. To perform this task, an infrared (IR) light is used for excitation. No fluorescence is directly excited by these IR photons. However, if two of them combine, they can excite the dye to its excited state, so that a photon is emitted when the molecule returns to its ground state. The probability of such two-photon events is very small and needs a very high intensity to trigger a detectable fluorescence. This condition on intensity is met only at the focal point of the objective (and requires very-high-power lasers). Compared to classical confocal microscopy, there is no need for a pinhole and the corresponding optics. All of the light coming from the excitation volume is collected. Furthermore, as the excitation volume is confined at the focus, there is no bleaching in the rest of the light cone, and three-dimensional images can be reconstructed with better accuracy, even for very diffusing sample

Let us end this section by mentioning another new and very powerful tool to investigate phenomena very close to surfaces and that is particularly well suited to the imaging of single molecules or other phenomena of low fluorescence. As its name suggests, total internal reflection fluorescence (TIRF) microscopy uses total internal reflection: in the right angular conditions, light is totally reflected from the interface and only a very small fraction of the energy penetrates in the sample. This penetration is limited to a distance of the order of a fraction of wavelength (a few hundred nanometers at most). This evanescent light can excite the fluorescence of the analyzed objects. The obvious advantage is the suppression of any background fluorescence coming from the part of the sample further away from the surface than the penetration length. The contrast is thus much better improved.

6.3.1.2   Electron Microscopy

Electron microscopes use an electron beam to probe objects. The transmission electron microscope works on a similar principle as the optical microscope using electrons in place of a light beam. Because of the much smaller wavelength of the electrons, the resolution obtained with these instruments is several orders of magnitude better than with optical microscopes (down to a fraction of nanometer). The optical path is exactly the one used to describe optical microscopes: The source of electrons is a heated filament, and the deflection of the electron beam corresponding to the deflection of light by glass lenses is obtained by magnetic fields. Contrary to the optical microscope, the resolution is never the theoretical resolution imposed by the Rayleigh formula (6.5) but is caused by aberrations unavoidable with magnetic lenses. Because of the interactions of the electrons with air, the whole set up is placed under vacuum. This limits the applications of electrons microscopes: no live cell or even hydrated sample can be imaged with such instruments.

Transmission electron microscopes (TEMs) image the density of electrons. The intrinsic absorption of every sample imposes us to work on thin samples. Either it is naturally the case (e.g., with molecules or membrane patches that are stuck on a thin carbon film), or thick samples have to be microtomed into thin slices before their observation.

In a transmission microscope, areas richer in electrons appear dark in the image. As the biological samples are composed of light elements (carbon, oxygen, hydrogen), heavy elements have to be added to increase the contrast. It is common to use chemical staining or gold nanoparticles functionalized with antibodies that can target particular proteins. They then appear as dark spots in the TEM images specifically located on the molecules of interest.

If TEM works in the transmission mode, the scanning electron microscope (SEM) works in the reflection mode. Again, the parallel with the optical path of an optical reflection microscope is tempting; the difference is that, in the present case, a SEM does not form the image of the reflected electron beam, it analyzes the electrons scattered by the surface where the incident beam has been focused. Again, the optics are magnetic lenses.

These electron beam techniques have a very high resolution up to the point where TEM-based techniques can resolve some protein structures. However, they need a sometimes tedious preparation of samples.

6.3.1.3 Atomic Force Microscopy

The atomic force microscope (AFM) works with a completely different principle [28]. Here, the surface to be analyzed is scanned under a fine stylus mounted on a flexible leaf spring (a cantilever) in the same way that a stylus probes the surface of a record in an old-fashioned phonograph (Figure 6.15). When one is interested in microparticles or macromolecules, the first step is to strongly adsorb them on this surface. The deflection of the cantilever is then a direct measurement of the topography of the surface. To get some order of magnitude, the radius of curvature at the apex of the tip is of the order of a few tens of nanometers, the cantilever spring constant is of the order of a few tens of milli-Newtons per meter. In most of the commercial instruments, not to say all of them, the detection of the position of the cantilever is performed optically by shining a laser beam on the back of the cantilever and measuring the reflected beam with a quadrant photodiode. The relative displacements of the sample versus the tip are performed by piezo electric actuators in the three directions of space.

In practice, the mode just described, where the vertical position of the sample is fixed and the force of the tip acting on it varies, is seldom used for two main reasons: First, by using the microscope this way, the force is higher on the ridges or the bumps of the surface and lower in the valleys. Like with any observation technique, applying a force is already a potentially perturbative process (the extreme case being scratching the surface), and having different forces on the surface may make the images very difficult to interpret. The second difficulty is more instrumental—getting true vertical distances from the measurement of the deflection of the cantilever would necessitate an accurate calibration of the detector for each experiment, which is practically unreliable.

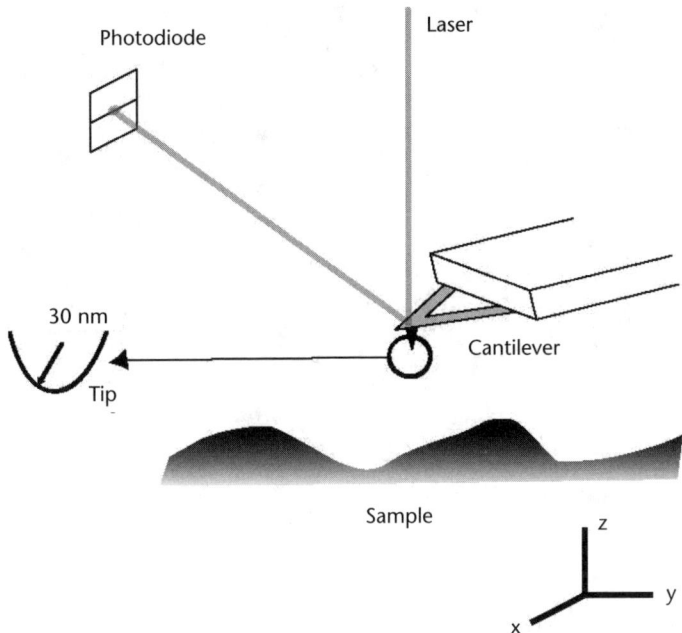

Figure 6.15   Principle of the AFM.

There is, however, a way to circumvent these difficulties; it follows a very general instrumentation strategy: The force (that is, the deflection of the cantilever) is kept at a fixed value, and the sample is dynamically moved up and down to keep this force constant during the scanning. This way, the two hurdles just mentioned vanish: on the one hand, the force applied on the sample is constant so the perturbation to the sample is the same everywhere, and, on the other hand, a constant force means a constant position of the cantilever extremity with respect to the laser. There is thus no more need for a calibration of the detector, since the read value stays constant. As the sample vertical position is monitored by a piezo-electric actuator easily calibrated in distance, the topography of the sample is directly given by these vertical displacements. The resolution is limited by the exact shape of the tip and is hard to define in the classical way, as the influence of this parameter depends on the size and shape of the imaged objects. It is certainly one of the strengths of the AFM to be able to image samples at a high resolution in a liquid and in particular in a buffer solution. In this respect, it is a major advantage compared to other techniques, such as electron microscopy. Because it is a simpler technique to use, AFM is often used also on dried samples.

It is clear from Figure 6.16 that AFM imaging is resolutive enough to access some protein structures. Although the resolution is somewhat poor compared to diffraction techniques and very partial, as only the exterior surface can be imaged since single molecules or complex ones can be imaged, there is no need for crystallization. This is an invaluable advantage in the case of membrane proteins. Furthermore, these images are taken in buffer solutions, sometimes directly on cells, thus in conditions very close to the functional environments of the proteins.

In biology, the samples one is interested in can have two characteristics that make them difficult to image with an AFM: They can be very soft, and they may need

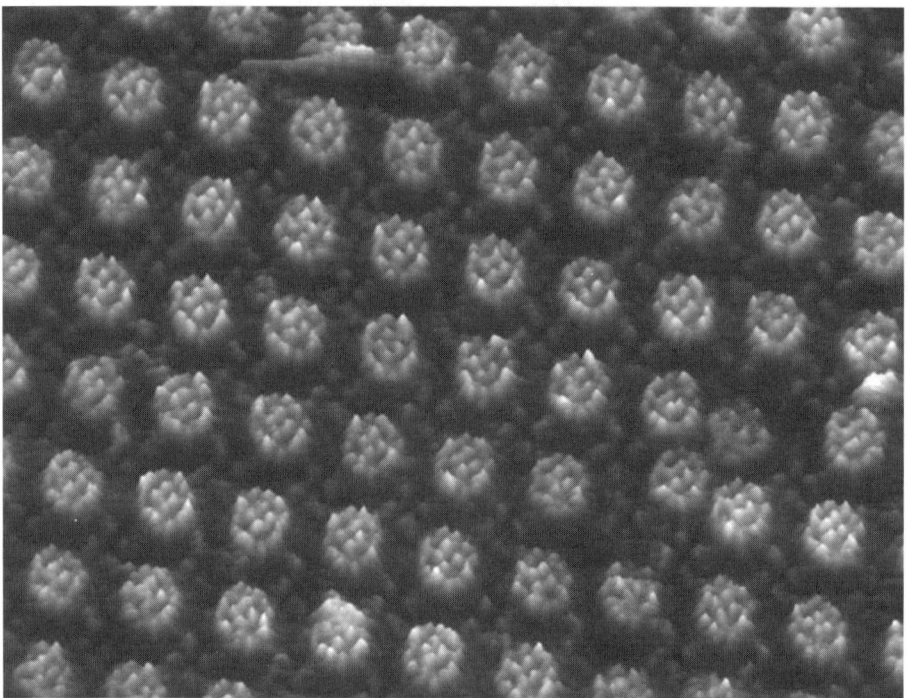

Figure 6.16 Image of bacterial aquaporin, a membrane protein still inserted in its native membrane [29]. Individual tetramers are clearly resolved. The lateral size of the image is 70 nm. (Courtesy of Dr. Simon Scheuring.)

to stay hydrated. Most of the time, it is both. The interaction of the stylus with the surface can then be too strong and the surface is scratched, the molecules of interest are damaged, or they are wiped out. To minimize these problems, other modes have been developed in which in the spring sustaining the tip oscillates and periodically "taps" the surface. The signal used for the feedback is then not the average position of the tip but the amplitude of oscillation measured by a lock-in detection. Although the tip still interacts rather strongly with the sample when it touches it, there is no transverse force (friction) applied to it, and a great deal of damage is avoided.

The AFM cantilever then behaves as a damped oscillator: It is characterized by a characteristic resonance frequency and a quality factor Q that witnesses the viscosity of the medium. In air (even more so in a vacuum) the quality factor is high (easily 100–1,000) meaning that the resonance is well defined. In water, Q is much lower (1–10) and consequently, because the resonance frequency is not well defined, the feedback is less sensitive and more damage is brought to the sample. New electronic methods aimed at electronically increasing the quality factor may become a good alternative to these problems.

With the microscopic sizes dealt with in this book, it is illustrative to describe the motion of a free cantilever subjected to thermal agitation. In that case, one can use the equipartition theorem on the potential elastic energy of this harmonic oscillator

$$\left\langle \frac{1}{2} m \omega_0^2 u^2 \right\rangle = \frac{1}{2} k_B T \qquad (6.6)$$

where m is the mass associated with the oscillator, u is its displacement, $\omega_0 = k/m$ is its resonance frequency, and k is the spring constant of the cantilever.

Equation (6.6) is used practically to calibrate the spring constant of the cantilever: When $k \sim 50$ mN/m, the thermal fluctuations are of the order of a few angstroms, which is readily measurable by the detector. Practically, a power spectral density is plotted and fitted with the theoretical lorentzian shape for a harmonic oscillator. The area under this curve A is then used to access the spring constant:

$$k = k_B T / A \tag{6.7}$$

6.3.2 Physical Characterization: Light Scattering

Initially used more to get molecular weight information out of polymers solutions in its static version, light scattering is also a powerful tool to directly measure hydrodynamic radii when it is in the form of dynamic light scattering.

The increase in computing capabilities and the reduction in size and cost of lasers have greatly popularized the use of these techniques.

6.3.2.1 Static Light Scattering [30]

It is a common observation that, when light hits a suspension, some of it is scattered along all directions. Rayleigh scattering describes quantitatively this scattering for particles smaller than the wavelength of the light. Here, we do not take into account the temporal fluctuations of the scattered light but average the signal over long times. The dynamic aspect will be treated later.

The theory proceeds by computing the interactions of the electric field associated with the incident light with the polarizability of the particle with which it interacts. The intensity $I(\theta)$ at an angle of incidence θ is then given by

$$I(\theta) \approx \frac{I_0 c \alpha^2}{r^2 \lambda^4 M} \left(1 + \cos^2 \theta\right) \tag{6.8}$$

where α is the particle polarizability, I_0 is the incident beam intensity, c is the concentration in particles, r is the distance to the detector, λ is the wavelength, and M is the molecular weight.

This strong dependency of the intensity with the wavelength gives a qualitative explanation of the color of the sky: the particles present in the atmosphere scatter the short (blue) wavelengths more than the other colors.

As the polarizability varies linearly with the molecular weight, at a given angle, the intensity is thus also proportional to M

$$I(\theta) = k(\theta) \cdot c \cdot M \tag{6.9}$$

If we consider a mixture of polymers of different masses or a polydisperse sample, we have to sum over all of the contributions:

$$I(\theta) = K\sum_i c_i M_i = Kc\,\frac{\sum_i c_i M_i}{\sum_i c_i} = KcM_w \qquad (6.10)$$

where M_w is the weight averaged molecular weight.

This is an ideal formula, and, particularly for polymers, the classical analysis of these experiments proceeds by plotting I for various concentrations and various angles (Zimm plot).

For larger objects whose size becomes comparable with λ, the light can scatter from different places on the same object. This effect decreases the scattered light even more so for large angles, and one has to take into account a structure factor that can be analytically expressed only for a few simple geometries.

6.3.2.2 Dynamic Light Scattering [30, 31]

Also denoted photon correlation spectroscopy, dynamic light scattering (DLS) consists of measuring the scattered light *dynamically* at a fixed angle. As the molecules diffuse within the observation volume, the emitted light resulting from the scattering interferes. The analyses of these time fluctuations are then used to deduce a diffusion coefficient. This technique is well suited to particles in the range 10 nm–1 μm.

This analysis is performed by computing the correlation function $G(\tau)$. $G(\tau)$ can be accurately modeled and is found to be exponential

$$G(\tau) = \langle I(t)\cdot I(t+\tau)\rangle = A\left[1 + B\exp(-2\Gamma\tau)\right] \qquad (6.11)$$

where $\Gamma = Dq^2$, $q = (4\pi n/\lambda)\sin(\theta/2)$.

When several different particles characterized by different diffusion coefficients are present, a multiexponential is used to access the different diffusion constants. From these measurements, one gets the hydrodynamic radius R_h by inverting the Stokes-Einstein relationship:

$$R_h = \frac{kT}{6\pi\eta D} \qquad (6.12)$$

Thus, the dimension measured with this technique is really the radius of the equivalent sphere that would diffuse similarly. This can have severe consequences in the case of nonspherical particles: an increase in the length or in the diameter of a rod, for instance, contributes very differently to its hydrodynamic radius.

6.3.3 Biochemical Characterization

6.3.3.1 Surface Plasmon Resonance [32]

The surface plasmon resonance (SPR) technique is used to quantify the amount of material on a surface. In biotechnology, this technique is used to detect and measure in real time the kinetic parameters of the interaction between two or more molecules. One of the strengths of the technique is that it does not require labeling the molecules.

The SPR effect is based on the interaction of light with a metallic surface in the conditions of total internal reflection, which means that, in the absence of a metallic layer, the light propagating in the solid (a glass prism) is totally reflected. Under these conditions, an evanescent wave exists at the surface of the glass. The intensity of this nonpropagating field decreases normally to the surface over a typical distance of a fraction of a wavelength (Figure 6.17).

A thin metallic layer present on the glass surface will not qualitatively modify this picture: most of the light is totally reflected. However, the evanescent wave can couple with the free electron clouds of the metal to create a plasmon (a cloud of excited electrons). For this phenomenon to occur, the energy carried by the incident photons has to exactly match the energy of the plasmon. As these plasmons are confined within the metal layer, there are drastic conditions of angle and wavelength that yield an efficient coupling. When these requirements are fulfilled, energy is effectively transferred to the plasmons, which results in a minimum in the reflected intensity.

The important point here is that, since we are dealing with evanescent fields, the conditions for which this transfer is efficient are highly dependent on the immediate environment of the metallic layer—typically within a wavelength in depth (i.e., within a few hundred nanometers). A modification of the refractive index of the solution next to the metal surface then results in a change in these conditions. The instruments used for biological applications use a fixed wavelength and monitor the incidence angle corresponding to the minimum of the reflected intensity. For protein-protein interactions, this change in the index of refraction is proportional to the amount of material present on the surface. SPR analysis systems can thus compute in real time the excess on the surface and use these measurements to get the kinetic parameters of the studied biomolecular interaction.

Practically, one of these two components is immobilized on the gold surface using a coupling strategy based on the ones described in Section 6.2.3. This is a critical step often overlooked; it has to be kept in mind that the results obtained by this technique describe a particular situation where the ligand is anchored on, and thus

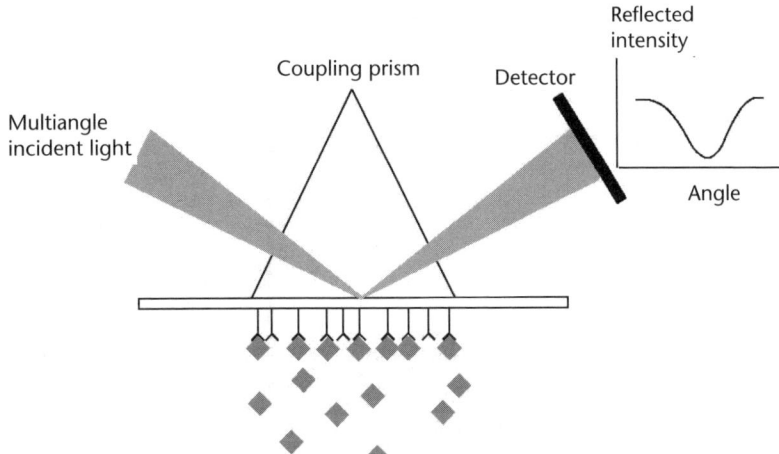

Figure 6.17   The surface plasmon resonance setup. The detector measures the angle of minimum reflection to access the surface excess.

influenced by, a solid surface. Even though this aspect can be minimized (e.g., by the use of long polymeric linkers or of biological gels), it is sometimes a severe limitation. In other studies, the orientation is favored and an NTA-based approach is preferred.

The analyte is then injected, and the kinetics of association is followed by quantifying the amount of material on the surface. The shape of the evolution of this surface excess can be modeled by classical kinetic equations:

$$\frac{d[LA]}{dt} = k_{on} \cdot [L] \cdot [A] - k_{off} \cdot [LA] \tag{6.13}$$

where L represents the ligand and A represents the analyte. k_{on} is the association constant of the complex, and k_{off} is the dissociation constant.

After a certain time, the system reaches a steady state described by equilibrium constants K_a and K_d that are given by:

$$K_a = k_{on}/k_{off} \text{ and } K_d = k_{off}/k_{on} \tag{6.14}$$

After this steady state, the analyte-free buffer is flown over the surface. As there is no more analyte in solution, the mass action law imposes desorption of the ligands. This step is described by kinetic equations very similar to the ones describing the association. Finally, a dissociating agent is injected to remove the remaining ligands and to regenerate the surface (Figure 6.18).

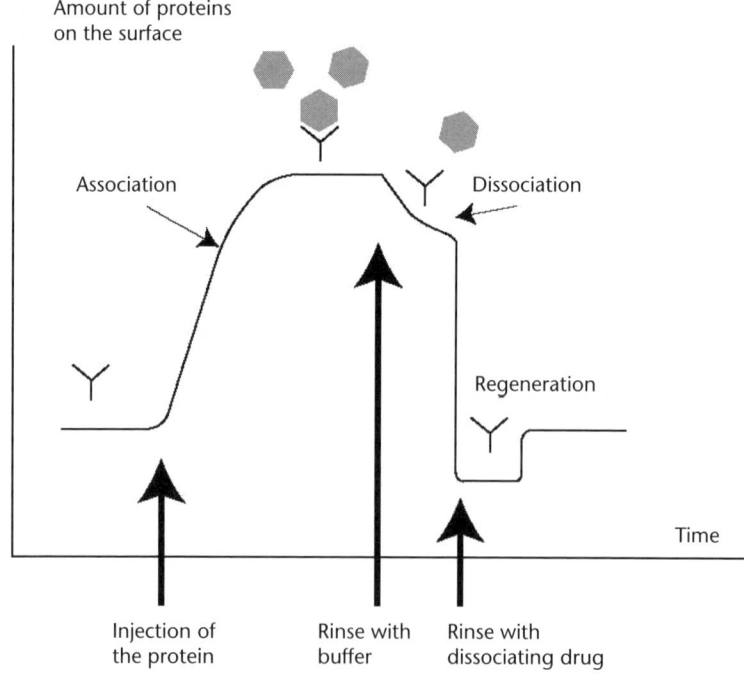

Figure 6.18 Typical sensorgram obtained by SPR illustrating the three steps of association, dissociation, and regeneration.

6.3.3.2   Flow Cytometry–Based Techniques [33]

Flow cytometry consists in flowing cells (although this principle can be used with any microparticle) one by one in front of a detector in order to measure their physical or chemical properties. According to this measurement, the population can be analyzed or even sorted in real time by addressing them toward the right container (Figure 6.19).

The most popular of these systems is the fluorescence activated cell sorting (FACS), in which case the signal measured by the detector is fluorescence. The cells are sorted according to their measured laser excited fluorescence. The cells also scatter some of this light (see Section 6.3.2.1), giving additional information on their dimensions. After these measurements, the water stream transporting the particles breaks into droplets in a very controlled way upon the application of a vibration of the nozzle. Each of these droplets contains one cell, and, according to the fluorescence measurement performed earlier, the drop is electrically charged with a positive or negative charge.

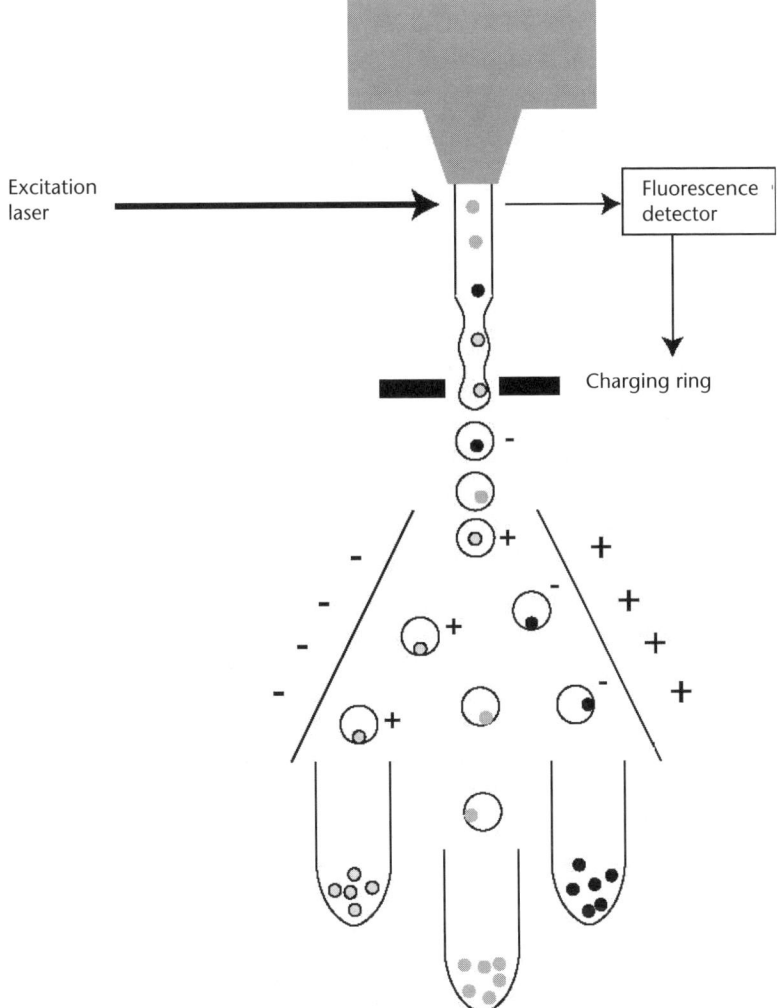

Figure 6.19   FACS can sort cells according to their fluorescence properties.

A static electric field is then applied transversally to this stream of drops and deflects them in one direction or the other depending on their electric charge. Sometimes, several colors can be excited by a single excitation wavelength, so different aspects of the cell content can be probed simultaneously.

For instance, it is possible with these instruments to sort cells by their protein content, provided the protein of choice is fluorescently labeled using immunofluorescence. Along the same line, assays based on GFP fluorescence (Section 6.2.1.2) can be used to analyze protein expressions of certain cells. In these situations, the cell can be kept alive and cultured after this sorting.

Different aspects of cell phenotype or genotype can be tested with assays based on DNA fluorescence. Live/dead assays, for instance, are routinely performed. They are based on the membrane permeability: When cells die, their membranes are disrupted and dyes can freely penetrate them. Dyes that fluoresce only in the presence of nucleic acids are used for this purpose. They can penetrate dead cells' membranes to bind to the nucleus DNA, so dead cells become fluorescent while live cells, whose membranes are intact, are not.

Drugs can drastically modify the biochemistry of cells, and their influence can thus be monitored by FACS. Some of these drugs modify the cell cycles, and a measurement of the DNA content of each cell of a population then gives a *signature* of this particular treatment. These measurements of DNA content are useful in other situations, such as the expression of a particular gene. To perform these measurements, cells are first permeabilized to allow the entry of the dye to the nucleus and thus the measurement of the DNA content.

6.4   Micromanipulation

6.4.1   Force Measurements

Micromanipulation using techniques derived from the classic mechanical micromanipulators are still very important. A derived strategy is to use the AFM described earlier to measure forces between individual objects (e.g., proteins). The principle is extremely simple: the two interacting proteins are immobilized on the tip of the instrument for one of them and on a facing solid surface for the other one. The tip and the surface contact, and the force necessary to separate them, are measured by the deflection of the cantilever. Using this technique, antigen-antibodies interactions have been measured, and the stretching of several biomolecules, including DNA and proteins, has also been characterized (Figure 6.20). Probing such interaction energies (close to kT) necessitates a particular theoretical treatment. In particular, the lifetime of a bond under a force is affected by this force [33, 34]. This model has been adapted to the problem of the measurement of rupture forces between single proteins, and it has been shown that this force F varies logarithmically with the velocity of separation [35]

$$F = \left(k_B\, T / x_\beta\right) \cdot Ln\left[r_f\, x_\beta / \left(k_{off}\, k_B\, T\right)\right] \qquad (6.15)$$

where r_f is the so-called loading rate (product of the cantilever spring constant by the velocity of separation), and x_β is a characteristic length of the bond.

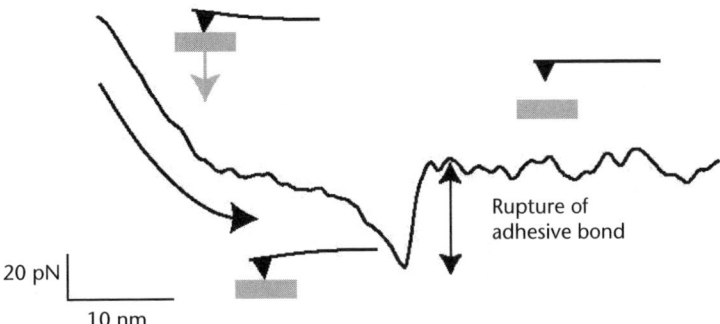

Figure 6.20 Separation of a surface from the AFM tip in the presence of an adhesive interaction. In this particular case, adhesion proteins are grafted on the tip and on the surface. When the force acting on the cantilever is too high, it snaps back to its equilibrium position. The nonlinear part of the curve represents the stretching of a single PEG linker [34].

6.4.2 Optical Tweezers

Optical tweezers (OTs) use a highly focused optical beam to trap particles at the focus. When a laser is focused through a high numerical aperture microscope objective, it identifies a well-defined light "cage" in which not only is the intensity maximal but the gradients in light intensity are also extremely strong [36]. We detail in Chapter 8 how a spatial gradient of an electric field can be used to trap particles of a different polarizability than the one of the surrounding medium (dielectrophoresis). The physical principle is exactly the same in the present situation: The light intensity within a laser beam is not uniform but is maximum in the center. In fact, the light distribution is Gaussian. The light intensity gradients then naturally drive particles of index of refraction higher than the solution toward the center of the beam. If the beam is tightly focused, the same effect drives the particles transversally toward the focus. The net effect is a trap localized close to the focal point—close, but not exactly at this point, because there is another force: the scattering force that tends to "push" the particle away.

These traps can immobilize and transports particles with forces in the range of a few tens of pico-Newtons, which is well suited to many practical situations. The lasers used for applications in biology are usually in the IR spectrum in a range for which there is no absorption of energy (and thus no heating).

OTs have been used with many different objects: viruses, bacteria, and organelles within cells, but the best known application is the trapping of the micron-sized beads that are used as handles on biopolymers or on molecular force-generating systems, such as biological motors [37]. Indeed, not only can OTs trap particles, they can also measure forces applied to them: As the object is pulled away by an external force, its position in the trap varies. Recent developments in optical microscopy have made it possible to track displacements of a few nanometers. The exact position of the bead then witnesses the force applied to it. In another configuration, the light intensity can be cranked up to balance for this external force so that the position of the bead remains unchanged. Both approaches need a calibration of the trap, usually by moving the particle in the fluid (or vice versa) at a known velocity and using Stokes-Einstein friction to measure the applied force.

We have discussed so far the use of these OTs as a mean to handle a single parti-cle. An interesting development is the possibility to create arrays of traps in which many particles can be manipulated at will. The simplest way to perform this task is by defining two or more positions for the focus and having the beam rapidly switch between these positions. Another approach is to define a holographic array where the energy landscape to which the particles are submitted (e.g., an array of traps) is defined in the Fourier plane [38].

6.4.3 Flow-Based Techniques

Flows can be used not only to transport particles or molecules but also to manipu-late them. For instance, elongational flows, where the velocity increases linearly in the direction of the flow, can be induced in microfluidic chambers. Single DNA mol-ecules stretched in this type of flow can be observed by fluorescence microscopy and their shape compared with existing theoretical models [39] (Figure 6.21).

Flows have also been used to align molecules: Elongated rigid objects such as microtubules orient naturally in the direction of the flow, and a receding meniscus can be used to perfectly align DNA molecules in the direction of drying. This *molec-ular combing* is performed on a modified surface and in the right pH conditions in order to have one end of the DNA stick to the surface [40].

Flow chambers are used to quantify the interactions of beads or cells with a sur-face. In the laminar flow conditions imposed by the geometry, the established Poiseuille flow imposes a constant shear rate near the solid surface. The force acting on the flowing object near the wall can then be computed and is found to be [41]

$$F \approx \eta a^2 Q \tag{6.16}$$

where Q is the flow rate, a is the radius of the particle, and η is the viscosity. The pro-portionality constant is imposed by the channel geometry and can be analytically

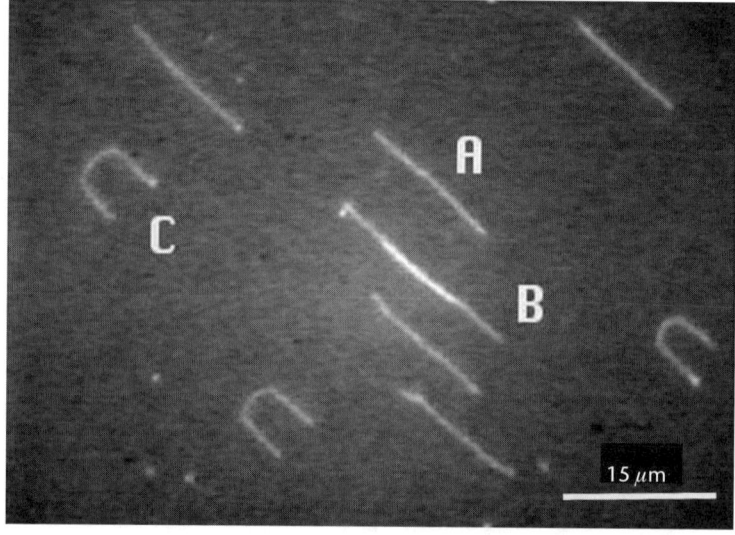

Figure 6.21 Fluorescently labeled DNA stuck on a solid surface after molecular combing. (*From:* [42]. © 1996 Institut Pasteur. Reprinted with permission.)

calculated. In the framework of the theory of Bell mentioned earlier (6.15), the duration of arrests of particles interacting with a protein bound on the solid surface is related to the dynamic characteristics of the bond and to a characteristic length [41, 43].

References

[1] Flory, P., *Principles of Polymer Chemistry* Ithaca, NY: Cornell University Press, 1971.

[2] Nakanishi, H., "Flory Approach for Polymers in the Stiff Limit," *J. Phys.*, Vol. 78, 1987, pp. 979–984.

[3] Antognozzi, M., et al., "Comparison Between Shear Force and Tapping Mode AFM—High Resolution Imaging of DNA," *Single Mol.*, Vol. 3, 2002, pp. 105–110.

[4] Lockart, D., and E. A. Winzeler, "DNA Array: Genomics, Gene Expression and DNA Arrays," *Nature*, Vol. 405, 2000, pp. 827–836.

[5] Pollack, J. R., et al., "Genome-Wide Analysis of DNA Copy-Number Changes Using cDNA Microarrays," *Nature Genet*, Vol. 23, 1999, pp. 41–46.

[6] Mathews, D. H., et al., "Expanded Sequence Dependence of Thermodynamic Parameters Improves Prediction of RNA Secondary Structure," *J. Mol. Biol.*, Vol. 288, 1999, pp. 911–940.

[7] http://www.rosalindfranklin.edu/cms/biochem/Walters/CABR/rna_folding.html.

[8] Sigel, R. K. O., et al., "Metal Ion Binding Sites in a Group II Intron Core," *Nature Structural Biology*, Vol. 7, 2000, pp. 1111–1116.

[9] Shea, J. E., and C. L. Brooks, "From Folding Theories to Folding Proteins: A Review and Assessment of Simulation Studies of Protein Folding and Unfolding," *Annu. Rev. Phys. Chem.*, Vol. 52, 2001, pp. 499–535.

[10] Song, L., et al., "Structure of Staphylococcal Alpha-Hemolysin: A Heptameric Transmembrane Pore," *Science*, Vol. 274, 1996, pp. 1859–1866.

[11] Zhu, H., and M. Snyder, "Protein Chip Technology," *Curr. Opin. Chem. Biol.*, Vol. 7, 2003, pp. 55–63.

[12] Hosch, S., et al., "Malignant Potential and Cytogenetic Characteristics of Occult Disseminated Tumor Cells in Esophageal Cancer," *Cancer Res.*, Vol. 60, 2000, pp. 6836–6840.

[13] Stenger, D. A., et al., "Detection of Physiologically Active Compounds Using Cell-Based Biosensors," *Trends in Biotechnology*, Vol. 19, 2001, pp. 304–309.

[14] Wu, R. Z., S. N. Bailey, and D. M. Sabatini, "Cell-Biological Applications of Transfected-Cell Microarrays," *Trends in Cell Biology*, Vol. 12, 2002, pp. 485–488.

[15] Berg, H. C., *Random Walks in Biology*, Princeton, NJ: Princeton University Press, 1993.

[16] Kruhlak, M. J., et al., "Reduced Mobility of the Alternate Splicing Factor (ASF) Through the Nucleoplasm and Steady State Speckle Compartments," *J. Cell Biol.*, Vol. 150, 2000, pp. 41–51.

[17] Stryer, L., and R. P. Haugland, "Energy Transfer a Spectroscopic Ruler," *Proc. Nat. Acad. Sci.*, Vol. 58, 1967, pp. 719–726.

[18] Dubertret, B., et al., "In Vivo Imaging of Quantum Dots Encapsulated in Phospholipid Micelles," *Science*, Vol. 298, 2002, pp. 1759–1762.

[19] Israelachvili, J., *Intermolecular and Surface Forces*, New York: Academic Press, 1992.

[20] Caffrey, M., "Membrane Protein Crystallization," *J. Structural Biol.*, Vol. 142, 2003, pp. 108–132.

[21] Ulman, A., *An Introduction to Ultrathin Organic Films from Langmuir-Blodgett to Self-Assembly*, San Diego, CA: Academic Press, 1992.

[22] Jeon, S. I., et al., "Protein Surface Interactions in the Presence of Polyethylene Oxide. 1. Simplified Theory," *Coll. Interf. Sci.*, Vol. 142, 1991, pp. 149–166.

[23] Hermanson, G. T., A. Mallia, and P. K. Smith, *Immobilized Affinity Ligand Techniques*, New York: Academic Press, 1992.

[24] du Roure, O., et al., "Functionalizing Surfaces with Nickel Ions for the Grafting of Proteins," *Langmuir*, Vol. 19, 2003, pp. 4138–45.

[25] Denk, W., and W. W. Webb, "Optical Measurements of Picometer Displacements," *Appl. Opt.*, Vol. 29, 1990, pp. 2387–2391.

[26] http://microscopy.fsu.edu.

[27] Diaspro, A., (ed.), *Confocal and Two-Photon Microscopy: Foundations, Applications, and Advances*, New York: John Wiley and Sons, 2002.

[28] Jena, B., and J. K. Horber, *Atomic Force Microscopy in Cell Biology*, New York: Academic Press, 2002.

[29] Scheuring, S., et al. "High Resolution AFM Topographs of the Eschrichia Coli Water Channel Aquaporin Z," *EMBO J.*, Vol. 18, 1999, pp. 4981–4987.

[30] Johnson, C. S., and D. A. Gabriel, *Laser Light Scattering*, Boca Raton, FL: CRC Press, 1981.

[31] Berne, B. J., and R. Pecora, *Dynamic Light Scattering*, New York: John Wiley and Sons, 1976.

[32] Cooper, M. A., "Optical Biosensors in Drug Discovery," *Nat. Review Drug Discov.*, Vol. 1, 2002, pp. 515–528.

[33] Bonner, W. A., H. R. Hulett, and R. G. Sweet, "Fluorescence Activated Cell Sorting," *Rev. Sci. Instr.*, Vol. 43, 1972, pp. 404–409.

[34] Bell, C. J., "Model for the Specific Adhesion of Cells to Cells," *Science*, Vol. 200, 1978, pp. 618–627.

[35] Evans, E., and K. Ritchie, "Dynamic Strength of Molecular Adhesion Bonds," *Biophys. J.*, Vol. 72, 1997, pp. 1541–1555.

[36] Ashkin, A., "Optical Trapping and Manipulation of Neutral Particles Using Lasers," *Proc. Nat. Acad. Sci.*, Vol. 94, 1997, pp. 4853–4860.

[37] Bustamante, C., et al., "Single Molecule Studies of DNA Mechanics," *Current Opinion in Structural Biology*, Vol. 10, 2000, pp. 279–285.

[38] Dufresnes, E. R., et al., "Computer Generated Holographic Optical Tweezers Arrays," *Rev. Sci. Instr.*, Vol. 72, 2001, pp. 810–816.

[39] Perkins, T. T., D. E. Smith, and S. Chu, "Single Polymer Dynamics in an Elongational Flow," *Science*, Vol. 276, 1997, pp. 2016–2021.

[40] Bensimon, D., et al., "Stretching DNA with a Receding Meniscus: Experiments and Models," *Phys. Rev. Lett.*, Vol. 76, 1995, pp. 4754–4757.

[41] Pierres, A., A. M. Benoliel, and P. Bongrand, "Measuring Formation and Dissociation of Single Bonds Interactions Between Biological Surfaces," *Curr. Opin. Coll. Interf. Sci.*, Vol. 3, 1998, pp. 525–533.

[42] http://www.pasteur.fr/recherche/unites/biophyadn/f-Fcombing.html.

[43] Alon, R., D. A. Hammer, and T. A. Springer, "Lifetime of the P-Selectin-Carbohydrate Bond and Its Response to Tensile Force in Hydrodynamic Flow," *Nature*, Vol. 374, 1995, pp. 539–542.

Selected Bibliography

Alberts, B., et al., *Molecular Biology of the Cell*, New York: Garland, 1989.

"Collection of Review Papers on Functional Genomics," *Nature,* Vol. 405, 2000.

CHAPTER 7

Magnetic Particles in Biotechnology

7.1 Introduction

In many biotechnological applications, the use of a carrier fluid to transport biological objects lacks specificity: For example, it is not always possible to bring by microfluidics transport a biological target to a specific location inside the biochip. A second complementary carrier is often needed. To this extent, magnetic beads are one of the most important categories of microparticles. Between 1990 and 2002, they were developed mostly for in vitro applications, principally for biodiagnostics and biorecognition, but also for purification and separation operations. More recently, their use has reached the domain of in vivo applications, such as cancer treatment.

In this chapter, we present first the nature of magnetic beads, their magnetic characteristics, and the force that can be applied on these beads. We then give examples of trajectory calculation for applications such as separation columns and magnetic field flow fractionation (MFFF). Finally, we show how assembly of magnetic beads has been used to build new biological tools, and we focus on chains of magnetic beads, ferrofluids, and magnetic membranes.

7.1.1 The Principle of Functional Magnetic Beads

At first sight it might seem strange to consider magnetic actuation of biological microsystems because DNA, proteins, antibodies, cells, and bacteria (except just one kind, but that is just anecdotic)[1] are not magnetic. However, the principle of functionalization has totally changed the approach: as soon as it became possible to bind DNA strands—or other biological or biochemical macromolecules—onto magnetic microparticles, these microparticles could be used to displace and manipulate complex biological molecules [1]. The principle of functionalization is schematized in Figure 7.1.

The principle is to find a chemical linker between the bead surface and the target in order to attach the target to the bead; there are many types of functionalizations, depending on both the surface of the particle and the target. For example, it has been found that the chemical group streptavidin-biotin is a good linker for the capture of DNA. It is a very complex task to find an adequate functional coating of the bead. To facilitate the task, prefunctionalized beads are currently sold by specialized suppliers.

---

1. The bacteria *magnetospirillum magnetoacticum* has magnetic microreceptors to use the Earth's magnetic field for orientation.

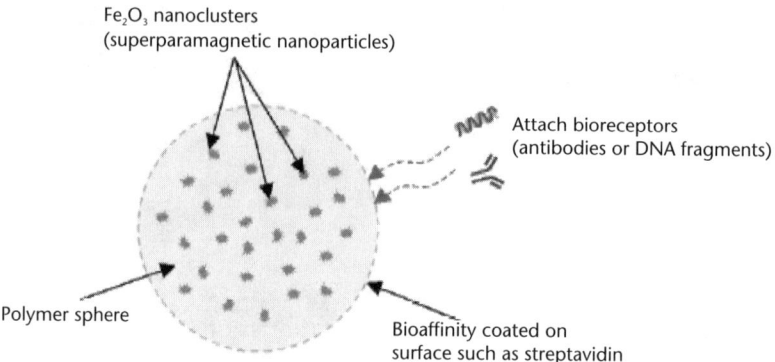

Figure 7.1 Schematic view of the principle of a functional magnetic microparticle. The bead is constituted by Fe_2O_3 nanoclusters embedded in a polymer sphere. The magnetic nanoclusters (or nanograins) have a size of 5 nm. A bead containing 20 nanoclusters—like here—has a diameter of about 150 nm. The surface is coated with streptavidin.

7.1.2 Composition and Fabrication of Magnetic Beads

Magnetic beads of 50 nm to 2 μm are available; the choice depends on the targets and the coating. The smaller beads are used to displace small-sized targets; for example, 50-nm Miltenyi magnetic beads are well suited to manipulate 32 basis pair (bp) DNA strands. For larger targets, larger beads or a larger concentration of small beads have to be used (in this case, more than one bead is attached to a single target).

The beads are fabricated to be superparamagnetic (i.e., they have a magnetization only if an external magnetic field is applied, and they totally lose their magnetization if the external magnetic field is removed). These beads are obtained by embedding paramagnetic nanograins (magnetic domains) of iron oxide Fe_2O_3 or Fe_3O_4 (of about 5 nm in size) in a biologically compatible matrix of latex or polystyrene. Generally, one wants to avoid having a remanent magnetic field, because this remanent magnetization does not allow the dispersion of the beads by Brownian motion when the external magnetic field is switched off, resulting in unwanted aggregates in the carrier fluid. Because large-sized beads (1–2 μm) contain more magnetic material, they experience a larger magnetic force, so that they can displace larger targets. For example, a 100-nm magnetic bead only contains about 13 to 15 paramagnetic nanograins of 5 nm.

Figure 7.2 shows perfectly spherical magnetic beads, but sphericity cannot always be achieved in the fabrication process. Usually large-sized beads have a more spherical shape than the smaller ones. In a general way, spherical beads are easier to manipulate because they interact with the others in a simpler way, as will be shown later in this chapter.

With the development of diagnostic devices, it has been found that beads that are both magnetic and fluorescent are very advantageous. Magnetic beads of 50 to 100 nm are not easily seen under a microscope, so this is much easier if they are fluorescent. Many efforts have recently gone into the development of magnetic fluorescent beads; such beads have the advantage of combining two functions: displacement when an external magnetic field is applied, and detection when the beads are excited at a wavelength corresponding to the peak in the spectrum of

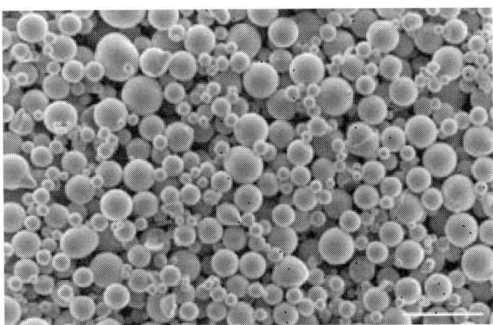

Figure 7.2   Microscope view of fluorescent magnetic beads of 200 nm.

fluorescence. Two different approaches to obtain combined magnetic/fluorescent effect have been followed. The first one consists of incorporating fluorescent markers inside the beads during their fabrication, but it was found rather difficult to obtain all of the convenient properties of magnetic beads (i.e., sphericity, biocompatiblility, monodispersion, and compactness). The other one consists of assembling a complex magnetic bead-target-fluorescent bead (Figure 7.3) or in binding the fluorescent particle directly to the magnetic bead.

The scheme of Figure 7.3 is widely used because it has the advantage of marking the target directly. Magnetic beads are used to bring the targets into a detection chamber, and they are usually removed after they have completed their task. It is thus advantageous to have the targets linked to a fluorophore for processes like detection.

Recently, the principle of a compound particle that is at the same time magnetic and fluorescent has been established by [2]. This principle is derived from the very simple example of Chapter 2. A spherical magnetic nanoparticle built around an iron-platinum (FePt) core, which is coated with a layer of cadmium and sulfur (CdS). This bead is heated to melt the outer layer of CdS. This liquid layer, because

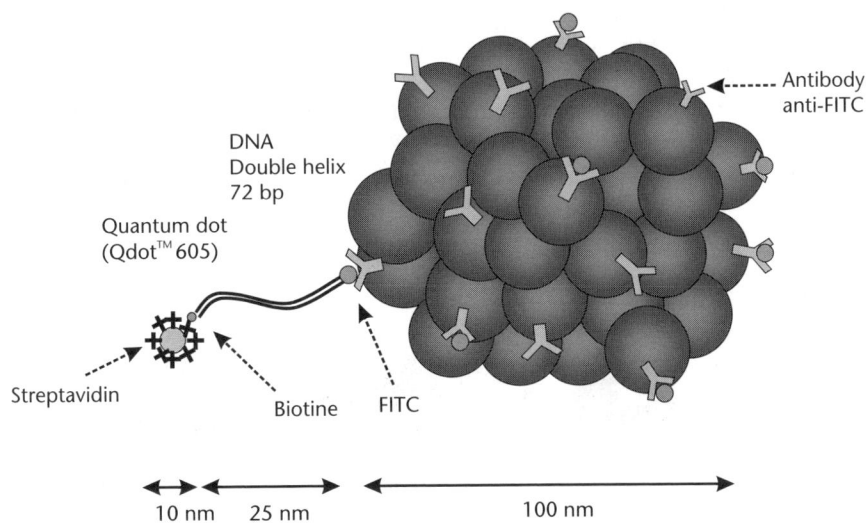

Figure 7.3   Schematic view of the association of a magnetic bead carrying 72-bp DNA and marked by a fluorescent quantum dot.

of its contact angle with the underlying solid, is not stable and migrates to take the shape of the Figure 7.4, where the liquid minimizes its surface energy.

The drawback of this construction is that the quantic efficiency of the particle is much less than that of a "free" quantum dot. When excited by a light source, an important part of the emitted light is absorbed by the magnetic sphere.

7.1.3 An Example of Displacement by Magnetic Beads for Biodetection

Microsystems for biorecognition or biodiagnostics require different operating steps, schematized in Figure 7.5. Suppose a fluid volume containing some target molecules (like DNA, proteins, or cells). Because direct detection is not effective for few targets in a large volume, it is necessary to concentrate the targets in a small chamber (detection chamber); at this point, functional magnetic beads are often used. They diffuse in the large volume and bind to the targets upon contact. After a binding time, magnetic force is used to concentrate the beads in a small chamber. The chemical linking of magnetic beads and targets can be broken if some conditions are changed in the chamber (e.g., an increase in the temperature). Finally, the magnetic beads are removed and the targets are concentrated in the detection chamber. A realization of such a microsystem is shown in Figure 7.6.

Note that in the realization of such a device, in order to achieve a satisfactory design, a careful modeling of the trajectories and concentration of the beads has been performed. Modeling of magnetic beads motion will be presented later in this chapter.

7.1.4 The Question of the Size of the Magnetic Beads

A recurring question is to decide what type of magnetic bead is best adapted to the problem one has to solve. Most of the time, the smaller beads are used, because, even if the magnetic traction exerted is weak, they have the property to be dispersed by Brownian motion as soon as the magnetic field is shut down (Figure 7.7). This is particularly useful, because, after the magnetic beads have done their job carrying target molecules, they must be separated from them (e.g., by thermal heating), before being removed out of the reaction chamber. A compact aggregate of magnetic beads cannot "free" the targets in the detection chamber.

The weak magnetic traction that can exert a small magnetic bead is not an important drawback: if the target is small (like a 32-bp DNA strand), the magnetic force need not be very important, and if the target is larger (a cell, for instance), more than one bead can be attached to the target and the magnetic traction is the resultant of all of the forces exerted by each bead.

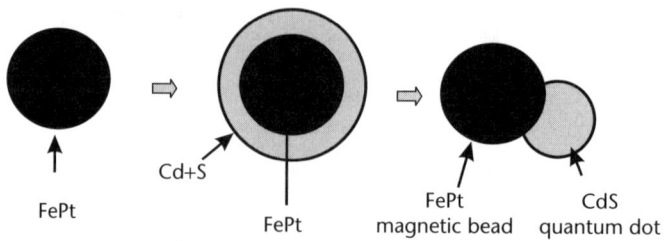

Figure 7.4 An example of fabrication of a bifunctional magnetic/fluorescent nanoparticle.

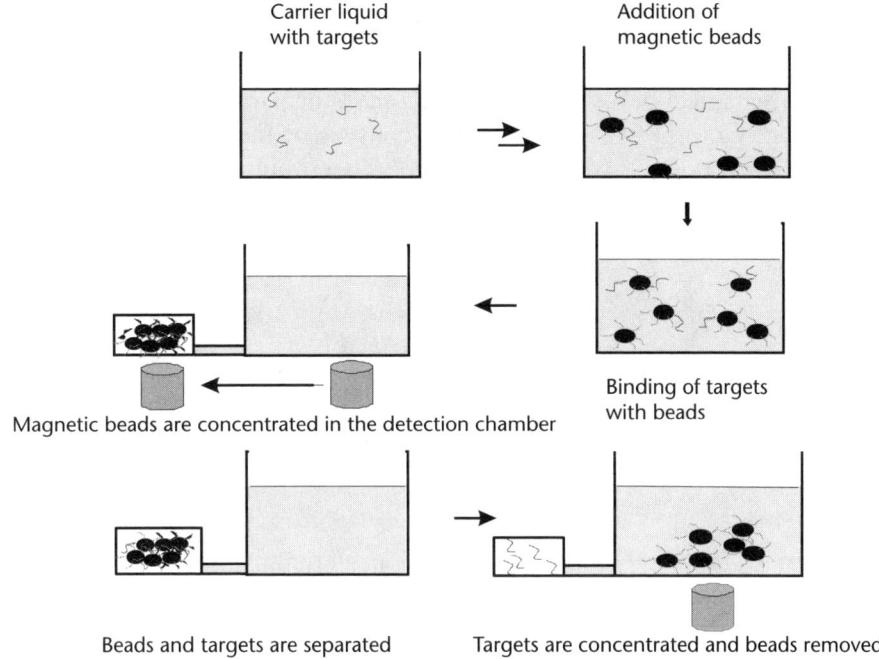

Figure 7.5 The principle of magnetic concentration of targets for biodiagnostics.

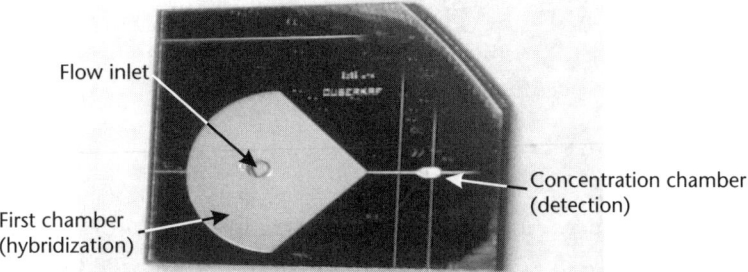

Figure 7.6 A DNA concentration microdevice with two chambers. The first chamber is the white sector at the left of the figure; the fluid and magnetic beads enter vertically through an orifice in the middle. After the time necessary for hydridization on the beads, the beads are attracted by a magnet into the concentration chamber, located at the right. (Courtesy of F. Ginot and R. Compagnolo, LETI/BioMérieux.)

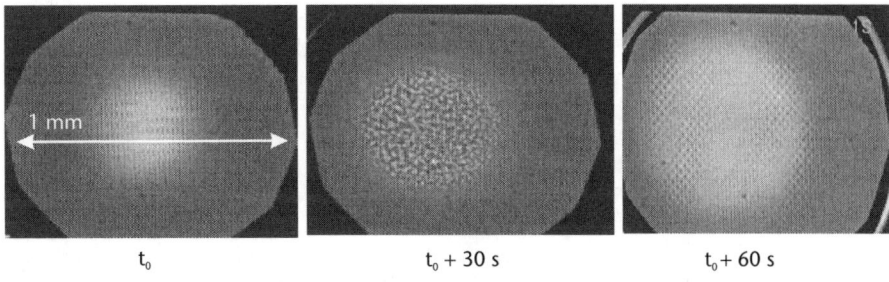

Figure 7.7 Dispersion under Brownian effect of small-sized magnetic beads in a liquid drop. (Courtesy of D. Massé, CEA/LETI.)

7.2   Characterization of Magnetic Beads

The mechanical behavior of magnetic beads depends on their magnetic properties. Characterization of magnetic beads consists of the determination of the average magnetic properties of the beads. On a general point of view, Maxwell's equations determine the electromagnetic behavior of any material. The magnetic induction in Maxwell's equation is related to the magnetic field by [3]

$$\vec{B} = \mu_0 \left(\vec{H} + \vec{M} \right) \tag{7.1}$$

where the magnetization is given by

$$\vec{M} = \chi \vec{H} + \vec{M}_r \tag{7.2}$$

where $\vec{M}_r$ is the remanent magnetization, so that

$$\vec{B} = \mu_0 \mu_r \vec{H} + \mu_0 \vec{M}_r$$

with $\mu_r = 1 + \chi$.

This shows that the information we need is contained in the relation between the magnetization $\vec{M}$ of the bead and the applied external magnetic field $\vec{H}$. This relation is usually determined experimentally by making use of a superconducting quantum interference device (SQUID) or a Hall probe.

7.2.1   Paramagnetic Beads

Magnetic micro- and nanoparticles used in biotechnology are nearly always paramagnetic—even superparamagnetic—because the magnetic force should vanish when the external field is switched off. If not, the beads would agglomerate, and it would be impossible to have them dispersed in the carrier fluid.

Paramagnetic media follows Langevin's law [4]

$$\frac{M}{M_s} = \coth\left(\frac{3\chi H}{M_s}\right) - \frac{1}{\dfrac{3\chi H}{M_s}} \tag{7.3}$$

Note that at low magnetic field, $\coth\left(\dfrac{3\chi H}{M_s}\right) - \dfrac{1}{\dfrac{3\chi H}{M_s}} = \dfrac{\chi H}{M_s}$ and (7.3) reduce to

the usual expression $M = \chi H$. For large magnetic field, $\dfrac{M}{M_s} = 1 - \dfrac{1}{\dfrac{3\chi H}{M_s}}$, and $M = M_s$

which states that saturation is reached. The diagram $M(H)$ is plotted in Figure 7.8.

If we assume that the paramagnetic beads are monodispersed (all of the beads are identical), Langevin's law may be applied to each bead.

7.2.2   Ferromagnetic Microparticles

The situation is more complex for ferromagnetic beads, because ferromagnetic objects keep a remanent magnetization when the external field vanishes. There is generally no analytical function for the magnetization, and one generally tries to fit the experimental curve by piecewise continuous polynomials.

7.3   Magnetic Force

A general expression of the magnetic energy of interaction of a particle immersed in a magnetic field $\vec{H}$ is [5, 6]:

$$U_m = -\frac{1}{2}\mu_0 \int \vec{M} \cdot \vec{H} dv \tag{7.4}$$

where $\vec{M}$ is the magnetization of the particle in the applied magnetic field $\vec{H}$, and U_m is the magnetic energy. The integration is taken over the particle volume. The magnetic force exerted by the magnetic field on the particle is the gradient of the interaction energy

$$\vec{F}_m = -\nabla U_m \tag{7.5}$$

Because of the very small size of the magnetic particles, the integration in (7.4) may be replaced by the value of the fields at the center of the particle, multiplied by the volume of the particle v_p

$$U_m = -\frac{1}{2}\mu_0 v_p \vec{M} \cdot \vec{H} \tag{7.6}$$

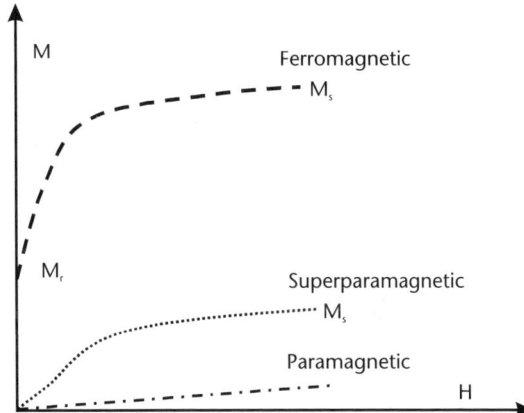

Figure 7.8   Relation $M(H)$ for the different types of materials.

7.3.1 Paramagnetic Microparticles

Most of the time, particles used in biotechnologies are paramagnetic (one exception will be discussed in Section 7.12). The magnetization is then aligned with the external field [7]:

$$\vec{M} = \frac{\chi}{1 + D_m \chi} \vec{H}$$

where D_m is the demagnetization coefficient ($D_m = 1/3$ for a sphere). Because $\chi \ll 1$ for the considered paramagnetic particles, magnetization can be approximated by

$$\vec{M} = \chi \vec{H}$$

and the energy of interaction is

$$U_m = -\frac{1}{2} \mu_0 v_p (\chi_p - \chi_f) |H|^2 \tag{7.7}$$

where χ_p is the magnetic susceptibility of the particle, and χ_f is that of the carrier fluid. Thus, the magnetic force on a paramagnetic microparticle is given by the gradient of (7.7)

$$\vec{F}_m = \frac{1}{2} \mu_0 v_p (\chi_p - \chi_f) \nabla |H|^2 \tag{7.8}$$

In the literature, another equivalent expression is often found [7, 8]. Indeed, if no electric field is associated to the magnetic field (or if the electric field is constant), Maxwell's equation implies that

$$\nabla \times \vec{H} = 0$$

and (7.8) can be rewritten as

$$\vec{F}_m = \mu_0 v_p (\chi_p - \chi_f)(\vec{H} \cdot \nabla)\vec{H} \tag{7.9}$$

7.3.2 Ferromagnetic Microparticles

In the case of ferromagnetic particles, we start from (7.5) and (7.6) [9] to obtain

$$\vec{F}_m = \frac{\mu_0}{2} v_p \nabla(\vec{M} \cdot \vec{H}) = \frac{\mu_0}{2} v_p \nabla((\vec{M}_p - M_f) \cdot \vec{H}) \tag{7.10}$$

7.4 Deterministic Trajectory

At a macroscopic scale, kinematics theory relates the mass acceleration of a body to the resultant of the external forces that act upon it. This is the well-known Newton's theorem.

$$m\frac{d\vec{V}}{dt} = \sum \vec{F}_e \tag{7.11}$$

where m is the mass of the particle, $\vec{V}$ is the velocity, and $\vec{F}_e$ is the external forces.

At a macroscopic scale, Brownian motion (random hitting by other molecules) is completely negligible. At a microscopic scale, the effects of the Brownian agitation are more visible, and Newton's formula should be replaced by Langevin's law [10]

$$m\frac{d\vec{V}}{dt} = \sum \vec{F}_e + R(t) \tag{7.12}$$

where $R(t)$ is a white noise due to the Brownian effect. However, it is often the case that an "average" trajectory can be calculated simply by using Newton's equation, especially if the size of the beads is larger than 1 μm or if the forces that act on the particle dominate the Brownian motion. In such a case, because Brownian motion can be considered as a white noise, the real positions of the beads are close to the "average" trajectory, with a nearly symmetrical dispersion. This "average" trajectory is often sufficient to predict the behavior of the microsystem and to design the relevant component.

Usually, for microfluidics systems using magnetic beads, three types of forces are present: gravity, hydrodynamic drag, and magnetic forces. In this case, Newton's equation can be written under the form

$$m\frac{d\vec{V}_p}{dt} = \vec{F}_{mag} + \vec{F}_{hyd} + \vec{F}_{grav} \tag{7.13}$$

where $\vec{V}_p$ is the velocity of the particle. The hydrodynamic drag is derived from the velocity field according to the equation

$$\vec{F}_{hyd} = -C_D\left(\vec{V}_p - \vec{V}_f\right) = -6\pi\eta r_h\left(\vec{V}_p - \vec{V}_f\right) \tag{7.14}$$

where C_D is the drag coefficient, η is the dynamic viscosity of the carrier fluid, r_h is the hydrodynamic diameter of the particle, and V_f is the velocity of the carrier fluid. It is assumed here that the velocity field of the carrier fluid is not affected by the presence of the beads, which is the general case, except if the volume concentration of the beads is important, leading to aggregation of the beads. Under this assumption, the velocity field of the carrier fluid must be calculated before attempting the calculation of the particles trajectories, using classical hydrodynamics equations (i.e., Navier-Stokes equations). A classical situation in microfluidics is the Poiseuille flow between two plates or in a rounded capillary. In such a case, the velocity field is obtained under a closed form.

Again, if the concentration of magnetic beads is not too important, the external magnetic field is not affected significantly by the presence of the magnetic beads. In such a case, the magnetic field must be calculated before attempting the calculation of the trajectories. Under this assumption, the magnetic force on a particle of volume v_p is given by

$$\vec{F}_{mag} = \mu_0 v_p \Delta\chi \nabla\left(\frac{1}{2}H^2\right) \tag{7.15}$$

where $\Delta\chi$ is the difference of magnetic susceptibility between a particle and the fluid, and H is the magnetic field. Equation (7.15) indicates that the magnetic force is aligned with the direction of the gradient of the square of the magnetic field. Finally, the gravity term is simply given by

$$\vec{F}_{grav} = g v_p \Delta\rho \hat{y} \tag{7.16}$$

where g is the acceleration of gravity, v_p is the volume of the particle, $\hat{y}$ is the vertical unit vector (oriented downward), and $\Delta\rho$ is the difference between the volumic mass of the particle and that of the liquid. After substitution of (7.14), (7.15), and (7.16) in (7.13), one obtains the equation for the particles velocity

$$m\frac{d\vec{V}_p}{dt} = \mu_0 v_p \Delta\chi \nabla\left(\frac{1}{2}H^2\right) - 6\pi\eta r_h\left(\vec{V}_p - \vec{V}_f\right) + g v_p \Delta\rho \hat{y} \tag{7.17}$$

x and y coordinates of the particle at a given time are linked to the velocity by the relations

$$\frac{dx}{dt} = V_{p,x}$$
$$\frac{dy}{dt} = V_{p,y} \tag{7.18}$$

Equations (7.17) and (7.18) define the particle trajectory. It is very seldom that they can be solved analytically, but, as we will see, it is always interesting to spend some time investigating whether an analytical solution may exist—even at the price of some simplification. An example will be given in Section 7.9. Most of the time, a numerical approach is required. Different methods can be used, like Runge Kutta or predictor-corrector. We indicate in Section 7.9 a very simple first-order predictor-corrector method, which is very efficient when the velocity of the carrier fluid is sufficiently low.

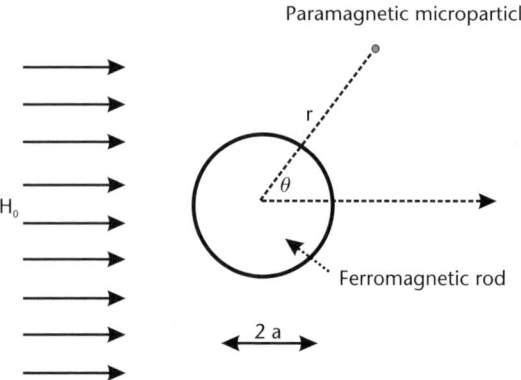

Figure 7.9   Schematic view of a ferromagnetic rod in an external magnetic field.

7.5   Example of a Ferromagnetic Rod

A very didactic example of trajectories of magnetic microparticles is that of the ferromagnetic rod. In this particular case, a closed-form solution exists for the magnetic field, and the trajectories may be calculated easily [7, 10]. Thus, it is a good test for the verification of numerical models. Moreover, it bears the principle of magnetic separators: we shall see that there are magnetic attraction zones and repulsion zones around the rod, and particles are expelled from the repulsion zones toward the attraction zones and concentrate on some regions on the rod surface. Assembly of ferromagnetic rods located in a large external magnetic field are called high gradient magnetic separators (HGMS) and are widely used in chemical and biological processes to remove magnetic particles from a carrier fluid [11].

7.5.1   Governing Equations

Suppose a ferromagnetic rod of radius a, surrounded by a carrier fluid containing paramagnetic microparticles and placed in a uniform external magnetic field $\vec{H}_0$ (Figure 7.9).

In this problem, there is no electric current, and the use of the magnetic scalar potential ϕ is sufficient to solve the problem. Using the relation

$$\vec{H} = -grad(\phi)$$

The potential ϕ is then solution of the following equation

$$div\left(-\mu_0\mu_r gr\vec{a}d(\phi)\right) + div\left(\mu_0\vec{M}_r\right) = 0 \qquad (7.19)$$

where μ_r is the relative magnetic permeability of the different materials of the computational domain. The remanent magnetization may be supposed uniform in the rod, and (7.19) reduces to

$$div\left(-\mu_0\mu_r gr\vec{a}d(\phi)\right) = 0 \qquad (7.20)$$

Equation (7.20) might not be linear if the external magnetic field is large enough to saturate parts of the ferromagnetic regions. Note that ϕ is continuous on the whole computational domain, even at the boundaries between different materials. The external field $\vec{H}_0$ is a Neumann boundary condition for (7.20). After solving for ϕ, $\vec{H}$ is obtained by $\vec{H} = -gr\vec{a}d(\phi)$.

Note that $\vec{H}$ is different from $\vec{H}_0$ in the vicinity of the rod: $\vec{H}$ is the sum of the external field $\vec{H}_0$ and the induced field $\vec{H}_i$. The magnetic force field on the paramagnetic microparticles is then obtained from $\vec{H}$ by calculating

$$\vec{F}_m = \mu_0 v_p\left(\chi_p - \chi_f\right)\left(\vec{H}\cdot\nabla\right)\vec{H} = \frac{\mu_0}{2}v_p\left(\chi_p - \chi_f\right)\nabla H^2 \qquad (7.21)$$

and the value of the magnetic force field is then plugged into the trajectory equation

$$m\frac{d\vec{V}_p}{dt} = \mu_0 v_p \Delta\chi \nabla\left(\frac{1}{2}H^2\right) - 6\pi\eta r_b\left(\vec{V}_p - \vec{V}_f\right) + g v_p \Delta\rho\hat{y}$$

This is the general case. However, in the special case of the ferromagnetic wire, the calculation of the particles' trajectories may be further detailed because the magnetic field may be calculated analytically [12].

7.5.2   Analytical Solution for the Magnetic Field

The total magnetic field in the vicinity of the rod is the sum of the external magnetic field plus the induced field due to the presence of the rod. It can be shown [9] that, at a distance r from the rod center and at angle θ from the external magnetic field H_0, the magnetic potential is

$$\phi_{ext} = \left(-H_0 r + \frac{1}{2}Ma^2 r^{-1}\right)\cos\theta \tag{7.22}$$

At this stage, two different cases may be distinguished: at very high external field, the rod is magnetically saturated and the total field $\vec{H}$ is given by

$$H_r^{ext} = \cos\theta\left(H_0 + \frac{A}{r^2}\right)$$
$$H_\theta^{ext} = \sin\theta\left(-H_0 + \frac{A}{r^2}\right) \tag{7.23}$$

where $A = \frac{1}{2}M_s a^2$. The magnetic force is then

$$F_m = -2\mu_0\left(\chi_p - \chi_f\right)v_p \frac{A}{r^3}\begin{pmatrix}\dfrac{A}{r^2} + H_0\cos2\theta\\[2mm] H_0\sin2\theta\end{pmatrix} \tag{7.24}$$

At sufficiently low magnetic field, the magnetization is linear homogeneous, and (7.22) leads to

$$H_r^{ext} = H_0\cos\theta\left(1 + a^2\Lambda r^{-2}\right)$$
$$H_\theta^{ext} = H_0\sin\theta\left(-1 + a^2\Lambda r^{-2}\right) \tag{7.25}$$

with $\Lambda = \dfrac{(\mu_w - \mu_f)}{(\mu_w + \mu_f)}$ and μ_w, μ_f are, respectively, the magnetic permeability of the rod and the fluid. The magnetic force is then

$$F_m = -2\mu_0\left(\chi_p - \chi_r\right)v_p \frac{a^2}{r^3} H_0^2 \Lambda \begin{pmatrix} \Lambda\left[\dfrac{a^2}{r^2}\right] + \cos 2\theta \\ \sin 2\theta \end{pmatrix} \tag{7.26}$$

Note the similarity between the two expressions of the magnetic forces (7.24) and (7.26). In both cases, in the vicinity of the rod, there are two attraction zones aligned with the external field (for θ close to zero or to π) and two repulsion zones in the direction perpendicular to the external field (for θ close to $\pi/2$ and $-\pi/2$). By taking the ratio between radial and azimuthal components of the force in either expression (7.24) or (7.26), it can be deduced that attraction forces are larger than repulsion forces. We show the magnetic field force in the vicinity of the rod in Figure 7.10.

7.5.3 Trajectories (Carrier Fluid at Rest)

We investigate first the case where the fluid is at rest. Using the algorithm for the calculation of trajectories described in Section 7.9 and the magnetic forces calculated in the preceding section, we find that the microparticles migrate from the two repulsion zones toward the two attraction zones, as shown in Figure 7.11. Theoretically, the repulsion and attraction regions extend to infinity, but the magnetic force decreases very rapidly from the rod surface, and particles located far from the rod will not move significantly.

7.5.4 Trajectories (Carrier Fluid Convection)

Suppose now that the carrier fluid is slowly moving around the rod. In this case, the flow is laminar and can be calculated either by solving the Navier-Stokes equations (Figure 7.12) or by using Lamb's equation [13]

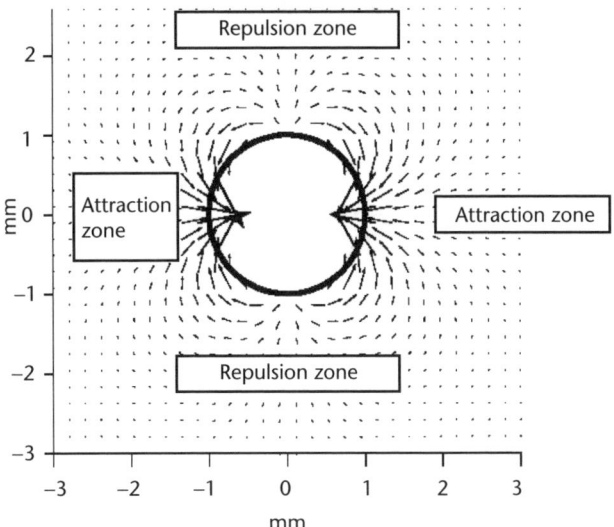

Figure 7.10 Magnetic force field around a cylindrical rod placed in a uniform magnetic field showing two attraction zones and two repulsion zones. The rod is perpendicular to the plane of the figure, and the external magnetic field is directed horizontally from left to right. (*From:* [12]. © 2000 EDP Science. Reprinted with permission.)

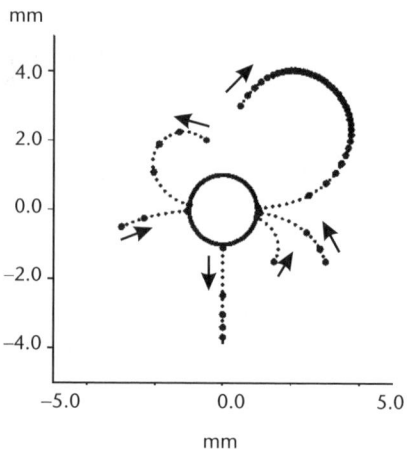

Figure 7.11 Trajectories of paramagnetic microparticles in the vicinity of the rod if the carrier fluid is at rest. (*From:* [12]. © 2000 EDP Science. Reprinted with permission.)

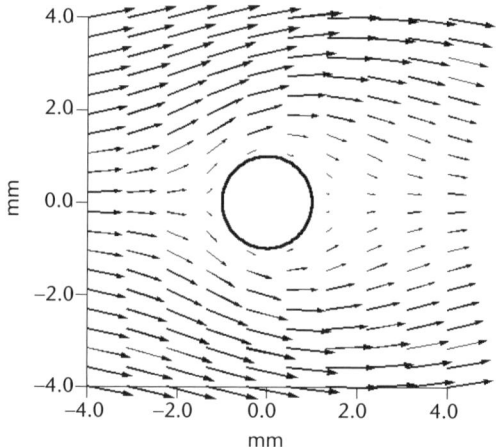

Figure 7.12 Velocity field of a laminar flow around a 2-mm diameter fiber. Velocity at infinity is 1 mm/sec.

$$u_r = -C\left[\ln(Re) - 0.5\left(1 - \frac{1}{Re^2}\right)\right]\sin\theta$$

$$u_\theta = -C\left[\ln(Re) + 0.5\left(1 - \frac{1}{Re^2}\right)\right]\cos\theta$$

(7.27)

with

$$C = \frac{u_{f0}}{\ln\left(\dfrac{7.4}{Re}\right)}$$

$$Re = \frac{2a\rho_f u_{f0}}{\mu_f}$$

The velocity field is represented in Figure 7.12.

Calculated trajectories for fluid velocities of 0.1 mm/sec and 1 mm/sec are shown in Figure 7.13 for a 1-mm radius fiber and 1-μm Dynal beads. The larger the carrier fluid velocity, the less the probability of capture by the rod.

7.6 Magnetic Repulsion

In the previous section, it was observed that the magnetic force field in the vicinity of a cylindrical rod placed in a uniform external magnetic field is highly nonuniform. The force is directed toward the rod in the regions aligned with the external field (from the rod center) and directed away from the rod in the regions perpendicular to the external field (from rod center). This observation can be extended to any shape of ferromagnetic body. We can even imagine more than one body in the external field, as shown in Figure 7.14, and obtain a confinement zone for the magnetic particles in a region located between the ferromagnetic bodies.

Note that the particles are not completely at rest in the confinement zone: they are repelled from the walls, and they move along the central axis to exit on both sides, as shown in Figure 7.15, from the calculation of the magnetic force field.

A very simple experimental set up may be realized to illustrate the principle of magnetic repulsion [14] (Figure 7.16). Place two ferromagnetic wires on a glass support perpendicular to two magnets. Between the magnets, the magnetic field is approximately uniform. If a drop containing paramagnetic particles is deposited on the plate between the wires, then the microparticles move rapidly toward the central axis; later, they migrate slowly along this central axis.

7.7 The Example of a Separation Column

It is important to be able to remove magnetic micro- and nanoparticles from a carrier liquid. This process may be performed in HGMS. In this section, we investigate

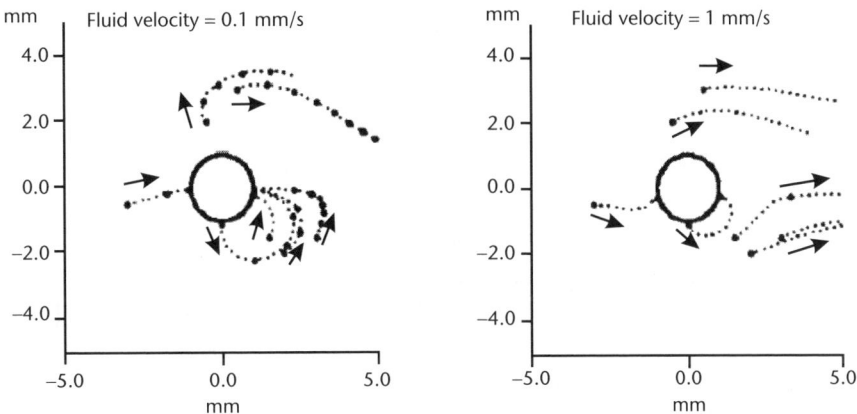

Figure 7.13 Trajectories of paramagnetic beads in the vicinity of a ferromagnetic rod under the action of a convective transport and a magnetic force field: fluid velocity at infinity 0.1 mm/sec (left); fluid velocity at infinity 1 mm/sec (right). (*From:* [12]. © 2000 EDP Science. Reprinted with permission.)

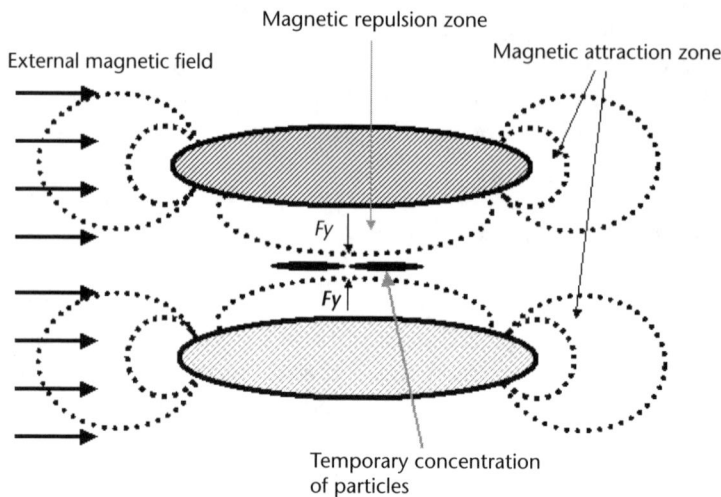

Figure 7.14   Principle of magnetic repulsion: repulsion zone due to two ferromagnetic objects in an external magnetic field.

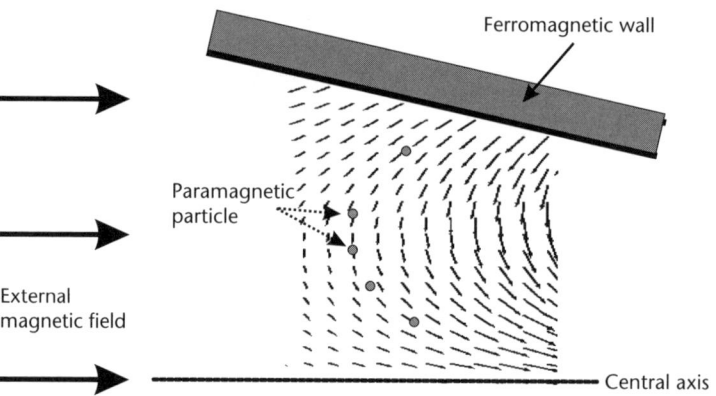

Figure 7.15   Force field created by two oblique symmetric ferromagnetic plates in a uniform external magnetic field. (*From:* [14]. © 2002 IEEE. Reprinted with permission.)

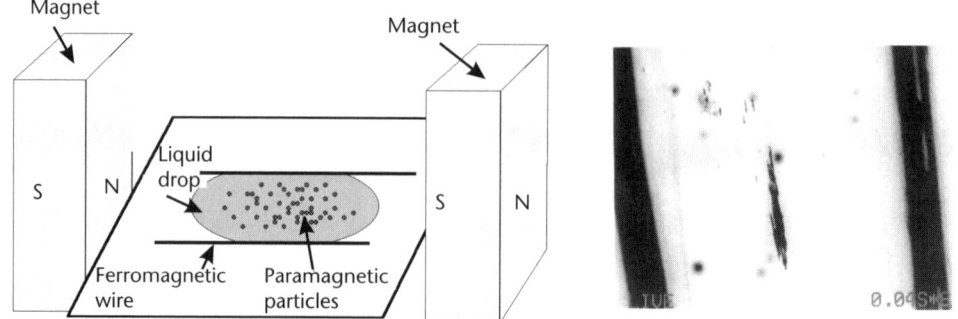

Figure 7.16   Experimental verification of the magnetic repulsion principle. Paramagnetic microparticles (1 μm) repulsed from two ferromagnetic parallel wires.

the behavior of magnetic beads submitted to a nonuniform magnetic field in the vicinity of an assembly of ferromagnetic rods or wires (Figure 7.17) and convected by a carrier fluid. Theoretically, this situation is just a generalization of the example of the ferromagnetic rod. The basic equations are the same [15].

The first step consists of calculating the magnetic field by using (7.20); then the magnetic force field is deduced from (7.21). Figures 7.18 and 7.19 show the magnetic vector field and magnetic forces in the assembly of wires. Magnitudes of these vectors are plotted in Figures 7.20 and 7.21. The situation is similar to the case of a single wire, but due to the packing of the rods, the repulsive forces are somewhat smaller.

The second step consists of solving the Navier-Stokes equations for the flow in the bundle (Figure 7.22, left). As expected, the flow is accelerated in the horizontal gaps between the rods.

The third step is the calculation of the trajectories following (7.17). Particles are principally captured on the surface of the rods facing the upcoming velocities (Figure 7.22, right). Few particles are captured in the attraction zones behind the rods; none are captured on the sides, where there are repulsive forces plus an acceleration of the flow. Note that we have supposed that the captured microparticles are not numerous enough to modify the magnetic field and the fluid flow circulation.

7.8   Concentration Approach

In this section, we present another approach to the behavior of magnetic particles. Particles are no more considered discrete entities but rather a continuous component in the carrier fluid. The presence of magnetic beads is defined by their volume concentration c equal to the number of particles per unit volume (or moles per unit volume). In the International System of units, c is expressed in particles/m^3. Because they use very small volumes of liquids, biologists and chemists use often other units: particles/μl or mole/μl.

Concentration repartition is calculated by solving the mass conservation equation for the particles in the liquid, taking into account convection due to the flow and to the magnetic forces [16]:

$$\frac{\partial c}{\partial t} = \nabla \cdot \left(D \nabla c - c \left(\vec{V}_f + u \vec{F}_{mag} \right) \right) = D \Delta c - \vec{V}_f \nabla c - \nabla \cdot \left(u \vec{F}_{mag} c \right) \tag{7.28}$$

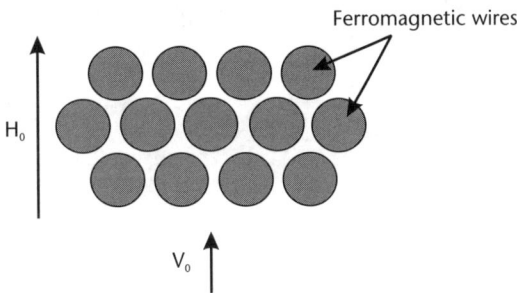

Figure 7.17   Assembly of parallel ferromagnetic wires in a uniform external magnetic field; the carrier fluid circulates between the wires.

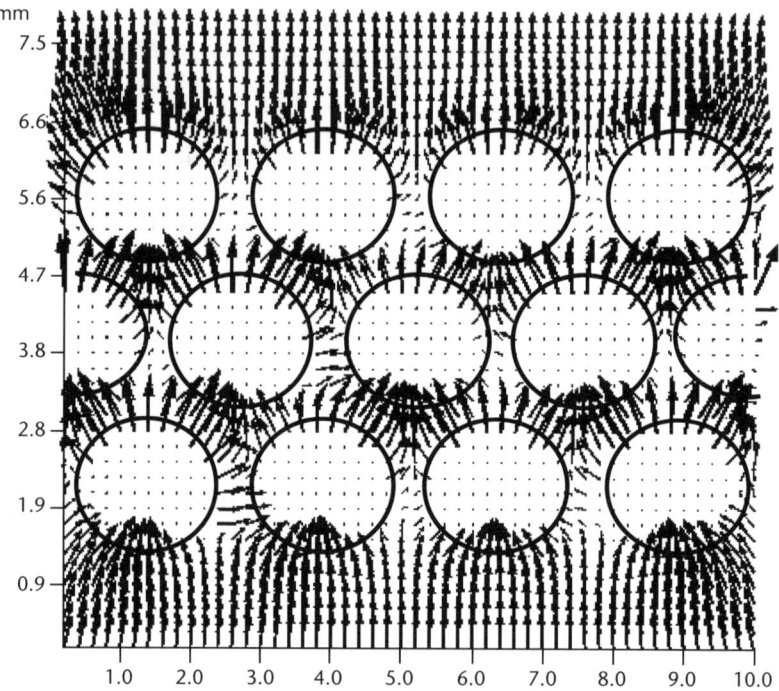

Figure 7.18 Magnetic field in an assembly of ferromagnetic wires. The magnetic field is similar to that around a single rod, but there are magnetic interactions between the rods. The magnetic field inside a rod is very small due to the magnetic permeability of the rod.

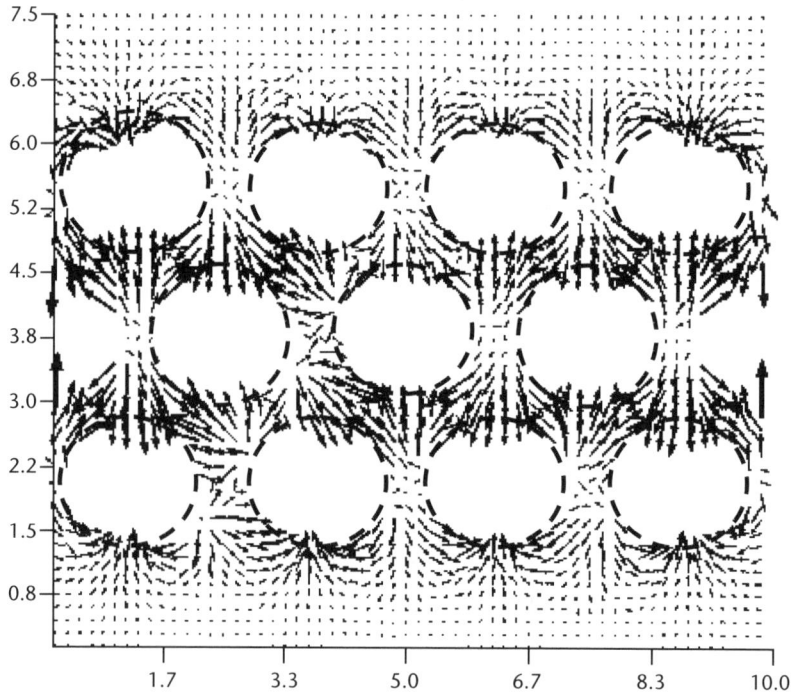

Figure 7.19 Magnetic forces. The magnetic force field is deduced from the magnetic field. Magnetic attraction forces can be clearly seen aligned with the external magnetic field. Repulsion forces are weak, due to the small gaps between the rods.

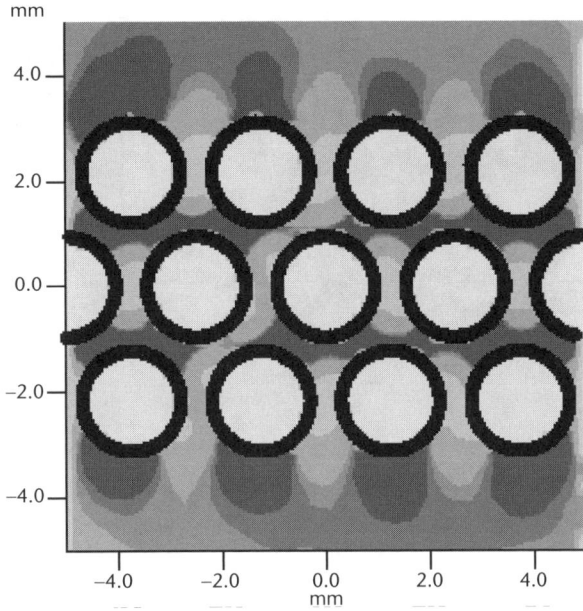

Figure 7.20 Magnitude of the magnetic field. Inside the bundle, the magnetic field is maximum in regions linking the rods at an angle of 45°.

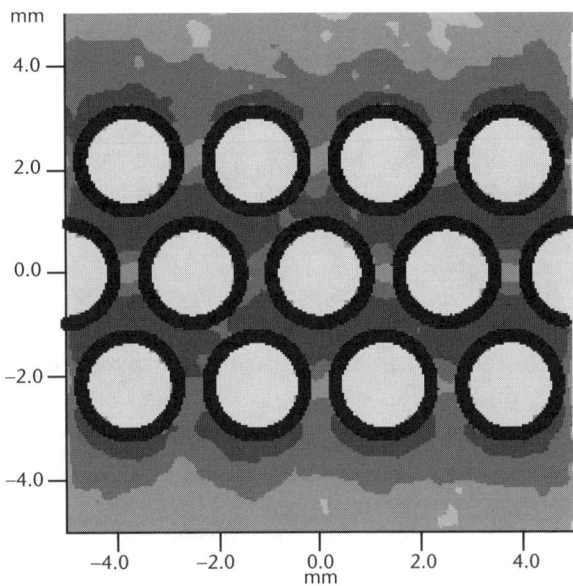

Figure 7.21 Magnitude of the magnetic forces: magnetic forces are minimum in the regions linking the rods along the x-direction.

In (7.28), D is the diffusion coefficient (m²/s) defined by Einstein's equation

$$D = \frac{k_B T}{6\pi\eta r_b} \qquad (7.29)$$

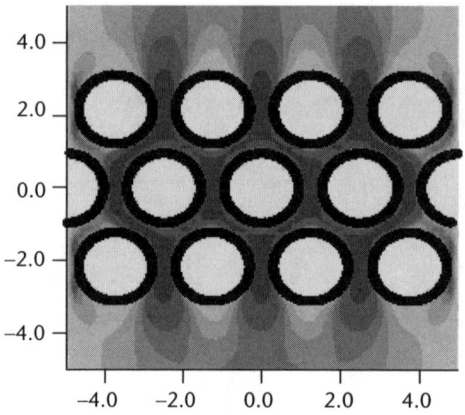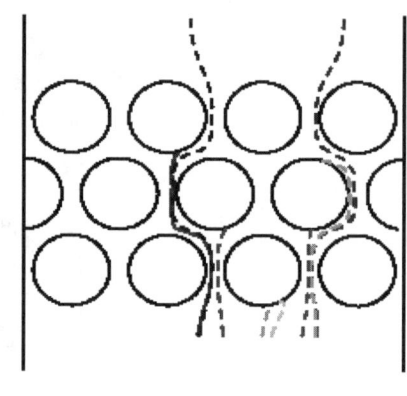

Figure 7.22 Left: Norm of the carrier fluid velocities. Right: Trajectories of particles through the bundle. Magnetic particles are primarily captured on the side of the wires facing the external magnetic field, some are captured behind, and none are captured on the sides.

where k_B is the Boltzmann constant and T is the Kelvin temperature. The mobility u is defined by

$$u = \frac{1}{6\pi\eta r_b} \qquad (7.30)$$

Note that u is the inverse of the drag coefficient C_D. In (7.28), the first term on the right-hand side is the diffusion term, which takes into account the Brownian motion; the second term is the advection due to the motion of the carrier fluid; and the third term is the convection due to the magnetic forces.

As for the calculation of particle trajectories, (7.28) requires the previous calculation of the velocity field of the carrier liquid and the magnetic force field.

Generally, two types of numerical approaches can be used to solve (7.28): finite elements method or finite differences/volumes method. In microfluidics, the boundaries of the domain have a very important impact, and the finite elements method is well adapted for such problems. However, in a simple geometry (rectangular or axisymmetric), finite differences method is very easy to use, and one can write its own numerical program to solve such problems.

Note that the problem is a weak coupled problem. It must be solved in three steps: first, compute the carrier fluid velocities; second, compute the magnetic force field; third, solve for the concentration distribution in the domain. Using the same numerical frame to solve successively the three equations is the most straightforward solution.

Note that boundary conditions must be specified to solve (7.28). At channel inlet, boundary conditions for the concentration are of the type $c = c_0$ or $c = 0$, depending on the location of particulate injection; and the condition $\dfrac{\partial c}{\partial n} = 0$ at the channel outlet. The solid walls are impermeable to fluid and particle transport, so that the corresponding boundary conditions are

$$\vec{J} \cdot \vec{n} = -D \ grad \ c \cdot \vec{n} + c\vec{v} \cdot \vec{n} + cu\vec{F}_{mag} \cdot \vec{n} = -D \ grad \ c \cdot \vec{n} + cuF_{mag} = 0 \qquad (7.31)$$

where $\vec{J}$ is the flux of particles.

7.9   The Example of Magnetic Field Flow Fractionation (MFFF)

In this paragraph, we illustrate how to calculate particle trajectory and concentration by taking the example of MFFF.

In biology and biotechnolgy, there is a constant need to separate particles, depending on their mass, electric charge, or magnetic properties. For example, this is the case for purification of proteins or for obtaining monodisperse magnetic beads. A very common method for separation of particles is FFF.

In a typical FFF device, the carrier fluid flows horizontally in a channel, and the particles experience a horizontal drag force; depending of the type of separation that is searched (mass, electric charge, magnetic properties), a relevant force field (gravity, electric field, magnetic field) is set up to act perpendicularly to the flow. Trajectories of the different particles differ in the FFF force field. Similar particles have similar trajectories and gather at the same location on the channel lower wall. An MFFF is sketched in Figure 7.23.

Much work has been done in the domain of sedimentation field flow fractionation (SdFFF) [17]; however, recently, new applications for biological processes—like cell separation—have required the use of submicronic paramagnetic particles. These are not much influenced by gravity, but MFFF is a well-suited method to separate these particles according to their size and magnetic permeability [18, 19].

The velocity field is well known in the case of a laminar flow between two horizontal parallel plates (the Reynolds number is much less than 1). From hydrodynamics considerations, it is well known that the velocity profile across any section is parabolic and given by

$$\vec{V}_f = \frac{3\vec{V}_0}{2}\left(1 - \frac{y^2}{d^2}\right) \qquad (7.32)$$

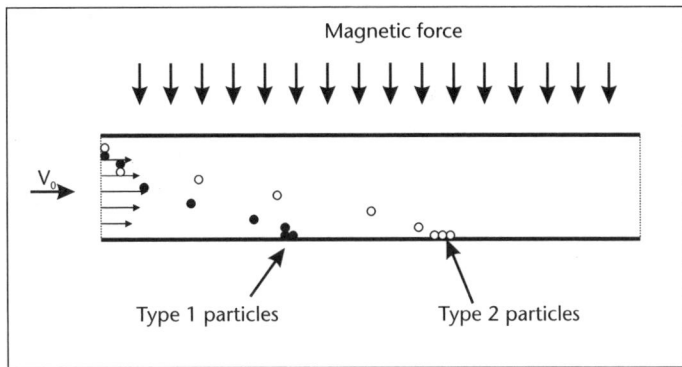

Figure 7.23   Schematic view of MFFF. The upper solid wall is called the depletion wall, and the lower wall is the accumulation wall.

where d is half the distance between the channel walls and V_0 is the average velocity.

7.9.1 Trajectories

Due to the design of the MFFF device, with a vertical external force field and a horizontal hydrodynamic drag, (7.17) can be decomposed on the x and y directions, and we obtain the following system for the particle velocity

$$\frac{dV_{p,x}}{dt} = -c_1\left(V_{p,x} - V_{f,x}\right)$$
$$\frac{dV_{p,y}}{dt} = -c_1 V_{p,y} + c_2$$

(7.33)

where c_1 and c_2 are given by

$$c_1 = \frac{6\pi\eta r_b}{m}$$

$$c_1 = \frac{v_p \Delta\chi \nabla\left(\frac{1}{2}H^2\right)\cdot\hat{y}}{m} + \frac{g v_p \Delta\rho}{m}$$

Equations (7.33) form a noncoupled, first-order, differential system. To solve numerically for such a system is classical; almost any mathematical software possesses an algorithm to solve such a system. However, (7.33) admits a closed-form solution for the velocity if we assume that c_1 and c_2 are constant [20]; this is not very restrictive since c_1 is constant in a homogeneous fluid, and c_2 is also constant if the magnetic gradient is uniform—which is the case when the dimension of the magnet is sufficiently large compared to the dimension of the channel. The closed form solution is then

$$V_{p,x} = V_{p,x0}e^{-c_1 t} + V_{fx}\left[1 - e^{-c_1 t}\right]$$
$$V_{p,y} = V_{p,y0}e^{-c_1 t} + \frac{c_2}{c_1}\left[1 - e^{-c_1 t}\right]$$

(7.34)

where the subscript zero corresponds to the initial values at $t = 0$. We have thus derived an analytical expression for the particle velocity that depends on the values of the magnetic force, drag coefficient, and gravity. In (7.34), the applied forces appear through the ratio c_2/c_1

$$\frac{c_2}{c_1} = \frac{v_p \Delta\chi \nabla\left(\frac{1}{2}H^2\right)\cdot\hat{y} + g v_p \Delta\rho}{6\pi\eta r}$$

(7.35)

It is seen in (7.35) that c_2/c_1 represent the ratio between the applied external (vertical) forces and the hydrodynamic (horizontal) drag force (c_2/c_1 has the dimension of a velocity). This ratio determines the trajectories. Two particles experiencing the same ratio c_2/c_1 and starting for the same point at inlet will follow the same trajectory—to some bias due to the Brownian motion.

We now advance a step further and search for a solution for the trajectory. We have to solve the first-order differential system

$$\frac{dx_p}{dt} = V_{p,x0}e^{-c_1t} + \frac{3V_0}{2}\left(1 - \frac{y_p^2}{d^2}\right)\left[1 - e^{-c_1t}\right]$$

$$\frac{dy_p}{dt} = V_{p,x0}e^{-c_1t} + \frac{c_2}{c_1}\left[1 - e^{-c_1t}\right]$$

(7.36)

where x_p and y_p are the coordinates of the particle at time t. This time, system (7.36) is coupled because the y-coordinate appears in the first x-equation—which means that the trajectory of the particle will not be linear.

7.9.1.1   Case of Analytical Solution to the Trajectory

If we examine in detail the system, it can be seen that if the velocity of the particle is zero at the inlet, $V_{p,x,0} = V_{p,y,0} = 0$—this is the case if the particles start at the top, along the upper (depletion) wall, time can be eliminated from (7.36), and the differential equation has the following particular form

$$\frac{dx_p}{dy_p} = \frac{3V_0}{2}\left(1 - \frac{y_p^2}{d^2}\right)\frac{c_1}{c_2}$$

(7.37)

Equation (7.37) is easily solved (taking into account $x_0 = 0$, $y_0 = -d$), and we find

$$x_p = -\frac{V_0}{2d^2}\frac{c_1}{c_2}\left[y_p\left(y_p^2 - 3d^2\right) - 2d^3\right]$$

(7.38)

This is the equation of a cubic; the corresponding trajectory has been plotted in Figure 7.25; one verifies that $x_p = 0$ for $y_p = -d$. By setting $y_p = d$ in (7.38), one finds the distance from inlet at which a particle meets the accumulation wall.

$$x_d = 2dV_0\frac{c_1}{c_2}$$

(7.39)

It can be readily checked that the larger the magnetic force, the smaller the distance x_d, and the larger the hydrodynamic drag, the larger the distance x_d.

Equation (7.39) can be used to define the separation efficiency of the MFFF. If the particles are injected at the top of the channel through a nozzle, the separation distance between the accumulation sites on the accumulation plate for two different types of particles (1 and 2) is given by (7.39). Provided that the particles are sufficiently small to neglect the gravity force before the magnetic force, one obtains a very simple relation [20]

$$\frac{L_2}{L_1} = \frac{\Delta\chi_1}{\Delta\chi_2}\frac{r_1^2}{r_2^2} \tag{7.40}$$

where L is the distance from the channel entrance. Usually, smaller paramagnetic particles have lower magnetic susceptibility, so that we can expect an efficient separation between two populations of magnetic beads (i.e., the two types of magnetic beads are gathering in two distinct packets on the accumulation wall).

Equation (7.40) for separation efficiency of an MFFF device shows the advantage of analytical methods when possible. They can produce relationships between parameters that are not always obvious when using numerical solutions. It is strongly recommended to always search first for analytical solutions—even to the price of simplifications—before turning to numerical methods.

7.9.1.2   General Case: Numerical Approach

System (7.34) determines the velocity field of the particles in the carrier fluid. The issue is now to solve system (7.36) for particle trajectory. More or less sophisticated methods can be used. But if we take advantage of the very slow velocity of the carrier fluid, a very simple predictor-corrector method can be set up.

Suppose that the particle has the coordinates x_i and y_i at time t_i.

The first step of the numerical scheme is to find a predictor point at time t_{i+1}. This predictor point is obtained by making use of the velocity $\vec{V}_{p,i} = (V_{p,x,i}, V_{p,y,i})$ at time t_i

$$\begin{aligned}\tilde{x}_{i+1} &= x_i + \Delta t V_{p,x,i}\\ \tilde{y}_{i+1} &= y_i + \Delta t V_{p,y,i}\end{aligned} \tag{7.41}$$

In reality, the particle velocity is not constantly $\vec{V}_{p,i}$ during the time interval $[t_i, t_{i+1}]$. Because we have found the predictor point $(\tilde{x}_{i+1}, \tilde{y}_{i+1})$, we now know the velocity at this point $\tilde{V}_{p,i+1} = (\tilde{V}_{p,x,i+1}, \tilde{V}_{p,y,i+1})$, and a more accurate velocity in the time interval $[t_i, t_{i+1}]$ is $\left(\dfrac{\tilde{V}_{p,x,i+1} + V_{p,x,i}}{2}, \dfrac{\tilde{V}_{p,y,i+1} + V_{p,y,i}}{2}\right)$.

The second step is then the following correction

$$\begin{aligned}x_{i+1} &= x_i + \Delta t\,\frac{\tilde{V}_{p,x,i+1} + V_{p,x,i}}{2}\\ y_{i+1} &= y_i + \Delta t\,\frac{\tilde{V}_{p,y,i+1} + V_{p,y,i}}{2}\end{aligned} \tag{7.42}$$

This two-step predictor-corrector method can be schematized graphically (Figure 7.24). The particle is located at the point M_i at time t_i, the predictor point is P_{i+1}, and the corrected point is M_{i+1}. The distance between these two points is the first-order error. The larger the carrier fluid velocity, the larger the distance $[P_{i+1},$

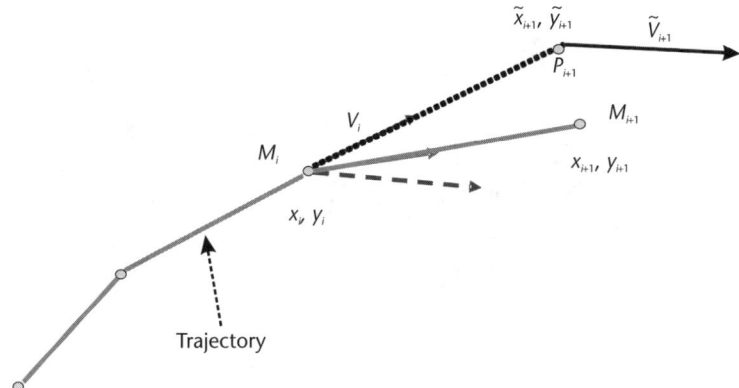

Figure 7.24 Graphical scheme of the first-order predictor-corrector method. The location of particle at time i is M_i, and the predicted location at time $i + 1$ is M_{i+1}.

M_{i+1}]. A second-order method using the same principle can be easily set up to verify that the precision is satisfactory.

7.9.2   Concentration of Magnetic Beads

The starting point is (7.28), where the velocity field is given by (7.32) and the magnetic force by (7.21)

$$\frac{\partial c}{\partial t} = D\Delta c - \frac{3\bar{V}_0}{2}\left(1 - \frac{y^2}{d^2}\right)\nabla c - uv_p \Delta\chi \nabla\left(\frac{1}{2}H^2\right)\cdot\nabla c \tag{7.43}$$

In (7.43), two terms on the right-hand side contribute to the convection of the particles: the drag force of the velocity field and the magnetic force term. Equation (7.43) must be solved by finite elements or finite differences/volumes numerical schemes.

7.9.3   Results and Comparison

We examine the case of an MFFF channel of 100-μm width and 1-mm length. The fluid carrier is water, and its average flow velocity is 0.1 mm/sec. Particles have a hydraulic radius of 1.4 μm and a magnetic susceptibility of 0.2; their diffusivity is 1.53 μm^2/sec, and the magnetic force is 3.45 pN.

In this particular case, the magnetic force is assumed constant. The velocity field is obtained under a closed form by (7.34) and the trajectory by (7.38)—if the injection point is located at the top plate—or else by a predictor-corrector scheme (7.42).

Figure 7.25 shows contour plots of the particles' concentration compared to the calculated trajectories. The location of the injection is either at the top of the channel or at one-third of the vertical height. As expected, the concentration contours are centered on the trajectories: Brownian motion slightly disperses the particles from their determinist trajectory.

In a second step, the same method was then applied to the case of the separation of a colloid mixture containing two different types of submicronic paramagnetic

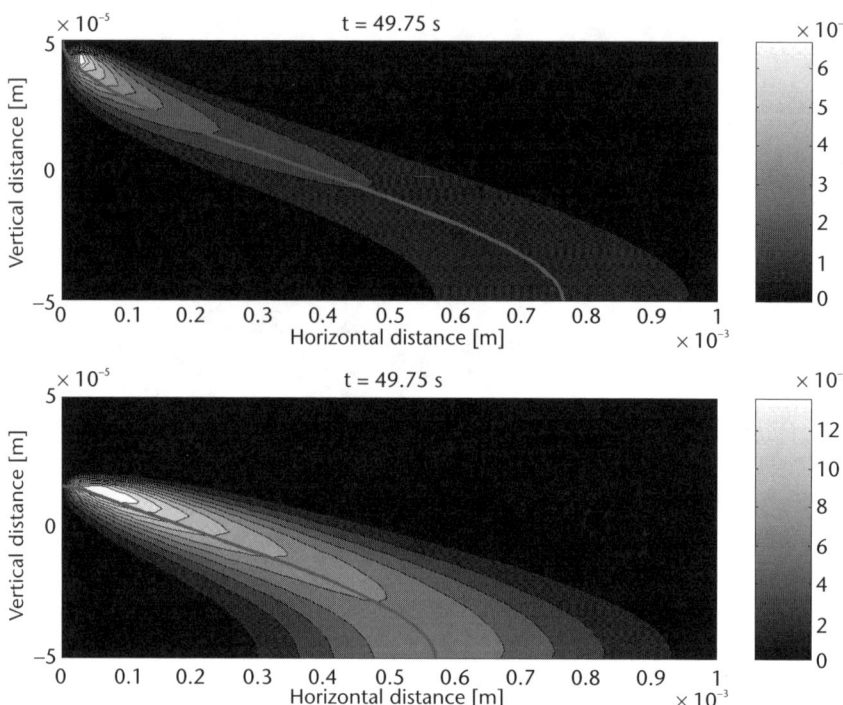

Figure 7.25 Comparison between calculation of trajectory and concentration contours, showing that the *determinist trajectory* is the center line of the concentration contours. In the figure on top, the trajectory is the cubic curve given by (7.18); bottom: the trajectory is obtained by a first-order predictor-corrector scheme. The channel width is 100 μm, and the average fluid velocity 0.1 mm/s. Particles have a radius of 1.4 μm and a magnetic susceptibility of 0.2. They are submitted by the magnetic gradient to a force of 3.45 pN. (*From:* [20]. © 2001 NSTI. Reprinted with permission.)

particles (Figure 7.26) differing by their magnetic susceptibility (0.2 and 0.8). We verify that the characteristic relation (7.40) applies even for trajectories not starting from the depletion (upper) wall.

7.10 Assembly of Magnetic Beads—Magnetic Bead Chains

Superparamagnetic beads in an external magnetic fields are similar to small induced magnetic dipoles [21, 22]. Their magnetic moment is given by

$$\vec{m} = \frac{1}{2}\frac{\mu_r - 1}{\mu_r + 2}\pi a^3 \vec{H} \tag{7.44}$$

where a is the dipolar distance. This induced magnetization creates the induced magnetic field

$$\vec{H}_i = \frac{1}{4\pi r^3}\left(3\vec{i}\left(\vec{i}\cdot\vec{m}\right) - \vec{m}\right)$$

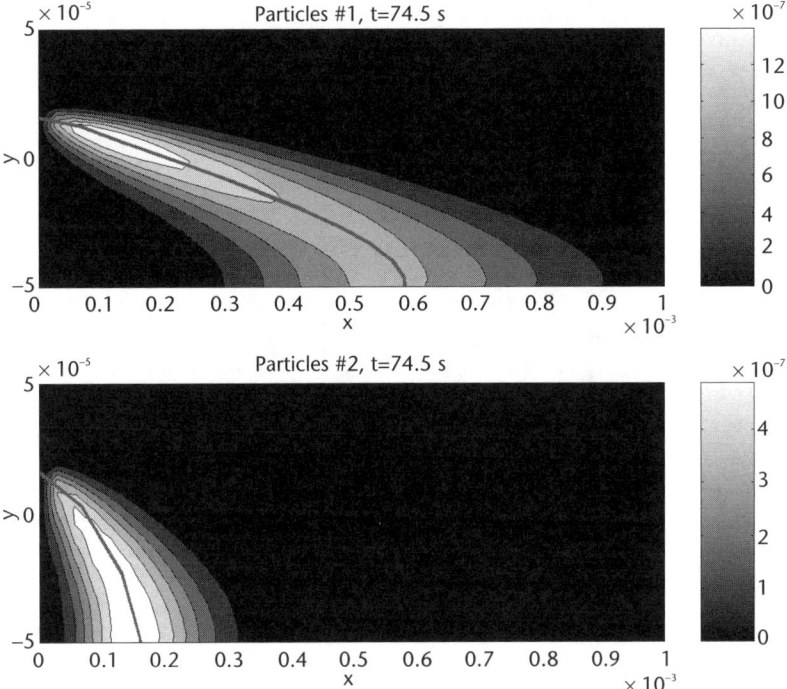

Figure 7.26   Two different families of trajectories for two different types of beads. In both cases, the starting point is the same, but the particles on the figure at the top have a magnetic susceptibility of 0.2, whereas those in the lower figure have a susceptibility of 0.8. One can see that (7.40) is also verified in the case of the numerical calculation of the trajectories. (*From:* [20]. © 2001 NSTI. Reprinted with permission.)

at a point defined by the vector $\vec{ri}$ (from the dipole center). In the total field $\vec{H}$, the induced magnetic field of a microparticle of radius a is

$$\vec{H}_i = \frac{1}{4}\frac{\mu_r - 1}{\mu_r + 2}\frac{a^3}{r^3}\vec{H} \qquad (7.45)$$

Take two of the same microbeads—of radius a—referred to by indices 1 and 2. The magnetic field *at the center of each bead* is

$$\vec{H}_1 = \vec{H}_0 + \vec{H}_{i,12}$$
$$\vec{H}_2 = \vec{H}_0 + \vec{H}_{i,21}$$

Using (7.45) with the corresponding indices, we find the coupled system

$$\vec{H}_1 = \vec{H}_0 + \frac{1}{4}\frac{\mu_r - 1}{\mu_r + 2}\frac{a^3}{r^3}\vec{H}_2$$

$$\vec{H}_2 = \vec{H}_0 + \frac{1}{4}\frac{\mu_r - 1}{\mu_r + 2}\frac{a^3}{r^3}\vec{H}_1 \qquad (7.46)$$

Let $\alpha = \dfrac{1}{4}\dfrac{\mu_r - 1}{\mu_r + 2}$, and we can solve (7.46) to find

$$\vec{H}_1 = \vec{H}_0\left(\frac{1}{1 - \alpha u^3}\right)$$

with $u = a/r$. For $r = a$ (when the two beads contact)

$$\vec{H}_1 = \vec{H}_0\left(\frac{1}{1 - \alpha}\right)$$

Using the formulation of the magnetic force, the force exerted by one sphere on the other is

$$F_1 = -6\mu_0\pi\frac{a^6}{r^4}H_0^2\frac{\alpha^2}{(1 - \alpha)^3}$$

When the two spheres contact each other, $r = a$ and

$$\vec{F}_1(contact) = -6\mu_0\pi a^2 H_0^2\frac{\alpha^2}{(1 - \alpha)^3} \tag{7.47}$$

Equation (7.47) shows that the binding force between two similar beads is proportional to the square of the bead radius and to the square of the magnetic field. In the same way that we have shown the existence of magnetic attraction and repulsion regions at a wire surface, it can be shown that the same exists at the surface of a spherical bead [21]: The two regions around the bead aligned with the external magnetic field are magnetically attractive, and the region around the *magnetic equator* of the bead is a repulsion zone (Figure 7.27).

Under the action of the Brownian motion, beads randomly contact each other; contacting beads will tend to stick together by their attraction regions and progressively will form a linear chain of beads aligned in the external field (Figure 7.28). These binding forces are more efficient for larger beads (1 μm) than for smaller beads (50 nm). Nano-sized beads—below 50-nm diameter—are often dispersed by the Brownian motion.

Magnetic chains are new tools in biotechnology (they can even be stabilized by polymer coating), and they have found an application for DNA separation: these chains are used to separate DNA segments in the same way as a gel. Let us first recall how DNA separation works in a gel: Under an electric field, the DNA segments are migrating under electrophoretic forces (see Chapter 8) at a different speed depending on their size (Figure 7.29).

The longer the strand of DNA, the more it encounters obstacles, and the more it is delayed in its motion. This technique has been widely used to decrypt the human genome. The major drawback is that a characteristic time for the separation is of the order of 24 hours. Some other solutions have been searched to obtain shorter

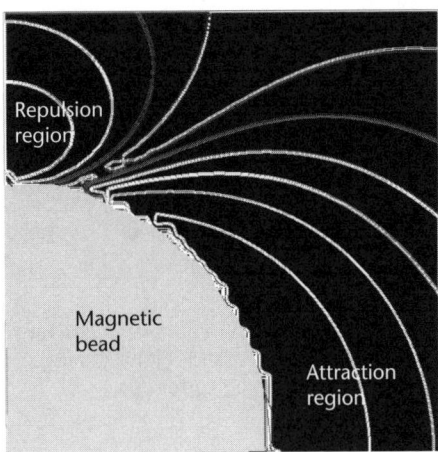

Figure 7.27 Magnetic attraction and repulsion regions around a spherical bead. The external magnetic field is oriented from left to right. Only one-quarter of the space has been represented.

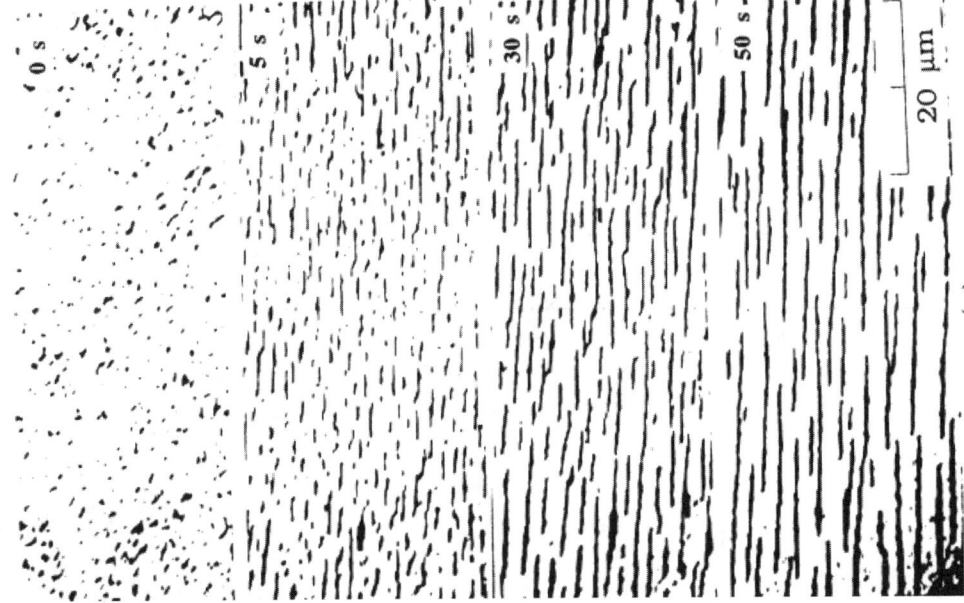

Figure 7.28 Chains of magnetic beads, aligned with the applied magnetic field, form in less than 1 minute.

separation times. The idea was to mimic the action of a gel by using lithography techniques to fabricate a lattice of micropillars. Due to the difficulty in fabricating this type of microcomponent, an interesting solution with magnetic beads has been set up [23, 24]. Magnetic beads (2.8-μm Dynal) initially dispersed in a flat microchannel limited between two plates form vertical columns (vertical chains) when an external magnetic field is applied vertically. The columns are naturally regularly spaced, and they perform the same function as gels during DNA electrophoresis. However, the separation duration is much shorter than that of gels

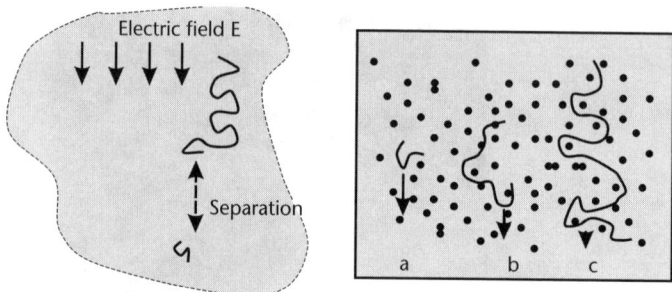

Figure 7.29 Left: Schematic view of DNA fragments of different sizes migrating at different speed in an agarose gel under a constant electric field. Right: A longer strand encounters more difficulties moving in the porosities of the gel than a shorter one.

(100 minutes instead of 24 hours). In this case, the assembly of magnetic beads is similar to a *calibrated* gel (Figure 7.30).

7.11 Magnetic Fluids

7.11.1 Introduction

We have seen that when the concentration of magnetic micro- or nanoparticles in a carrier fluid is sufficient and the magnetic field is uniform, magnetic chains will form. If the external field is not uniform, aggregates will form (Figure 7.31).

But what if the particles remain dispersed by the use of surfactants or electric repulsion? In such a case, one obtains a magnetic fluid—or a ferrofluid—which is a coherent fluid formed by a stable suspension of magnetic nanoparticles (grains of magnetite Fe_3O_4 or maghemite) in a carrier fluid (Figure 7.32, right). There are two types of ferrofluids, depending on their base (carrier fluid): the base may be organic and the particles are dispersed by surfactants, or the base may be polar and the

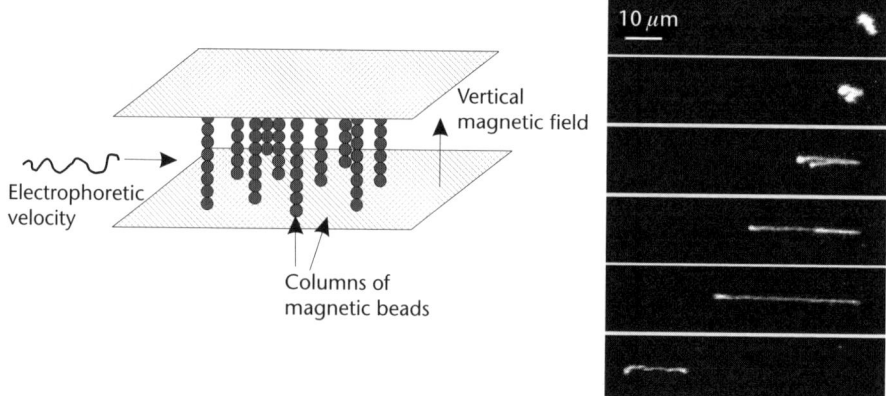

Figure 7.30 Left: Schematic view of a DNA separator constituted by magnetic beads aligned in vertical columns. Right: A long DNA strand moves from right to left; when meeting an obstacle, it stretches on both sides of the obstacle, and the longer side slowly drags out all of the DNA strand. DNA motion restarts until the next obstacle. (Courtesy of J. L. Viovy, Institut Curie.)

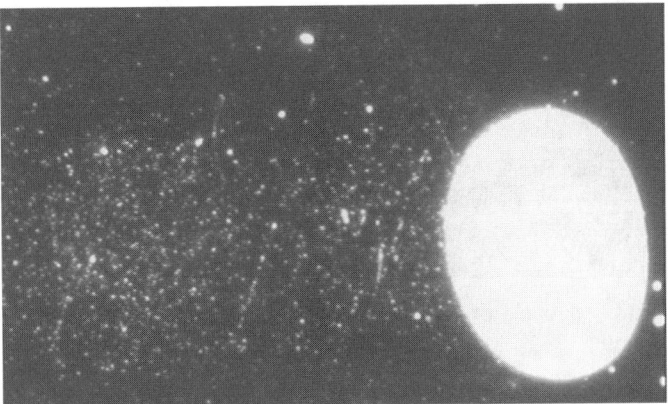

Figure 7.31   Image of an aggregate of magnetic microparticles (1-μm Dynal).

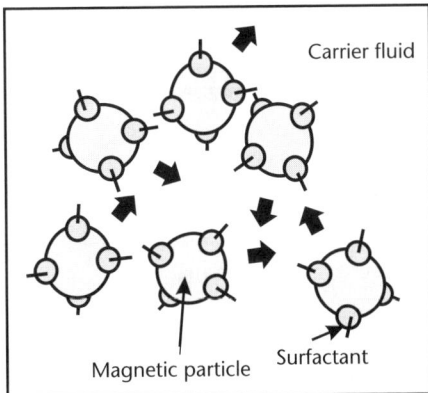

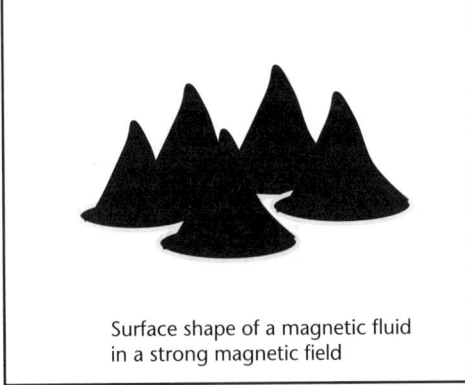

Surface shape of a magnetic fluid
in a strong magnetic field

Figure 7.32   Left: Schematic view of an organic-based ferrofluid. Right: View of ferrofluid drops attracted by a magnet.

particles are dispersed by electric charges (Figure 7.32, left). Note that there exist no natural magnetic fluids, because liquid metals are not magnetic, as their liquefaction temperature is above Curie temperature.

The advantage of magnetic fluids is that they can be actuated externally by a micromagnet or a microelectromagnet and that they can be inserted as *active plugs* in microsystems (Figure 7.33) to pump or regulate the flow [25].

7.11.2   Magnetic Force on a Plug of Ferrofluid

Suppose a ferrofluid plug in a capillary tube controlled by an external magnetic field (Figure 7.34).

The magnetic forces on the plug result in a magnetic pressure difference between the two edges of the plug according to Rosensweig's formula [26]. If index *a* stands for the advancing front and *r* stands for the receding front, we have

$$\Delta P_{mag} = \mu_0 \int_{H_r}^{H_a} M dH + \frac{\mu_0}{2}\left(M_{n,a}^2 - M_{n,r}^2 \right) \qquad (7.48)$$

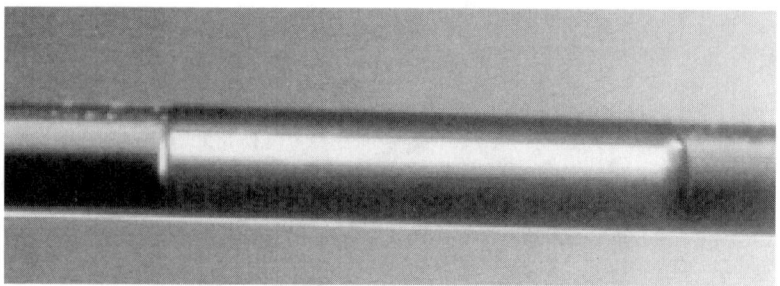

Figure 7.33   Ferrofluid plug in a capillary tube (diameter of 200 μm).

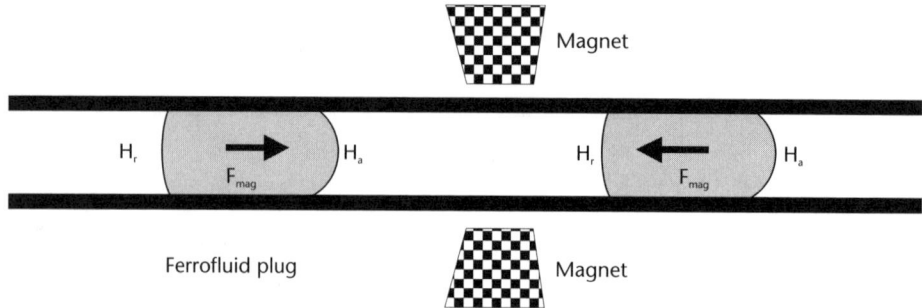

Figure 7.34   Schematic view of a plug approaching a magnet. The fluid flow is directed from left to right. The magnetic force on the plug is always directed toward the magnet.

where H is the magnetic field along the capillary axis, M is the magnetic moment, and M_n is the magnetic moment normal to the plug interface with water. Now if we note that a ferrofluid behaves like a paramagnetic media, in the limit of low magnetization $M = \chi H$ and after substitution in (7.48), one obtains

$$\Delta P_{magnetic} = \frac{1}{2}\frac{\mu}{\mu_0}\left(\mu - \mu_0\right)\left(H_a^2 - H_r^2\right) \tag{7.49}$$

Note that an exact derivation may be done for any magnetic field by integrating the full Langevin's formulation [27]:

$$\frac{M}{M_s} = \coth\left(\frac{3\chi H}{M_s}\right) - \frac{1}{\dfrac{3\chi H}{M_s}} \tag{7.50}$$

resulting in the expression

$$\Delta P_{mag} = \frac{\mu_0 M_s^2}{3\chi}\left[Ln\left(\frac{sh\left(\dfrac{3\chi H}{M_s}\right)}{\dfrac{3\chi H}{M_s}}\right)\right]_{Hr}^{H_a} + \frac{\mu_0 M_s^2}{2}\left[\left(\coth\left(\frac{3\chi H}{M_s}\right)\right)^2\right]_{Hr}^{H_a} \tag{7.51}$$

This result shows that at low magnetic field, the difference of the squares of the magnetic field at the interfaces determines the magnetic force on the plug. The interface that is nearest to the magnet is dominant in (7.49) or (7.51), and the force on the ferrofluid plug is always directed toward the magnet. If the force is sufficient, the magnet will block the plug, and the flow in the capillary will stop.

In conclusion, one has to be cautious with the use of magnetic liquids; they are not biocompatible (e.g., they block PCR), and they often must be separated from the biofluid by a secondary plug, usually made of oil, for biocompatibility.

7.12 Magnetic Micromembranes

In this chapter, we are dealing with magnetic micro- and nanoparticles, and it seems interesting to present a very useful application of magnetic microparticles in biotechnology. In this section, we show how magnetic nanoparticles can be assembled inside a polydimethyl-siloxane (PDMS) matrix to form a magnetic micromembrane. The advantage of such membranes is that they can be actuated externally (from outside the microdevice) by a time-varying magnetic field.

7.12.1 Principle

It has been found that flexible, elastomer micromembranes could be used in microsystems to separate a biofluid from another fluid or gas. Increasing the pressure (e.g., by acoustic waves) or the temperature of the auxiliary fluid results in the pressurization or agitation of the biofluid [28].

Another actuation of elastomer micromembranes can be achieved if the membrane has magnetic elements attached to or embedded in it. Magnetic actuation has been obtained first by fixing permalloy microplates to an elastomer micromembrane [29]; more recently, magnetic microparticles have been embedded in a PDMS matrix to obtain biocompatible, smooth-surfaced, and totally deformable magnetic micromembranes [30].

The principle is to mix paramagnetic (carbonyl iron) or ferromagnetic (ferrite) nanoparticles with liquid PDMS in a mass ratio of 25–50%. This matrix is spread by spin coating and left to polymerize. Very uniform membranes can be obtained (Figure 7.35). An interesting property is that such membranes are nearly as deformable and flexible as pure PDMS membranes. Deflection of membranes is governed by two numbers: Young modulus (usually noted E and expressed in Pascals) and Poisson coefficient (usually noted v and without unity). Young modulus and Poisson coefficient for 100-μm-thick PDMS-only membranes are of the order of $8 \cdot 10^5$ Pa and 0.5, whereas that of PDMS membrane containing carbonyl iron in a mass ratio of 25% are of the order of $9.5 \cdot 10^5$ Pa and 0.55.

7.12.2 Deflection of Paramagnetic Micromembranes

Suppose now a circular paramagnetic micromembrane with a radius of $500 \, \mu$m that is $50 \, \mu$m thick; this membrane is clamped to a solid wall by its circular edge (Figure 7.36).

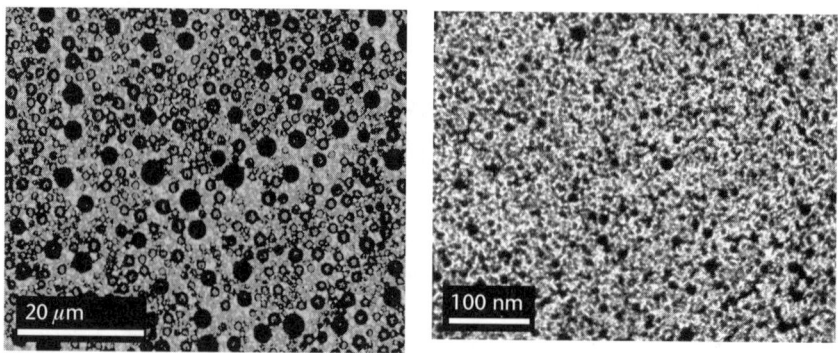

Figure 7.35 Two types of paramagnetic membranes. Left: 4-μm nanograins of carbonyl iron embedded in PDMS. Right: 10-nm nanograins extracted from ferrofluid embedded in PDMS. (Courtesy of F. Ricoul and A. Vigier, CEA/LETI.)

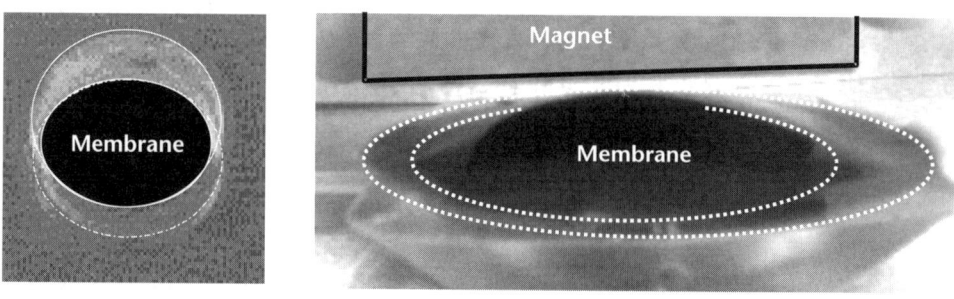

Figure 7.36 Magnetic micromembrane (right) clamped in a PDMS matrix and (left) deformed under the action of a magnet. (Courtesy of F. Ricoul and A. Vigier, CEA/LETI.)

An estimate of the deflection of such membranes can be done by using the linear theory—if the maximum deflection is less than about 0.2 times the membrane thickness [31]:

$$w_{max} \le \frac{h}{5} \tag{7.52}$$

and the maximum deflection is given by the relation

$$w_{max} = \frac{Pa^4}{64D} \tag{7.53}$$

where

$$D = \frac{E.h^3}{12\left(1 - v^2\right)} \tag{7.54}$$

a is the membrane radius, and P is the uniform applied pressure on the membrane. Using $E = 9 \cdot 10^5$ Pa and $v = 0.5$, D is obtained through (7.54): $D = 1/8 \cdot 10^{-7}$ and we find $w_{max} = 10 \, \mu$m for an applied magnetic pressure of 128 Pa. This deflection is just at the upper limit of the linear regime given by (7.52). It is immediate to see that the limit of the linear regime is given by

$$P \leq \frac{13Dh}{a^4} \tag{7.55}$$

At this stage, the remaining question is how to calculate the applied magnetic pressure. An approximate solution is to use the formula previously derived for a magnetic microparticle alone (7.8) and to make the summation over all magnetic particles located inside the membrane. By doing this, we neglect the field interaction between the magnetic particles. However, this approximation is expected to be reasonably good because (1) the particles are very small—the interaction forces are weak and very local—and (2) the applied external field is perpendicular to the membrane, whereas the interaction forces between the particles are contained in the plane of the membrane. Because the size of the membrane is generally small compared to the scale of the applied external magnetic field, we can assume a uniform magnetic gradient and we obtain

$$\vec{F}_{mag} \approx \mu_0 \sum_{grains} v_p \left(\vec{M} \cdot \nabla \right) \vec{H} = \mu_0 N v_p \left(\vec{M} \cdot \nabla \right) \vec{H} \tag{7.56}$$

where v_p is the volume of each magnetic microparticle, and N is the number of particles embedded in the membrane. Let f be the ratio between the volume of the magnetic material and the total volume of the membrane, then $fV = Nv_p$, and the magnetic force on the membrane is:

$$\vec{F}_{mag} \approx \mu_0 fV \left(\vec{M} \cdot \nabla \right) \vec{H} \tag{7.57}$$

where V is the total volume of the membrane. The magnetic pressure is then given by the expression

$$P_m = \frac{\vec{F}_{mag}}{S} \approx \mu_0 fh \left(\vec{M} \cdot \nabla \right) \vec{H} = \frac{\mu_0}{2} fh\chi_p \nabla H^2 \tag{7.58}$$

Using (7.53), we find that the membrane deflection is then linked to the gradient of the square of the magnetic field by

$$w_{max} \approx \frac{\mu_0 f\chi_p \left(1 - v^2 \right) a^4}{10Eh^2} \nabla H^2 \tag{7.59}$$

The membrane maximum deflection is proportional to the content in magnetic particles f, to the magnetic susceptibility of the particles χ_p, to the fourth power of the radius a^4, and to the gradient of the square of the magnetic field ∇H^2. It is inversely proportional to the Young modulus E and to the square of the membrane thickness h^2.

Because micromagnets or microelectromagnets do not deliver an important magnetic field—and consequently an important gradient—the efficiency of a magnetic micromembrane depends on the use of very magnetizable particles that can be packed in the PDMS matrix without increasing too much the Young modulus and Poisson coefficient. To this extent, it is likely that micromembranes containing nanoparticles will be more efficient than that containing metal microplates.

7.12.3 Oscillation of Magnetic Membranes

An application of magnetic micromembranes in biotechnology is the mixing of fluids. We analyze here briefly the principle of actuation of micromembranes by an oscillating (or pulsating) magnetic field.

7.12.3.1 Paramagnetic Membranes

In the previous section, we saw that the deflection of a paramagnetic membrane is proportional to the gradient of the square of the external magnetic field ∇H^2. Suppose now that the source of the external magnetic field is an electromagnet and that this magnetic field is periodically reversed. Changing $\vec{H}$ into $-\vec{H}$ leaves the term ∇H^2 unchanged. The magnetic force on a paramagnetic membrane is always attractive, and the deflection of the membrane is always directed toward the source of the external magnetic field.

Another way of looking at this phenomenon is to consider Langevin's law for paramagnetic media (Figure 7.37). When the field is reversed to its opposite ($\vec{H}$ is changed into $-\vec{H}$), the magnetization is changed to its opposite ($\vec{M}$ is changed into $-\vec{M}$), and the term $(\vec{M} \cdot \nabla)\vec{H}$ in the expression of the force remains identical.

7.12.3.2 Ferromagnetic Membranes

Ferromagnetic membranes show a different behavior when placed in an oscillating magnetic field. When the external field vanishes, during a change from $\vec{H}$ to $-\vec{H}$, there is still a remanent magnetization, as shown in Figure 7.38.

Oscillating magnetic fields are obtained by using miniaturized electromagnets, and the delivered magnetic fields are often weak, so that the magnetization of the

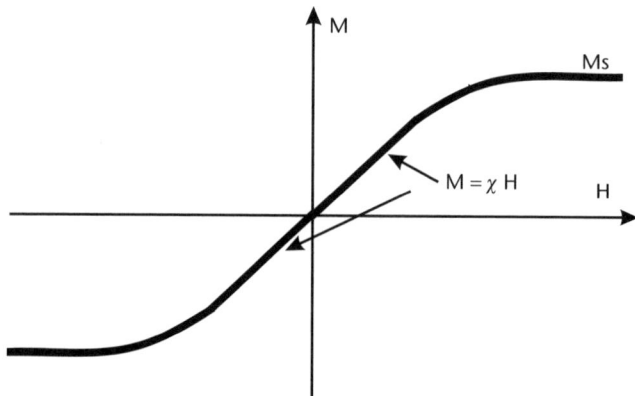

Figure 7.37 Langevin's law for paramagnetic material. Magnetization is aligned with the magnetic field.

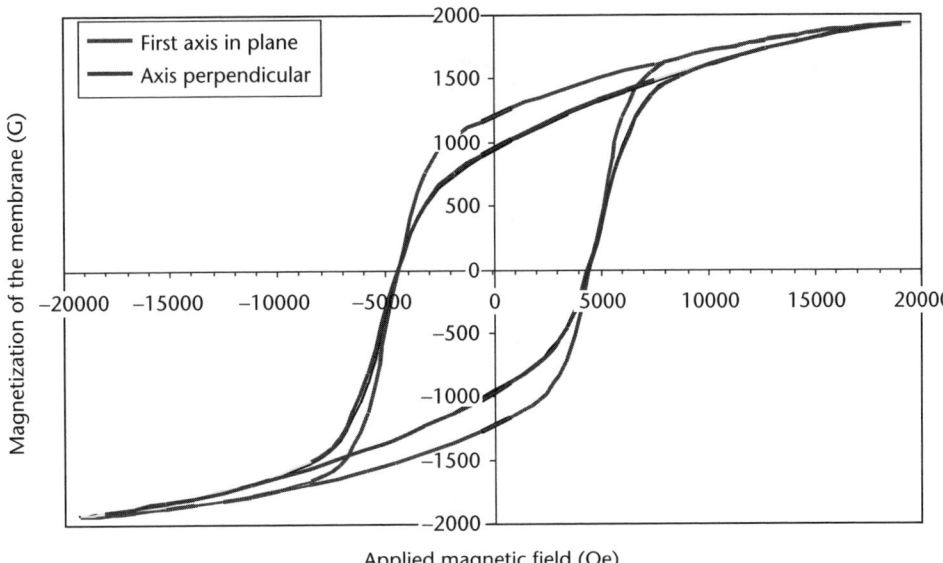

Figure 7.38 Magnetic moment of a ferromagnetic membrane, measured in the plane of the membrane and along the axis perpendicular to the membrane.

ferromagnetic particles remains close to the remanent magnetization $\vec{M}_r$. In this case, the expression of the force is approximately

$$\vec{F}_{mag} \approx \frac{\mu_0}{2} f \nabla\nabla \left(\vec{M}_r \cdot \vec{H} \right) \tag{7.60}$$

When the magnetic field is reversed, the force on the membrane is also reversed, and the membrane oscillates with the external field.

7.12.3.3 Example

In order to obtain a larger deflection, ferromagnetic micromembranes have been fixed perpendicularly to a solid wall and oscillate like a "fish tail" under the actuation of a pulsating magnetic field (Figure 7.39). It has been shown that such motion greatly enhances the mixing in the microchamber [30].

7.13 Conclusion

In the miniaturization trend of biotechnological microsystems, microflows are often becoming not selective or specific enough to correctly perform their expected role of carrier. Functional magnetic microbeads are then used as additional carriers, bringing the required selectivity. This is why magnetic beads have been found to be a very powerful complementary tool for manipulating biological objects, and they appear now to be one of the most useful tools in biotechnology.

Magnetic beads can bind to many different target molecules by the principle of functionalization, and they can unbind by elution, so they can be used as temporary

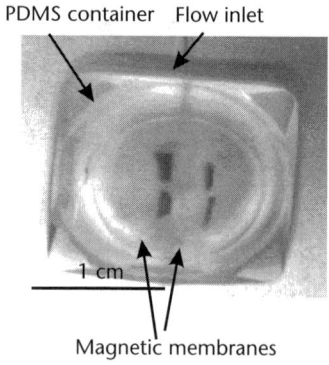

PDMS container   Flow inlet

1 cm

Magnetic membranes

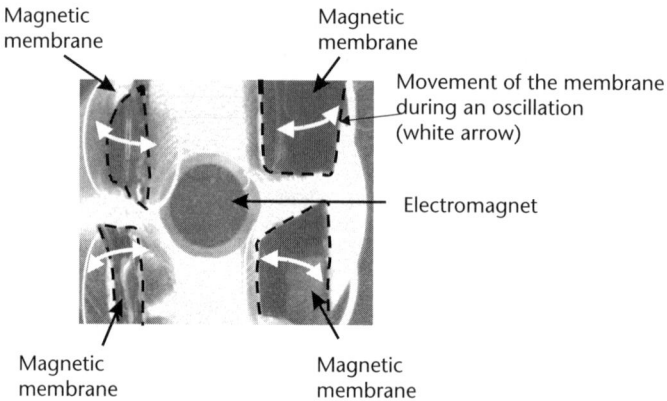

Magnetic
membrane

Magnetic
membrane

Movement of the membrane
during an oscillation
(white arrow)

Electromagnet

Magnetic
membrane

Magnetic
membrane

Figure 7.39   A view of the microchamber with four micromembranes, and the inlet capillary can be seen at the top of the picture (top); view of the membranes during the oscillations—the electromagnet is the round disk behind the membranes (bottom). (Courtesy of F. Ricoul and A. Vigier, CEA/LETI.)

carriers. The applied force on magnetic beads can be precisely controlled by the action of an external magnetic field, leading to a simplified design of the microsystem. Moreover, with the combination of magnetism and fluorescence, beads are easily tracked and detected under the microscope.

Magnetic beads can be assembled to form new biotechnological tools like magnetic chains, magnetic fluids, and magnetic membranes. Magnetic chains can be arranged to form sieving matrices to separate long DNA strands; magnetic fluids are used primarily to perform micropumping; and magnetic membranes are well suited to enhance mixing in microsystems.

References

[1]   Xie, X., et al., "Preparation and Application of Surface-Coated Superparamagnetic Nanobeads in the Isolation of Genomic DNA," *J. Magnetism and Magnetic Materials*, Vol. 277, No. 1-2, June 2004, pp. 16–23.

[2]   Gu, H., et al., "Facile One-Pot Synthesis of Bifunctional Heterodimers of Nanoparticles: A Conjugate of Quantum Dot and Magnetic Nanoparticles," *JACS*, Vol. 126, 2004, p. 5664.

[3]   Portis, A. M., *Electromagnetic Fields: Sources and Media*, New York: John Wiley and Sons, 1968.

[4]   Chantrell, R. W., J. Popplewell, and S. W. Charles, "Measurements of Particle Size Distribution Parameters in Ferrofluids," *IEEE Trans. on Magnetics*, Vol. MAG-14, No. 5, September 1978, pp. 975–977.

[5]   Oberteuffer, J. A., "Magnetic Separation: A Review of Principles, Devices, and Applications," *IEEE Trans. on Magnetics*, Vol. MAG-10, No. 2, June 1974, pp. 223–238.

[6]   Aharoni, A., "Traction Force on Paramagnetic Particles in Magnetic Separators," *IEEE Trans. on Magnetics*, Vol. MAG-12, No. 3, May 1976, pp. 234–235.

[7]   Lawson, W. F., W. H. Simmons, and R. P. Treat, "The Dynamics of a Particle Attracted by a Magnetized Wire," *J. Applied Physics*, Vol. 48, No. 8, August 1977, pp. 3213–3224.

[8]   Kelland, D. R., "Magnetic Separation of Nanoparticles," *IEEE Trans. on Magnetics*, Vol. 34, No. 4, July 1998, pp. 2123–2125.

[9]   Zimmels, Y., "Effect of Concentration and Characteristic Distributions on Electromagnetic Separation of Polydisperse Mixtures," *IEEE Trans. on Magnetics*, Vol. MAG-20, No. 4, July 1984, pp. 597–607.

[10] Schlick, T., *Molecular Modeling and Simulation*, New York: Springer, 2000.

[11] Gerber, R., "Theory of Particle Capture in Axial Filters for High Gradient Magnetic Separation," *J. Phys. D: Appl. Phys.*, Vol. 11, 1978, pp. 2119–2129.

[12] Pham, P., P. Massé, and J. Berthier, "Numerical Modeling of Superparamagnetic Sub-Micronic Particles Trajectories Under the Coupled Action of 3D Force Fields," *European Physical J.*, Vol. 12, 2000, pp. 211–216.

[13] Zebel, G., "Deposition of Aerosol Flowing Past a Cylindrical Fiber in a Uniform Electric Field," *J. Colloid Science*, Vol. 20, 1965, pp. 522–543.

[14] Berthier, J., et al., "Magnetic Confinement of Paramagnetic Micro and Nano-Particles Away from Solid Walls," *IEEE Trans. on Magnetics*, Vol. 38, No. 2, March 2002, pp. 913–916.

[15] Berthier, J., et al., "Numerical Modeling of Paramagnetic Microparticles Trajectories in a Densely Packed Ferromagnetic Wire Bundle," *CFEC Conference*, Milwaukee, WI, 2000.

[16] Davies, L. P., and R. Gerber, "2D Simulation of Ultra-Fine Particle Capture by a Single-Wire Magnetic Collector," *IEEE Trans. on Magnetics*, Vol. 26, No. 5, September 1990, pp. 1867–1869.

[17] Beckett, R., et al., "Measurement of Mass and Thickness of Adsorbed Films on Colloidal Particles by Sedimentation Field-Flow Fractionation," *Langmuir*, Vol. 7, No. 10, 1991, pp. 2040–2047.

[18] Rheinlaender, T., et al., "Different Methods for the Fractionation of Magnetic Fluids," *Colloid Polym. Sci.*, Vol. 278, 2000, pp. 259–263.

[19] Rheinlaender, T., et al., "Comparison of Size-Selective Techniques for the Fractionation of Magnetic Fluids," *J. Magnetism and Magnetic Materials*, Vol. 214, 2000, pp. 269–279.

[20] Berthier, J., P. Pham, and P. Massé, "Numerical Modeling of Magnetic Field Flow Fractionation in Microchannels: A Two-Fold Approach Using Particle Trajectories and Concentration," *MSM 2001 Conference*, Hilton Head Island, SC, March 19–21, 2001.

[21] Alward, J., and W. Imaino, "Magnetic Forces on Monocomponent Toner," *IEEE Trans. on Magnetics*, Vol. MAG-12, No. 2, 1986, pp. 128–133.

[22] Eisenstein, I., "Magnetic Separators: Traction Force Between Ferromagnetic and Paramagnetic Spheres," *IEEE Trans. on Magnetics*, Vol. MAG-13, No. 5, September 1976, pp. 1646–1648.

[23] Furst, E. M., et al., "Permanently Linked Monodispersed Paramagnetic Chains," *Langmuir*, Vol. 14, No. 26, 1998, pp. 7334–7336.

[24] Doyle, P. S., et al., "Self-Assembled Magnetic Matrices for DNA Separation Chip," *Science*, Vol. 295, March 2002, p. 2237.

[25] Perez-Castillejos, R., et al., "The Use of Ferrofluids in Micromechanics," *Sensors-and-Actuators-A-(Physical)*, Vol. A84, No. 1-2, August 1, 2000, pp. 176–180.

[26] Rosensweig, R. E., *Ferrohydrodynamics*, New York: Cambridge University Press, 1985.

[27] Berthier, J., and F. Ricoul, "Numerical Modeling of Ferrofluid Flow Instabilities in a Capillary Tube at the Vicinity of a Magnet," *Technical Proc. 2002 International Conference on Modeling and Simulation of Microsystems*, Vol. 1, p. 764.

[28] Cooney, C. G., and B. C. Towe, "A Thermopneumatic Dispensing Micropump," *Sensors and Actuators A*, Vol. 116, 2004, pp. 519–524.

[29] Khoo, M., and C. Liu, "Micro Magnetic Silicone Elastomer Membrane Actuator," *Sensors and Actuators A: Physical*, Vol. 89, No. 3, April 15, 2001, pp. 259–266.

[30] Berthier, J., and F. Ricoul, "Development of Magnetic Micromembranes for the Actuation of Fluids in Biological and Chemical Microsystems," *Applied Nanoscience*, Vol. 1, 2004.

[31] Timoshenko, S., and S. Woinowski-Krieger, *Theory of Plates and Shells*, 2nd ed., New York: McGraw-Hill, 1959.

CHAPTER 8

Micromanipulations and Separations Using Electric Fields

Because of their versatility, electric field–based techniques are the most widely used for bioanalyses, not only for the separation of biological objects through the well-known *electrophoresis*-derived techniques but also for the handling and micromanipulation of particles, cells, or even biomolecules through *dielectrophoresis*. We'll review in this chapter the principles of these techniques with their use, their limitations, and their foreseen applications. With the transfer of microfabrication techniques to this field and the advent of the lab-on-a-chip approach, particularly well suited to many of these experiments, new original tracks are now opened for potential developments. Although we certainly do not pretend to make an exhaustive list of these new routes, we will try to give a few examples illustrative of this approach.

8.1 Electrophoresis

Quite generically, objects dispersed in water bear a surface charge. The origin of this charge is diverse: It may come from the natural ionization of some of these colloids (e.g., proteins) that give them a net charge depending on the pH of the solution, but other phenomena such as the adsorption of charged ions or the intrinsic ionic nature of some of these particles may come into play.

However, it is important to remember that a charged particle should never be considered *naked* in the solvent. It is always surrounded by counter-ions, so that its global charge is strictly zero (electroneutrality). Because of the thermal agitation, this layer of counter-ions is diffuse.

Electrophoresis consists of applying to these charged objects a direct external electric field that sets them into motion. Their mobility is then indicative of some of their characteristics (charge, molecular weight, and so forth). This approach was pioneered by A. Tiselius [1], who got the Nobel Prize in 1948 for his work on protein electrophoresis.

In this part, we will first describe the structure of the counter-ions layer. Then, by increasing complexity, we will describe the electrophoresis of hard spheres, DNA, and finally proteins or other biological objects.

8.1.1 The Debye Layer

Let us first calculate the distribution of the counter-ions around a charged colloidal particle.

This classical calculation starts by writing the Poisson equation that relates the potential ψ to charge density ρ

$$\nabla^2 \psi = -\frac{\rho}{\varepsilon_0 \varepsilon_r} \tag{8.1}$$

where ε_0 is the vacuum permitivity ($\approx 8.85 \cdot 10^{-12}$ $J^{-1}C^2m^{-1}$) and ε_r is the relative permitivity (≈ 80 for water).

At equilibrium, ρ is given by summing the number of charges in the solution. The number n_i of ions of each species is given by a Boltzman distribution

$$n_i = n_{i0} \exp(-z_i e\psi / k_B T) \tag{8.2}$$

where z_i is the valency of the ions, e is the elementary charge (1 eV = $1.6 \cdot 10^{-19}$ C), k_B is the Boltzmann constant ($\sim 1.38 \cdot 10^{-23}$ J $\cdot$ K^{-1}), and T is the temperature; n_{i0} is the number of these ions far from the surface. The charge density is then given by

$$\rho = \sum_i n_i z_i e = n_0 \left(-ze \cdot \exp(-ze\psi / k_B T) + ze \cdot \exp(ze\psi / k_B T)\right) \tag{8.3}$$

in the case of symmetric electrolytes. The Gouy-Chapman model combines (8.1) and (8.3); the boundary conditions are given by the global electroneutrality condition and by the surface charge.

In a plane geometry, this equation can be solved analytically, but the case $e\psi <<$ $k_B T$ is instructive (Debye-Hückel approximation). The linearization of (8.3) leads to the classical expression

$$\psi = \psi_0 \cdot \exp(-\kappa \cdot x) \tag{8.4}$$

where

$$\kappa^2 = 2 \frac{z^2 e^2 n_0}{\varepsilon_0 \varepsilon_r k_B T} \tag{8.5}$$

κ^{-1} is an important length that describes the range of these electrostatic interactions—it is known as the Debye length. ψ_0 is the surface potential, and x is the distance from the surface. Still in the framework of the Debye-Hückel approximation, the potential in the case of spherical particles of radius a is [2]

$$\psi = \psi_0 a \frac{\exp(-\kappa(x-a))}{x} \tag{8.6}$$

Here, x is the distance from the center of the sphere.

It is good at this point to get an order of magnitude of the Debye length, which obviously depends strongly on the salinity: for a monovalent ion ($z = 1$) at a 0.1 M concentration, $\kappa^{-1} \sim 1$ nm; if $c \sim 10^{-7}$ M (which is the minimum that can be reached in ultrapure water where the salinity is determined by the H^+ and OH^- ions), $\kappa^{-1} \sim 1\mu$m. At a distance from the surface larger than κ^{-1}, the particles are neutral (Figure 8.1).

There are more elaborate models that describe this double layer more realistically. In particular, the Stern model considers a more complex profile of the potential: a proximal zone where the counter-ions are immobilized, and a diffuse layer where the earlier discussion is valid again. It is then common to introduce the ζ potential that gives the value of the potential at the point where the shear around the particle becomes significant (Figure 8.2).

Practically, ζ is the only parameter that can be experimentally determined either by electrophoresis for colloids or by other means: electro-osmosis for walls, sedimentation potential or streaming potential (the electric potential developed by the fall of the objects or by flowing water on them) for heavy or tormented particles [2]. Let us note at this point that, in the simple Debye-Hückel description, the ζ potential is located strictly on the surface of the particle ($\zeta = \psi_0$). For a spherical particle (8.6), the total charge q is then given by

$$q = \int_a^\infty 4\pi r^2 \rho dr = 4\pi\varepsilon_0\varepsilon_r a(1+\kappa a)\zeta \qquad (8.7)$$

Equation (8.7) gives a relationship between the surface charge density σ and ζ

$$\sigma = \varepsilon_0\varepsilon_r\zeta(1+\kappa a)/\alpha \qquad (8.8)$$

8.1.2 Electrophoresis of a Charged Particle

What are the consequences of an electric field E on a charged particle? Although it is tempting to express the resulting force acting on the particle in the form $F = qE$, this is generally too simplistic an approach. Given the complex distribution of charges described earlier, this is not surprising. Indeed, the electric field acts first on the counter-ions, which are then driven in the direction opposite to the particle, therefore increasing its hydrodynamic friction. When dealing with small particles, the dissipation is principally viscous. We call η the viscosity of the solvent.

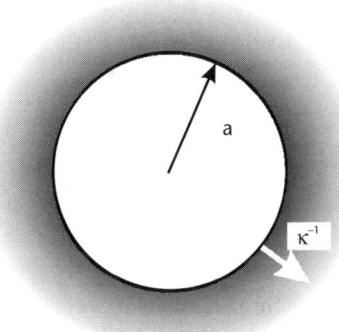

Figure 8.1 Distribution of charges around a spherical particle.

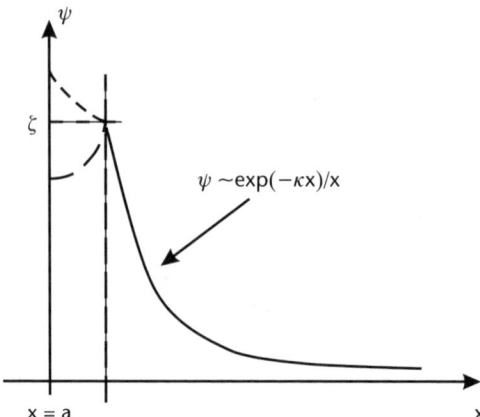

Figure 8.2 Profile of the double layer in the presence of a proximal immobile layer (Stern layer) for a spherical particle. The potential takes the analytical form of (8.4) at a certain distance from the surface of the particle.

For weak electric fields, the velocity V of the particle can then be considered linearly dependent on the electric field

$$V = \mu \cdot E \tag{8.9}$$

where μ is defined as the electrophoretic mobility.

Depending on the salinity and the radius of the particle, two extreme cases can be analytically considered.

If the charge density is high or for large particles, we have $\kappa a \gg 1$, and we can use the *Smoluchowski* calculation on plane surfaces similar to the one derived for electro-osmosis (see Section 8.3) [3, 4]. In this case, μ can be derived from the equilibrium of forces in the double layer, assuming that the charge profile is only marginally modified by the electric field, and we obtain

$$\mu = \frac{\varepsilon \zeta}{\eta} \tag{8.10}$$

In the framework of the Debye-Hückel approximation, plugging (8.7) in (8.10) yields

$$\mu = \frac{q}{4\pi\eta\kappa a^2} \tag{8.11}$$

The net result here is that the mobility depends only on the ζ potential of the particle—not on its size or on its shape. This result is a direct consequence of the confinement of the flow within the Debye layer. The physical mechanism underlying this motion is thus very different from the usual Stokes law, where the bead submitted to an external force such as gravity experiences a viscous drag.

We have so far focused on large particles or small Debye length; we can go to the other extreme case of a very small particle or a very small salinity (large Debye length). We then have $\kappa a \ll 1$, and the particle can be seen as a point-like object surrounded by a very diffuse cloud of counter-ions that extends many radii away from it. In that case, the flow lines extend far from the object over a characteristic length now close to the bead radius. Not surprisingly, given the earlier discussion, this case is then described by the more familiar Hückel calculation and, using (8.7), the mobility is accurately given by

$$\mu = q/6\pi\eta a = 2\varepsilon\zeta/3\eta \qquad (8.12)$$

Between these limiting results, the behavior of the particles is more complex and best described by numerical simulations [4]. Furthermore, assumptions of the previous calculations are often discovered to be incorrect. For instance, the Smoluchowski calculation assumes that the charge distribution around the particle is only marginally disturbed by its motion. However, this is generally not true, as the cloud of counter-ions is deformed by the electric field, and this shape in turn modifies the friction. The same becomes true if the velocity of the particle becomes very high [4].

In other words, out of the framework of the simplistic approximations presented earlier, electrophoresis becomes rapidly a very complicated phenomenon, and it should be kept in mind that the practical interpretation of these experiments is largely empirical.

One last word in this rapid theoretical introduction: We have considered a particle in an infinite medium at rest. However, the suspension is contained in a physical cell. In the general case, the walls of this cell are not passive, but the charge they bear responds to the external electric field by developing electro-osmotic flows (see Section 8.3) [2]. These induced flows superimpose a plug flow to the electrophoretic motion of the particles. They can dramatically affect the performances of the devices and should be well controlled.

Double-strand DNA is the molecule most commonly separated by electrophoresis. We will first focus on this case. However, there are many objects on which electrophoretic separation is used: proteins or single-strand DNA are molecular examples. Cells or organelles are other more macroscopic examples.

8.1.3 Electrophoresis of DNA

DNA is a polyelectrolyte (a polymer chain whose monomers bear a charge). On top of the complexity of charged objects we have just discussed, one must add the one inherent to polymers. We will shortly present it and infer from these results the main characteristics of DNA electrophoresis.

8.1.3.1 Polymer Chains in Solution

We saw in Chapter 6 that in dilute solutions, DNA chains adopt a coil configuration whose radius, called the radius of gyration R_g, is directly related to the size of the monomers b and their number N through the relation

$$R_g = b \cdot N^{1/2} \qquad (8.13)$$

To understand the behavior of a polyelectrolyte chain in an electric field, we can model it by a succession of charged beads linked together by springs (Rouse model) [5]. The friction of each bead with the solvent is $6\pi\eta a$. Without getting into the full calculation, it turns out that the coil is transparent to the hydrodynamic flow, and the friction of the chain in the liquid comes exclusively from the individual frictions of the beads [6]. The effective friction over the whole length of the chain is then proportional to the number of beads and thus to its length. As the electric force applied to the chain is also proportional to its length, the net result is that the mobility of a long polyelectrolyte is independent of its length (similar to the fact that it does not depend on the size or on the shape of the particle in the case of the electrophoresis of a hard sphere) [7].

Now we know that a polymer chain is not a succession of independent beads. However, more refined models, and in particular the ones derived from the Zimm description where the hydrodynamic interactions between beads are taken into account, show that the added terms induce only small deviations to this law [7].

Practically, we can conclude that no separation of long DNA will occur by simply applying an electric field to a solution of these molecules.

8.1.3.2 Gel Electrophoresis

To achieve a separation, it is then necessary to use a sieving matrix that slows down the largest molecules. Most commonly, this is a gel made from polyacrylamide or agarose, whose mesh size (from a few tens of nanometers to a few hundred nanometers) conditions the separation.

The migration of small molecules within the gel can be modeled by computing the free volume available to the particle [8, 9]. This volume depends on the gel characteristics as well as on the particle effective volume, and one gets ultimately the following expression for the mobility

$$\mu \sim 1/\log N \tag{8.14}$$

This equation is indeed well verified and routinely used in a limited range of molecular weights.

However, longer molecules deform in the gel and take the average orientation of the electric field (Figure 8.3). They can be modeled by reptating chains within an array of obstacles, and their mobility becomes independent of the molecular weight again [10].

Practically, the solution containing the mixture to be analyzed is deposited at one extremity of a gel slab. The whole slab is immersed in the buffer, and an electric field (up to ~100 V/cm) is applied via electrodes immersed in it. In these experiments, the electrodes are physically separated from the sample (although in electrical contact with it via the buffer solution); thus, the electrochemistry at their surface (electrolysis in particular) has no consequence. After applying the electric field for a certain time, the molecules are stained and appear as bands (Figure 8.4).

For molecules too large to be separated by direct gel electrophoresis, pulsed field electrophoresis in gel is the preferred technique [11]. In these experiments, the orientation of the electric field is periodically switched between two values. The angle between these two directions can be finite (crossed field electrophoresis) or the value

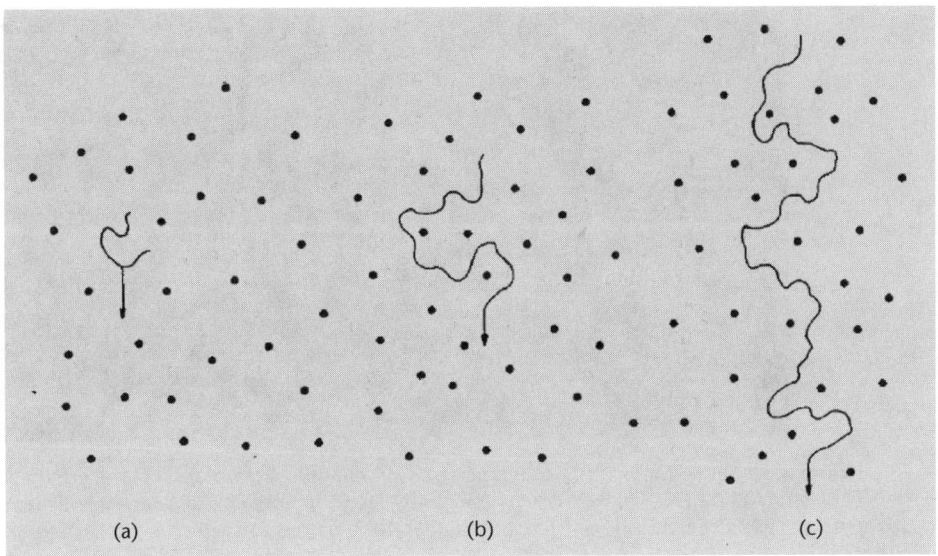

Figure 8.3 Electrophoretic migration of a DNA chain in a gel for three different DNA sizes: (a) Ogston regime; (b) and (c) reptating chains in the gel. (*From:* [12]. © 2000 American Physical Society. Reprinted with permission.)

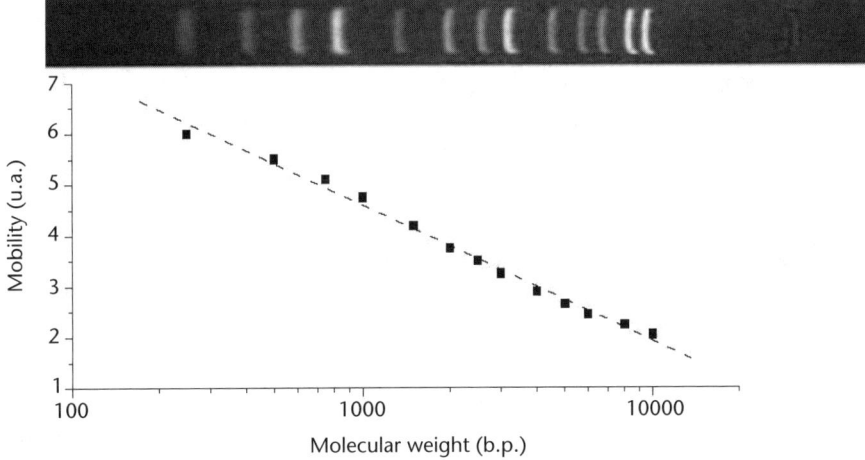

Figure 8.4 Electrophoregram of a *DNA ladder* showing the logarithmic dependence of the mobility on the molecular weight (8.14).

of 180° can be taken, so that the electric field changes its direction along the same orientation (field inversion electrophoresis). This approach is based on transient states, where the long molecules do not have time to fully extend in the direction of the electric field before its change of direction, whereas small molecules do (Figure 8.5). Therefore these small molecules can reach a steady state in which they are elongated but switch their direction with the field. Long molecules that spend most of their time orienting themselves along the field are considerably slowed down. This technique has been a breakthrough in the separation of long molecules (up to 10 Mbp) [13]. Extremely efficient, it is, however, also very time-consuming.

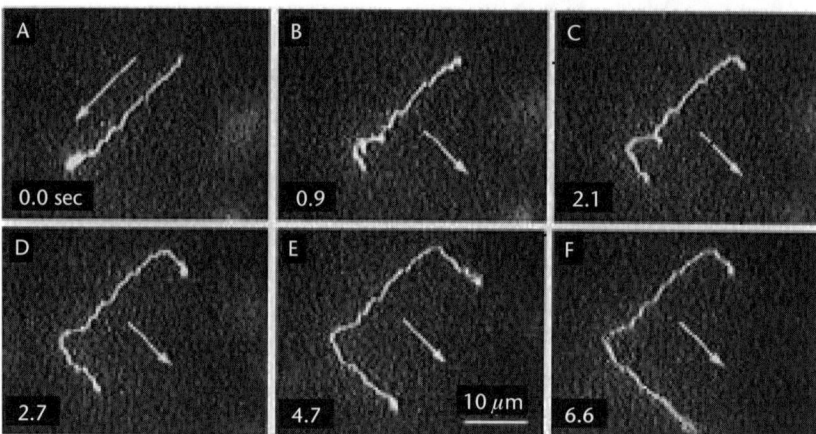

Figure 8.5   Single-molecule DNA image at a 120° change of orientation of the electric field (arrow) between (A) and (B). To allow for this single-molecule image, DNA is fluorescently labeled. (*From:* [14]. © 1996 Oxford University Press. Reprinted with permission.)

8.1.3.3   Capillary Electrophoresis, Electrophoresis on Chip

Capillary electrophoresis is a gel-free technique. In this case, the gel slab is replaced by a capillary whose typical dimensions are 100 μm in diameter and 50 cm in length. Because of the geometry, high electric fields can be used (up to 1 kV/cm). This strategy is much easier to automate and parallelize than the gels and has been massively used in *brute force* sequencing of various genomes, including the human genome. The sieving matrix in this case is a polymer solution that acts similarly to a gel. The main difficulty in using this approach is electro-osmosis, which superimposes to the electrophoretic motion and which is quite detrimental to the resolution. This effect is very dependent on the details of the chemistry of the surfaces, and it is necessary to "hide" them to the electric field (e.g., by chemically grafting or by adsorbing neutral polymers on it).

A strong tendency nowadays is to integrate these capillaries within "chips," where all of the dimensions can be reduced and that have on the same chip the injection, separation, and detection steps [15]. With this miniaturization of the device comes naturally the diminution of the volume used for the analysis, another extremely positive point. Furthermore, heat dissipation is even better than for standard capillaries, enabling higher fields to be used. As mentioned in the preceding chapters, the fabrication of these channels uses a technology stemming from the microelectronics industry. Many different materials can be used for the fabrication of the chips (e.g., plastics or plastic-silicon hybrids), although it should be stressed that silicon, the preferred material in microfabrication, is generally too conductive to be compatible with the high electric fields required for this particular application. When it arrives at full maturity (which can be forecast within the next years), this approach is expected to lead to disposable, inexpensive devices. However there are still hurdles to overcome: the use of new materials induces new constraints to fight electro-osmosis, and there seems to be a limit to the electric fields that can be used as values that are too high can have dramatic consequences, such as the aggregation of long DNA chains by electrokinetic effects [16].

8.1.3.4 New Exploratory Fields

Besides the lab on a chip electrophoresis approach, micro- and nanofabrication has enabled new ideas and new concepts aiming at the separation of DNA molecules and more and more proteins.

Artificial Gel

As early as 1992, there have been experiments where the gel commonly used as a separation sieving matrix was replaced by a solid-state microfabricated array of pillars [17]. The photolithography-derived techniques enable the fabrication of micron-sized posts separated by a few microns, dimensions close to the gel's pore size used for long DNA molecules. Combined with the ability to visualize single DNA molecules, these experiments have pioneered a whole school of experiments, where the behavior of the chains could be unambiguously correlated to their extension, hooking and releasing from the posts, and so forth. They constitute an invaluable help in the comprehension of the phenomena. Indeed, the first experiments of migration of DNA subjected to a low electric field in such arrays of posts have shown a lot of common features with their behavior in gels. Although this strategy uses well-known techniques derived from the microelectronics industry for the fabrication of the substrates, it has not yet come to the point where it is used for routine analysis. Similarly to gels where the range of accessible molecular weights is dictated by the gel's pore size and thus its degree of crosslinking, the efficiency of these devices is conditioned by the size and spacing of the posts, parameters that can be tuned during the fabrication step but that are not as readily accessible as for a gel. However, for research purposes, this approach has been very successful and is no doubt a tool that can and will be used in particular situations.

Indeed, after these proofs of concept experiments, pulsed-field electrophoresis has been performed in two-dimensional hexagonal arrays of pillars with extremely good results (see Figure 8.6). Because of the high regularity of the array, DNA chains move more regularly than in a gel, and sequences of electric field turned by 120° enable extremely fast separation of long molecules (up to a few hundred kilobasis pairs in a few tens of seconds, which is 100 to 1,000 faster than pulsed field gel electrophoresis) [18]. Again, it seems relatively straightforward to integrate

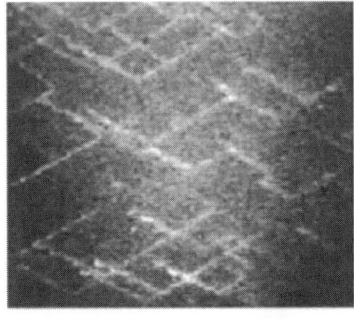

30 μm

Figure 8.6 Pulsed field electrophoresis in a hexagonal array of posts (diameter and spacing around 1 μm). Picture taken after a 120° change in orientation of the electric field. (*From:* [18]. © 2001 American Chemical Society. Reprinted with permission.)

such structures in on-chip separation devices. More recently, these monolithic posts have been replaced by pillars made of magnetic beads, which align in a regular hexagonal array upon the application of a magnetic field. This strategy has some obvious advantages (no microfabrication and easy replacement of the medium); however, it does not allow a range of variation in the spacing between posts as large as would be desirable [19].

Nanopores and Nanochannels

Not only can DNA and other biopolymer chains be forced through a microchannel or a microaperture, they can be threaded through a nanopore. In the present version of these experiments, single-strand DNA is driven by an electric field through a single biological pore, the α-hemolysin channel that has been previously isolated and inserted in a lipid membrane [20]. The structure of this proteic complex enables the translocation of one single-strand DNA at a time (see Figure 8.7). The electric field that drives the chain in the pore is also used for its detection: When the DNA is inside the pore, it effectively blocks the ionic electrical current increasing the electrical resistance. By analyzing the details of this time-resolved resistance, one can get the length of the molecule (by the duration of the pulse) and some crude information on its sequence (by the value of the resistance). It is hoped that, at some point, this technique will be used for ultrafast DNA sequencing. To achieve this goal, it is necessary, however, to develop a solid-state alternative to the α-hemolysin pore. Intense work is presently devoted to drill nanometer diameter holes in thin inorganic membranes with sophisticated nanotechnology. The next generation of these devices is expected to be more than simple pinholes. They should also carry in the same plane some kind of very local detection set up. This can be molecular, with the chemical grafting of molecules responding to the passing of some of the nucleotides (e.g., by fluorescence energy transfer) or, maybe more realistically, in the form of transverse electrodes embedded in the membrane, which will allow us to achieve a single base pair resolution through the measurement of a tunneling current.

Still in the nanoworld, nanochannels are the other field of application of this *nanoelectrophoresis*. In this case, double-strand DNA is forced into nanochannels or nanoconstrictions with the electric field and observed by fluorescence microscopy. Currently, several techniques can be used for this nanofabrication. The chains

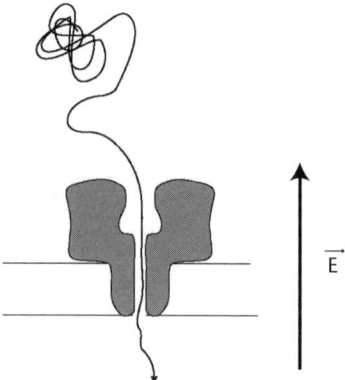

$\vec{E}$

Figure 8.7 Schematic of a DNA strand translocating a α-hemolysin protein embedded in a lipid bilayer. The drawing is not to scale. The structure of this protein is represented Figure 6.5.

can considerably stretch in these structures (up to 60% of their fully extended length), and it is observed that their extended length is proportional to the number of base pairs. By a simple averaging of this projected length, an extremely accurate determination of the molecular weight can be obtained [21]. Another foreseen application of this confinement is the precise localization of proteins bound to the chain. If the protein and the DNA are labeled with fluorescent dyes of different colors, a fluorescence image can reveal very rapidly and accurately the position of this protein along the chain.

In the same spirit, structures alternating micro- and nanoconstrictions (respectively, 1.5 μm and 90 nm) have been developed that induce different confinement regimes for the DNA molecules [22]. Upon the application of the electric field, molecules experience many transitions, from thick regions to regions smaller than the radius of gyration.

Although it was predicted that long molecules would be retarded by the constrictions by an entropic barrier, practically, their mobility turns out to be higher. A qualitative interpretation involves a large deformation of the coil. Strong interactions with the surface in these ultraconfined geometries are probably also to be taken into account [23].

8.1.4 Electrophoresis of Proteins

After the genomics era, and now that more and more genomes have been sequenced, proteomics that aims at identifying the proteins expressed in living organisms is the next big challenge [24]. From a chemical physics point of view, proteins are not as well defined as DNA, as their characteristics of charge and hydrophilicity varies a lot. Besides, the separation of these biomolecules takes a different meaning than that of DNA, where the problem is to discriminate between identical objects differing only by their molecular weights. For proteins, completely different molecules differing by their primary sequence and therefore their length, charge, hydrophilicity, and so forth have to be analyzed. The determination of the size is performed in polyacrylamide gels in the presence of a surfactant, and pH gradients are used to measure their charge. Two-dimensional gels combine these two techniques.

8.1.4.1 Sodium Dodecyl Sulfate Polyacrylamide Gel Electrophoresis

This technique is used for separating proteins according to their size. As they are globular charged objects taking various shapes, it is first necessary to unfold them. This step is performed in a denaturing solution of sodium dodecyl sulfate (SDS). This charged surfactant efficiently unfolds proteins so they can be considered classical flexible polyelectrolytes whose charge is imposed by the bounded SDS. Since, on average, all the proteins have the ability to fix a similar amount of SDS molecules (per unit length), all of the proteins of the mixture can be assimilated to the same polyelectrolyte, differing only in length, even though their primary sequence is different. Polyacrylamide gel electrophoresis (PAGE) of these SDS-protein complexes is then conceptually similar to gel electrophoresis of DNA and is heavily used, in particular the Ogston regime ($\mu \sim 1/\log(M)$) (8.14).

8.1.4.2 Isoelectric Focusing (IEF)

Because of the acidic and amine groups present in their structure, all proteins bear a net positive charge at low pH and negative charge at high pH. The pH value at which their charge is zero is called the isoelectric point. It is measured by creating a pH gradient in a gel and then measuring the point where the net charge of the proteins is strictly zero. When an electric field is applied to the gel, the mobility changes its sign at this isoelectric point. A given protein is thus electrofocused and accumulates at this point.

8.1.4.3 Two-Dimensional Electrophoresis

Electrophoresis can thus be used independently for molecular weight determination or for measurement of the isoelectric focusing (IEF). In these complex systems, it is common to combine these two determinations by making two-dimensional electrophoresis (2D PAGE) [25].

The principle is to make an IEF measurement over a pH gradient, say, along the horizontal direction. This step sorts the proteins with respect to their charge (Figure 8.8). The slab is then submitted to an SDS PAGE along the vertical axis that separates all of the proteins of the same spot (i.e., having the same charge) with respect to their molecular weight.

These analyses are performed routinely. Despite the complex patterns due to the high diversity of the proteins of a particular sample (Figure 8.9), they give good and reliable results (e.g., in following the evolution of the expression of a given protein in different cell environments via the intensity of its related spot in the electrophoregram).

8.1.5 Cell Electrophoresis

Getting again more complex than proteins, cells are difficult to separate with electrophoresis. The main reason is that, like for proteins, size is only one part of the

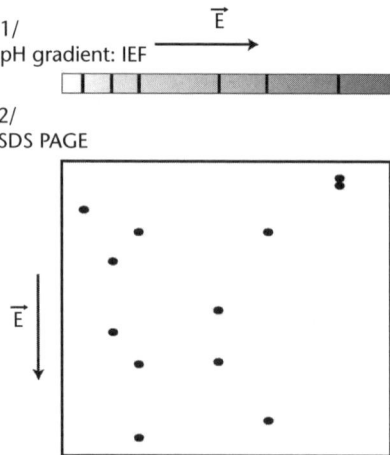

Figure 8.8 Principle of two-dimensional protein electrophoresis. The first step is an IEF determination (charge sorting); the second step an SDS PAGE (molecular weight sorting). In the final electrophoregram, each spot is a single protein.

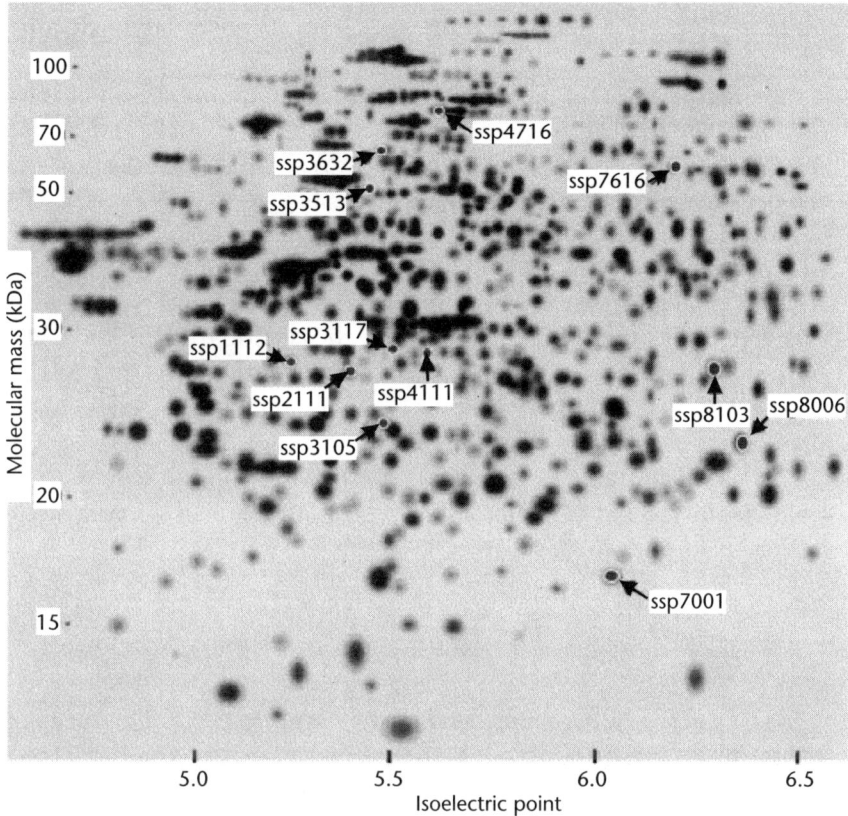

Figure 8.9   Raw two-dimensional SDS PAGE of human diploid fibroblast. The spots indicated by an arrow show significant difference in relative abundance between sick and normal fibroblasts. (*From:* [26]. © 1998 Elsevier. Reprinted with permission.)

information. The task is usually to separate cells that have a distinctive property from other cells. The equivalent of SDS has not been found for cells, so even a size separation does not give very good results. Besides, gels are not well suited for these large objects. Therefore, electrophoresis does not seem to be the technique adapted to this problem. Dielectrophoretic measurements, described in the next section, are more suitable.

8.2   Dielectrophoresis

In contrast with the previously described electrophoresis, dielectrophoresis is not a transport technique, except in some cases that we will review in the end of this part; it is rather a micromanipulation tool. Based on a contrast between the polarizabilities of the particle and the medium, it can be used on any kind of particle, charged or not.

8.2.1   Theoretical Basis: The Dielectrophoretic Force

Although the physics of dielectrophoresis (DEP) has been known and characterized for a long time, this effect has recently regained some interest in the biology-biotechnology community mainly because of its potential when coupled to micro-structures [27].

By definition, DEP is the motion induced by *nonuniform* electric fields and is due to a contrast of polarizabilities between the particle and its solvent. For the interested reader, some good textbooks provide a far more detailed description than the present short chapter [27–29].

Let us consider a particle in a solvent submitted to an electric field. The application of an electric field leads to a charge accumulation at the interface with the surrounding medium. This nonuniform charge distribution creates a dipole, and it is this very dipole that now interacts with the electric field. If the field is different on both sides of the particle, a net force acts upon it that drives it toward the high electric field areas if its polarizability is higher than that of the medium and in the other direction if it is smaller (Figure 8.10).

More quantitatively, the basic Maxwell equations tell us that a particle of polarizability α and of radius a experiences a force F in the presence of an external electric field E given by

$$F = \frac{2}{3}\pi a^3 \alpha \nabla |E|^2 \tag{8.15}$$

The particles we are interested in are *lossy dielectrics*. This means that, on top of the intrinsic permitivity of the particle, one has also to consider its conductivity and the energy dissipated via this ionic conduction. In this framework, expressing the polarizability in (8.15) leads to

$$F = 2\pi a^3 \varepsilon_0 \varepsilon_{r,l} Re(f_{CM}) \nabla E^2 \tag{8.16}$$

where ε_0 is the vacuum permitivity, $\varepsilon_{r,l}$ is the relative permitivity of the solvent, f_{CM} is the so-called Clausius-Mossoti factor, and $Re(f_{CM})$ is its real part

$$f_{CM} = \frac{\varepsilon_p^* - \varepsilon_l^*}{\varepsilon_p^* + \varepsilon_l^*} \tag{8.17}$$

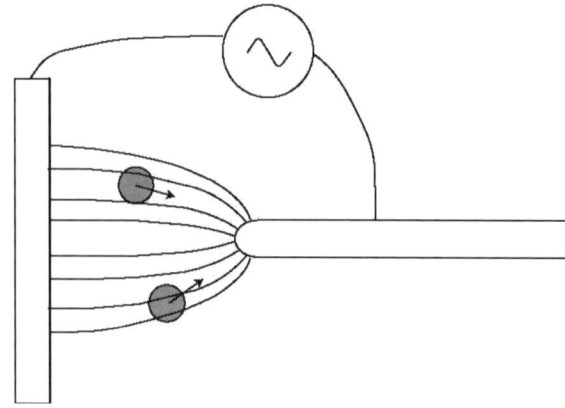

Figure 8.10 Polarizable particles attracted toward high electric field region in positive DEP regime.

where $\varepsilon_p{}^*$ and $\varepsilon_l{}^*$ are respectively the complex permitivities of the particle and the solvent

$$\varepsilon^* = \varepsilon_0 \varepsilon_r - j\frac{\sigma}{\omega} \qquad (8.18)$$

where ε_r is the relative permitivity, σ is the conductivity, and ω is the frequency of the electric field.

We also note $\varepsilon_l = \varepsilon_0 \varepsilon_{r,l}$ and $\varepsilon_p = \varepsilon_0 \varepsilon_{r,p}$.

Several consequences can be immediately derived from this expression:

- The direction of the force exerted on the particle depends on the sign of the real part of the Clausius Mosssoti factor: if $Re(f_{CM}) > 0$, the particle is attracted to the regions where the field is maximum. This is what is usually called positive dielectrophoresis. In the other case, it is repelled from these areas (actually, it is the solvent that is attracted to these areas, forcing the particles to be repelled from them). It is the negative DEP case where the particles appear to be driven toward the low electric field areas.

- Another interesting point is that $F \sim \nabla E^2$. This dependence with the square of the amplitude of the electric field implies that DEP arises with *dc* or *ac* fields. In the first case, however, electrophoresis will compete with DEP in the motion of the particles. The use of high frequency ac fields is particularly welcome because it suppresses electrolysis or, more generally, electrochemistry at the surface of the electrodes.

- The force depends on the gradient of the electric field intensity. Therefore, to transport particles over large distances, it would be necessary to maintain a large gradient over such distances. This requires large electric fields (and thus high voltages) in the macroscopic world. On the other hand, large local gradients are more easily created in microstructures (small scales). This explains why, with the development of microfabrication and its use in life science, this effect has been effectively rediscovered recently to manipulate, characterize, or sort particles. However, to transport them over large distances remains a challenge. As we will see later in the text, a traveling wave or a succession of elementary displacements triggered by gradients over small distances seems the best approach.

One last word before detailing the consequences of these expressions: the DEP traps are formed and conditioned by the geometry of the electrodes that are *in the solution* (except for the electrodeless DEP detailed further in this chapter). This is markedly different from electrophoresis, where the electrodes are placed outside of the channel (although in electrical contact with it). Working with high frequencies minimizes the electrochemistry at the surface; however, we will see later that it can't avoid electrohydrodynamic flows or a direct contact of metal electrodes with the objects to be manipulated. This last point can be very detrimental to the technique because of the interactions of metals with charged objects, in particular through the formation of an electric image that leads to irreversible sticking [28, 30]. Care has thus to be taken to coat these electrodes to minimize these effects [28, 30, 31].

8.2.2 The Clausius-Mossoti Factor

If we want to go one step further in the calculation of the force induced by the electric field, we have to expand the Clausius-Mossoti factor. This discussion is very different, depending on whether we deal with low frequencies or high frequencies.

At low frequencies (lower than 10 kHz), there is some dispersion of the dielectric constant of the particle. That is, ε_p is actually a function of the frequency of the electric field [32]. This dispersion is mainly due to the relaxation time of the polarization of the double layer surrounding the particle (this time depends on the radius of the particle and on the Debye length). The situation is quite complex, and no satisfactory model fully describes all of the observed phenomena [33].

At higher frequencies, the counter-ions of the double layer do not have enough time to move, and the particle is basically nondispersive. This means that ε_p and σ_p are independent on the frequency. The polarization is then only due to the contrast in dielectric constants between the particle and its surrounding medium. This interfacial polarization (Maxwell-Wagner effect [27]) is described by a unique relaxation time τ that depends only on the permitivities and conductivities of the particle and the medium.

We can then express the real part of the Clausius-Mossoti factor

$$Re(f_{CM}) = \frac{(\varepsilon_p - \varepsilon_l)\omega^2\tau^2}{(\varepsilon_p + 2\varepsilon_l)(1 + \omega^2\tau^2)} + \frac{(\sigma_p - \sigma_l)}{(\sigma_p + 2\sigma_l)(1 + \omega^2\tau^2)} \qquad (8.19)$$

where

$$\tau = \frac{\varepsilon_p + 2\varepsilon_l}{\sigma_p + 2\sigma_l} \qquad (8.20)$$

In this case, $Re\ (f_{CM})$ varies monotonously with the electric field frequency between the extreme values $\dfrac{\sigma_p - \sigma_l}{\sigma_p + 2\sigma_l}$ (for $\omega \to 0$) and $\dfrac{\varepsilon_p - \varepsilon_l}{\varepsilon_p + 2\varepsilon_l}$ ($\omega \to \infty$).

There remains, however, some discrepancies between this expression and the forces quantitatively measured in diluted colloidal suspensions. It becomes important to take into account the ionic double layer that effectively modifies the conductivity of the particles. Some authors have treated this problem by considering an infinitely thin conductive layer on the surface of the particle. Empirically, one then adds a surface conductivity λ_s to the particle conductivity σ_p. The total effective conductivity then becomes [27]

$$\sigma_p' = \sigma_p + 2\lambda_s/a \qquad (8.21)$$

The behavior of the particles is then well described by using σ_p' rather than σ_p. We have plotted on Figure 8.11 the expression of $Re(f_{CM})$ in this high-frequency regime. The low-frequency regime down to dc behavior cannot be described by this curve. Below typically 1–10 kHz, $Re(f_{CM})$ decreases, and the particles may even take a negative DEP behavior.

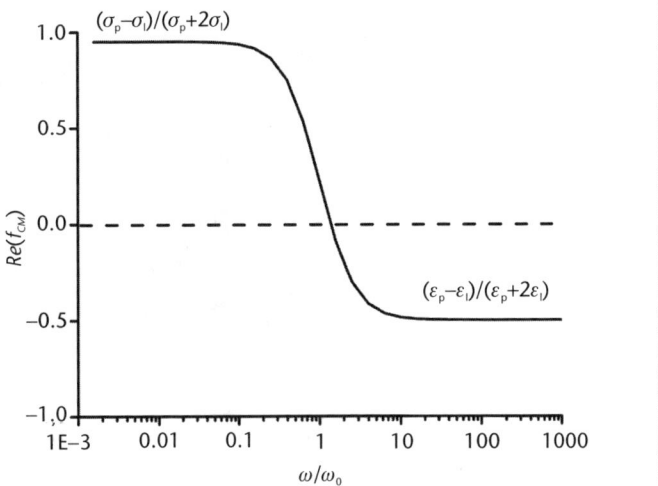

Figure 8.11   Evolution of the real part of the Clausius-Mossoti factor with the frequency. The point ω = ω_0 is the crossover frequency (see more details in Section 8.2.4.3).

8.2.3   Optimization of the Electric Field

8.2.3.1   Electrode Geometries

The exact three-dimensional landscape of the electric field intensity shapes the force acting on the particles. Depending on whether the final application is to trap them, to set them into motion or in rotation, or to combine trapping and another force field (such as an hydrodynamic flow), the constraints on the geometry of the electrodes are different. For instance, one of the early geometries used *castellated electrodes* that provided a good array of traps [34]. The exact calculation of the electric field spatial variations can be routinely performed via a finite elements analysis, and the optimization of their shapes and disposition in space becomes possible. Planar four electrodes geometry can give a precisely located trapping site in negative DEP, but, when possible, three-dimensional arrangements of two sets of four electrodes facing each other give the best trapping cages [35] (Figure 8.12).

8.2.3.2   Electrodeless DEP

Practically, DEP comes inevitably with electrohydrodynamic flows that we will detail later. In microstructures, even for modest applied voltages, electric fields can become extremely high and can easily trigger electrochemical instabilities at the surface of the electrodes. This drawback is even worse when using two-dimensional electrodes in a channel or chamber having some depth (although there are some geometries where the electrodes themselves are shaped in three dimensions: indeed, the instabilities are much reduced in this geometry—see Section 8.2.7.2).

We can avoid some of these problems by using the so called electrodeless DEP, and although it is not as flexible as the usual electrode-based methods, it offers enough advantages to be favorably considered in some circumstances [36–38].

This technique is based on insulating channels having constrictions. If one applies an electric field between the entrance and the exit of the channel, the field

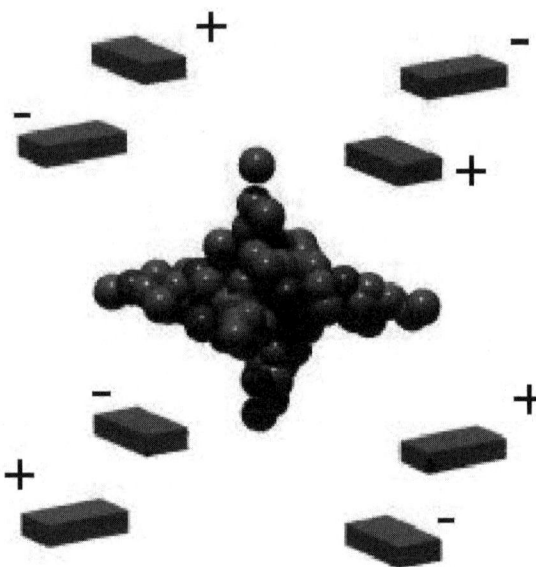

Figure 8.12 Numerical calculation of trapping of particles in an octopole geometry. (*From:* [39]. © 1996 Springer. Reprinted with permission.)

lines have to squeeze in the constrictions, therefore creating high gradients (Figure 8.13). It is thus an easy way to use the DEP effect only by structuring the channel. No electrochemistry is involved, and most of the instabilities disappear. This has a cost, however: using this geometry, only one electric field controls the displacements of the particles in the channel. Therefore, all of the tasks have to be performed sequentially, whereas by using electrodes of the correct geometry, many individual DEP wells can be controlled independently in a parallel way.

8.2.4   Characterization of Particles

8.2.4.1   Real-World Particles

Of course, particles that need to be manipulated or separated, particularly in biomicronano technology, are generally not solid homogeneous spheres. They can actually be extremely diverse and go from DNA molecules [40] to cells [41] or viruses [42]. They are usually modeled by an effective sphere, although multishell models may be more accurate to account for some of the particle characteristics at the price of a higher number of parameters. More than absolute quantitative measurements, studies on complex particles are often relative between slightly different systems.

8.2.4.2   Collection Rate

Now that we know how to apply a force to a particle, and actually design the energy well in which this particle is immersed, we can use this knowledge to characterize it. Historically, the experiments were performed in the positive regime, and the measurement was a collection rate that implicitly assumes that the rate at which the particles accumulate in this well is proportional to the force they experience [27].

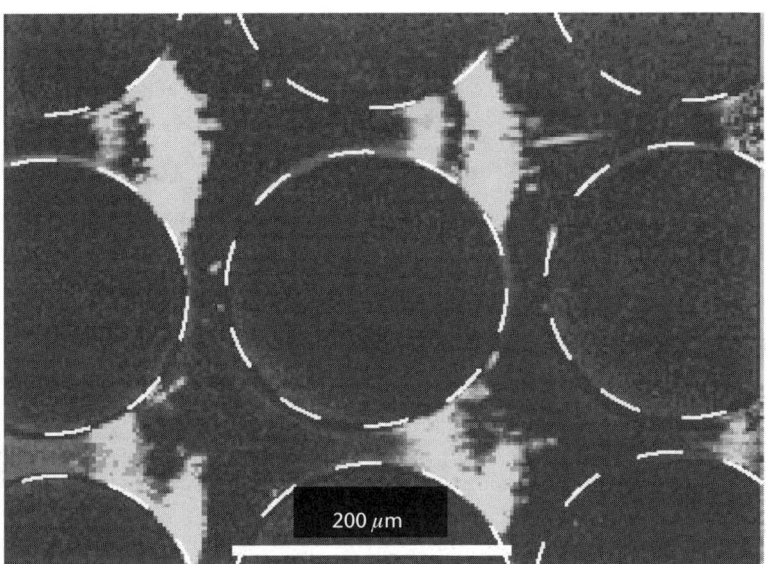

Figure 8.13 Electrodeless DEP of live and dead bacteria. The posts are made of glass, and the electric field is directed from left to right. Because of their different dielectric characteristics, dead bacteria are trapped at a different spot close to the constriction itself compared to live bacteria (in a wider region). (*From:* [38]. © 2004 American Chemical Society. Reprinted with permission.)

Making this measurement at different frequencies gives access to a spectrum of $Re(f_{CM})$ and thus to the various parameters it covers via multiparameters fits.

8.2.4.3 Levitation Height

However, it has been recognized more recently that these experiments in the positive DEP regime were problematic, because of the many experimental difficulties caused by electrohydrodynamic flows (see Section 8.2.6.2). An alternative is to measure the height at which the DEP force can counterbalance gravity and levitate a particle in the negative DEP regime [43].

This height is given by equilibrating the DEP force and gravity

$$Re(f_{CM})\nabla E^2 \approx \frac{(\rho_p - \rho_l)g}{\varepsilon_l} \tag{8.22}$$

Rather than static measurements, levitation is now usually combined with an external hydrodynamic flow, as we will see in Section 8.2.7.1.

8.2.4.4 The Crossover Frequency

From (8.19) and as illustrated on Figure 8.11, the behavior of the particles switches from positive to negative DEP when the frequency of the applied field varies. This frequency is called the crossover frequency and is a particularly convenient (although far from complete) way to characterize particles. For instance, a crossover frequency spectrum for different conditions, such as variable solvent conductivities, characterizes the dielectric constant and conductivity of the considered particle or complex bioparticles [29].

This frequency is easily calculated by solving the equation $Re(f_{CM}) = 0$, and its solution is then given by

$$f_0 = \frac{1}{2\pi} \sqrt{\frac{(\sigma'_p + 2\sigma_l)(\sigma'_p - \sigma_l)}{(\varepsilon_p + 2\varepsilon_l)(\varepsilon_p - \varepsilon_l)}} \qquad (8.23)$$

where σ'_p is given by (8.21).

A fit of such spectra gives an accurate description of some characteristics of the particle (Figure 8.14).

σ'_p is also a function of the particle's radius. Equation (8.23) has been tested for identical particles of various radii. The excellent agreement between theory and experiments is a good way to determine the surface conductivity of these particles (see Figure 8.15).

8.2.5 Electrorotation and Traveling Wave

We have seen how DEP could be used to apply a force on particles in a solvent and eventually to trap them. A similar physical principle can be used to set them in rotation. Consider a polarizable particle in a rotating electric field, such as the one that can be formed using four electrodes successively addressed. If the angular velocity of the field is sufficiently high, the induced-dipole orientation makes some angle with the electric field and, as a consequence, a torque acts on the particle and makes it rotate (see Figure 8.16). This torque experienced by the particle is then given by [27–29, 41]:

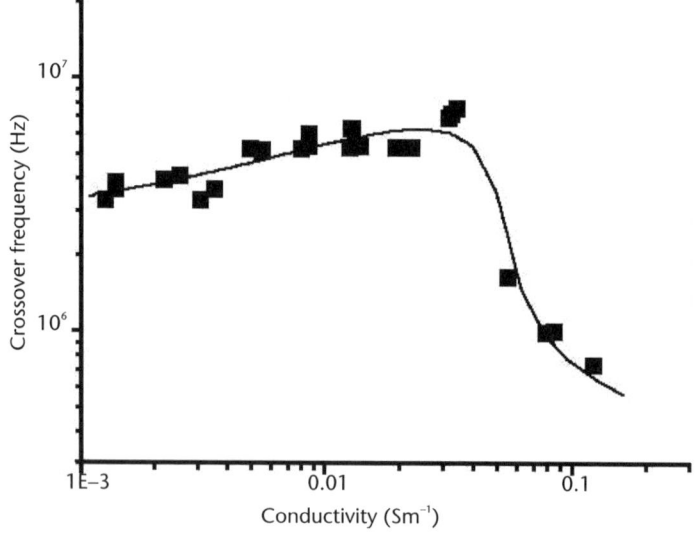

Figure 8.14 Fit of (8.23) (line) over experimental crossover frequencies measured at different medium conductivities (squares). Particles are 216-nm polystyrene latex beads. (*From:* [44]. © 2002 Elsevier. Reprinted with permission.)

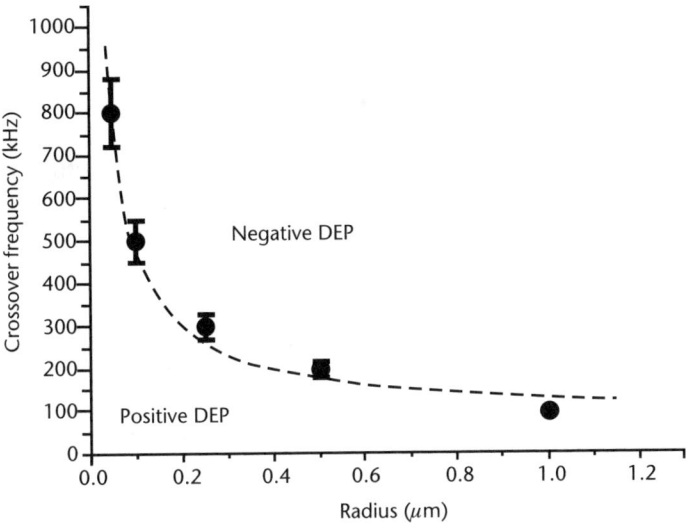

Figure 8.15 Crossover frequency between positive and negative DEP for polystyrene latex beads of different radii in pure water. The line is a best fit of (8.23).

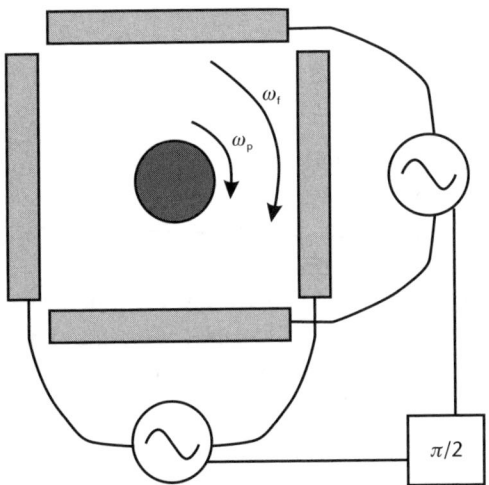

Figure 8.16 Physical principle of electrorotation.

$$\Gamma = -4\pi\varepsilon_0\varepsilon_{r,l}a^3\,\mathrm{Im}(f_{\mathrm{CM}})\cdot E^2 \tag{8.24}$$

with the same conventions as in the preceding part.

Although electrorotation presents some similarities with DEP, it also bears some fundamental differences in the sense that the direction of the ac electric field is not fixed but is continuously moving.

The full calculation based on (8.24) shows that the direction of the torque depends on the electrical characteristics of the particle and the medium but not on the frequency: There is no equivalent to the crossover frequency (no change of sign in the rotation direction). However, the amplitude of this rotation is not monotonic and, depending on its sign, exhibits a maximum or a minimum precisely at the crossover frequency.

Electrorotation is successfully used as an analytical tool to probe the dielectric properties of particles or cells. By varying the frequency, dielectrophoretic spectroscopy can be performed [41, 45]. It can also be useful in complementing DEP cages to stabilize them or to get more information out of these trapping measurements.

An extension of this calculation is the application of a similar sequence to electrodes aligned on a surface. Imagine this set up as the unfolding of the electrorotation set up. Instead of having a rotating electric field, we now deal with a traveling wave of electric field [46]. Similarly to the electrorotation set up, this implies successive addressing of many electrodes linearly arranged with phase shifted signals.

Traveling waves have been effectively used for the pumping of liquids [47] and to induce the motion of particles or cells [41, 45]. As with electrorotation, high-velocity switching between the electrodes leads to the asynchronous motion of the particles. This strategy can then been coupled with microfluidics and should lead to a directed motion in a channel. However, practically, a DEP force acting on the particles often comes into play and, to avoid sticking of the particles on the electrodes, care should be taken to work in negative DEP conditions.

8.2.6 Instabilities

8.2.6.1 Pearl Chaining

We have seen how the external electric field can induce a dipole in a particle. The interaction between this dipole and the electric field then drives the particle toward the high field region (for positive DEP). The story, however, does not stop here. If there is another particle in the vicinity of the first one, the electric field experienced by this second object is not the externally applied field, but a field modified by the presence of the first particle (of course, the field experienced by the first particle is also modified by the presence of the second one) (Figure 8.17). This leads to an arrangement of the particles in *pearl chains*, where they touch each other and where

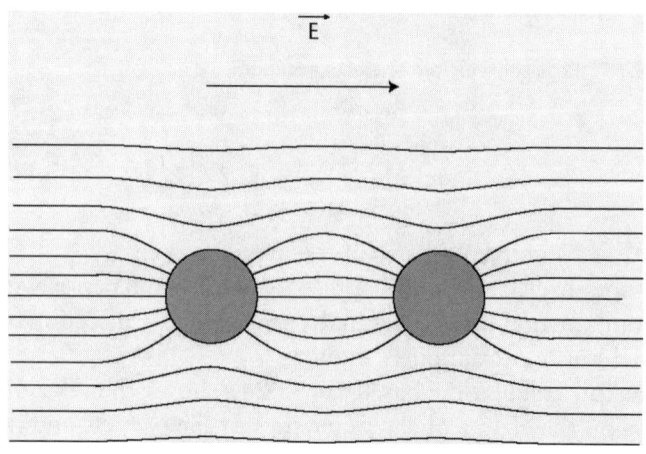

Figure 8.17 Interaction between particles. Each particle (in this case they are more polarizable than the medium) deflects the field lines and thus the effective electric field for the other particle.

this chain follows the fields lines (the whole assembly acts like a large dipole) [28, 48].

This effect effectively increases the trapping efficiency [49]. The critical value of the electric field necessary to observe pearl chaining has been computed by Pohl [27]. The calculation leads to

$$E_{crit} \approx \frac{\varepsilon_p + 2\varepsilon_l}{\varepsilon_p - \varepsilon_l} \sqrt{\frac{kT}{2\pi\varepsilon_0 \varepsilon_{r,l} a^3}} \qquad (8.25)$$

A detailed review on pearl-chaining effects can be found in [28].

Decreasing the concentration of particles in the solution is not a good way to avoid pearl chaining: As the high electric field region (close to the electrodes) collect the particles, their local concentration increases considerably in these areas to reach values for which pearl chaining is observed, even with extremely dilute initial solutions (Figure 8.18).

8.2.6.2 Electrohydrodynamic Instabilities

Most of the time, positive DEP brings hydrodynamic instabilities that can be strong enough to impair an efficient trapping or to modify the position of trapping. As a matter of fact, it has often been observed that the position of stable trapping is not at the point of maximal field but slightly away from it: on the electrodes themselves or on their edges. These effects can be qualitatively understood by taking into account field-induced flows in the solution.

Most of the time, thermal gradients in the solution are the strongest contribution to these effects [47]. The dissipated heat is given by $Q \sim \sigma E^2$. Thus, the same field nonuniformities that give rise to the DEP effect cause local thermal gradients that in turn are the source of convective flows. In negative DEP regime, these flows actually stabilize the trapping. However, they tend to destabilize positive DEP trapping. Temperature rise has other consequences, such as modifying the conductivgities or the permitivities, which are both a function of temperature (see [50] for a detailed review).

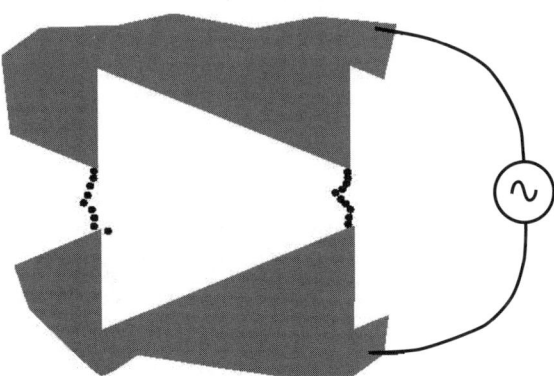

Figure 8.18 Experimental observation of pearl chaining of 0.5-μm latex particles between pointed electrodes. The distortion in the center is caused by electrohydrodynamic flows (see Section 8.2.6.2).

In addition to these thermal effects, the electric field directly interacts with the double layer present at the surface of the electrodes. Effects similar to electro-osmosis will manifest themselves at time scales larger than the time necessary to establish this double layer. Thus, for sufficiently small frequencies (that depend on the characteristics of the medium but that can be roughly estimated at 10–500 kHz), this effect, sometimes complicated by charge injection [51], manifests itself with a much higher intensity than the DEP itself. As a matter of fact, low-frequency observations where the trapping site is at the center of the electrodes or precisely on their edges can be explained by a combination of flows and DEP.

8.2.7 DEP-Based Separations

8.2.7.1 Combining DEP and a Hydrodynamic Flow

Contrarily to electrophoresis, and except for traveling-wave DEP, DEP is thus a trapping and an analytical tool. Therefore, to achieve any kind of separation with this concept, another flow should be added. In that sense, DEP-based separations are very similar to affinity chromatography, where a mixture is brought in contact with a matrix by a flow in a column. Some of the components of this mixture that have more affinity to the matrix are then trapped on it, while the other components can flow with the solvent. The trapped species is then eluted a second time by changing the solvent (for instance, its pH) or by adding a molecule that has an even higher affinity for the matrix.

DEP is the equivalent of a "smart" matrix in the sense that the affinity of particles for the traps can be tuned externally by changing the electric field characteristics. However, it still needs an eluant to flow the particles on the electrodes.

The simplest idea that comes to mind is to flow the particles on the energized electrodes. By adjusting the characteristics of the electric field, one component of the mixture is trapped on the electrodes while the others flow with the liquid. There are quite a few examples of this approach that use different geometries for the electrodes or sequences of flow in one or two directions [52–55].

Another way of achieving a good separation of particles is to combine DEP and gravity in the FFF technique. This is the coupling between DEP levitation and a Poiseuille flow, well controlled by a difference of pressure in a microfluidic channel. The height at which the particles levitate is given by (8.22) that yields after calculations [54, 56]:

$$h_{eq} = h_0 + A \ln\left(\frac{Re(f_{CM})}{(\rho_p - \rho_l)g}\right) \tag{8.26}$$

A and h_0 are empirical parameters specific to the electrode geometry.

As the Poiseuille flow is characterized by a parabolic velocity profile, the fluid velocity is different for each height, and particles of different dielectrophoretic characteristics—different $Re(f_{CM})$—will travel at different speeds, which results ultimately in different elution times. Negative DEP is particularly well adapted to this purpose, as it actively "pushes" the particles at different heights within the flow.

8.2.7.2   Dielectrophoretic Ratchets

A good illustration of the use of DEP is the so-called dielectrophoretic ratchet. This experiment can be declined in two versions [the Brownian ratchet or the shifted ratchets similar to the one sketched in Figure 8.19(a)]. They are briefly described here.

Brownian Ratchet

The theoretical models on which these particular experiments are based are described in [57]. Reference [58] provides a very complete review on these *force-free motion* phenomena. Let us imagine Brownian particles in a potential similar to the one described in Figure 8.19(a). They are trapped in the minima of this potential, and, if the potential barriers are large enough, which is supposed here, their concentration profiles are very narrow and centered close to these minima [Figure 8.19(b)]. Now, let us switch off this potential (i.e., we now impose to the particles a flat potential), the particles are going to freely diffuse, and, as a consequence, the concentration peaks will broaden [Figure 8.19(c)]. After a time t_{off}, we cycle back to the saw-tooth potential [Figure 8.19(d)]. The particles are again trapped in the minima of the potential. If t_{off} is sufficiently large, a nonnegligible fraction of them have diffused over a distance larger than the small side of the pattern of the potential; on the other hand, because of the asymmetry of this pattern, the fraction of the particles that have diffused over a distance larger than the large side of this pattern is much smaller. As a result, a fraction of these particles will be trapped in the minima next to the ones they previously occupied, and, with the conventions of Figure 8.19, more of them will shift to the left than to right [shadowed area under the concentration peak in Figure 8.19(c)]. By reiterating this process a large number of times, one

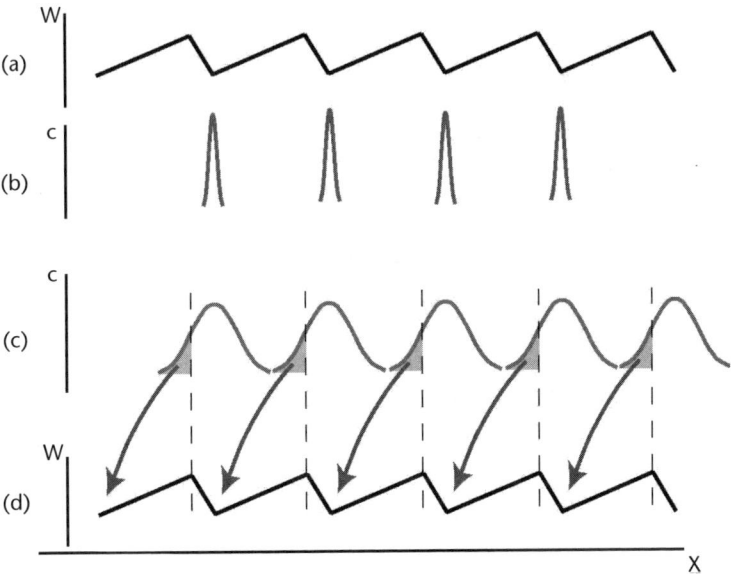

Figure 8.19   (a) Energetic potential and (b) concentration profile of the particles. The height of the energy barriers is much higher than thermal agitation. After a diffusion step of duration t_{off}, (c) the particles diffuse and (d) are trapped again when the potential is switched on. W = potential; c = concentration.

can then set these particles into motion over potentially large distances, whereas the only gradients present in the system are local. Hence, we have solved here one of the main limitations of DEP by making it a transport technique and not only an analysis technique. Furthermore, we have an adjustable "knob": t_{off}, the time during which we let the particles diffuse, is a control parameter for these experiments.

The macroscopic motion of the particles is expected to vary exponentially with their diffusion coefficient and thus with their size or their molecular weight, which is very promising for separating particles of different sizes [57].

This saw-tooth potential can be of DEP nature [31, 59, 60]. In particular, it can be created by setting an ac voltage between an electrode whose corrugations presents the "good" properties of periodicity and asymmetry and a planar one [31]. Qualitatively, by a simple *tip effect*, the electric field is of much higher intensity on the ridges than it is in the valleys. As a consequence, the asymmetry of the electrode reflects itself on the intensity of the electric field and thus on the energetic potential. This is confirmed by a careful computation of the electric field using finite elements. When experiencing the electric field at high enough frequencies, the beads are confined in the valleys (negative DEP).

The macroscopic mean velocity V of a particle at a given t_{off} is quantitatively described by the simple model exposed earlier (no adjustable parameter). Figure 8.20 plots the velocities of two different latex beads as a function of t_{off}. These curves exhibit a maximum whose position at a given t_{off} is very dependent on the size of the particles. In consequence, this is an extremely promising, although quite slow, technique.

Shifted Ratchets

The previous idea (Brownian ratchet) relies on diffusion. It implies small velocities, a disadvantage that can be corrected if we use two potentials similar to the one

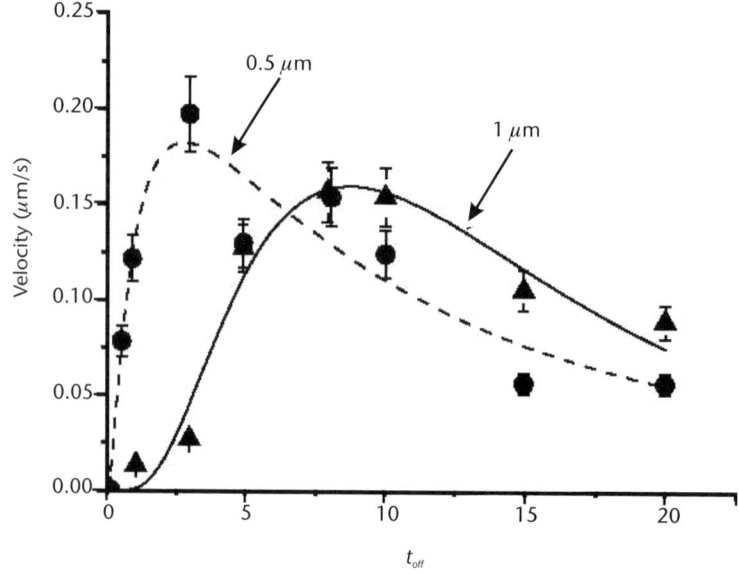

Figure 8.20 Macroscopic velocities as a function of t_{off} (diffusion step) for two different sizes of latex beads. In this particular case, setting $t_{off} = 3$ seconds leads to a factor of 10 between the velocities.

described in the preceding part [Figure 8.19(a)]. Here, these potentials are *shifted* by a fraction of their common period and addressed successively [61].

When the commutation time is too small, the particle cannot escape the corresponding trap, and the macroscopic velocity is zero. However, when both times are long enough, the particles have enough time to move by one total period per time cycle. This velocity V_{opt}, is the optimal velocity. We can rephrase this statement in terms of mobilities instead of residence times: For identical residence times, particles will have either a zero velocity or an optimal velocity according to their mobilities. In other words, this device is a *filter* according to the mobility of the particles. Moreover, the mobility threshold of this filter can be chosen by tuning the two residence times. On a separation point of view, this filter-like situation is obviously a great improvement compared to conventional techniques, as the velocity of some of the considered particles is exactly zero, and even subtle differences in mobilities should be usable for a separation.

With the use of two-dimensional electrodes sputtered on gold, a DEP trapping with the desired characteristics can be obtained by applying an ac voltage between the two electrodes of one slide. To get successively two of these potentials, two of these plates are stacked with their gold sides facing each other (Figure 8.21).

This situation emphasizes some of the points listed earlier: Negative dielectrophoresis is difficult to deal with in this particular geometry, as the regions of weakest electric field are located outside of the central channel, therefore expelling the particles from it. It is a strikingly different situation from the one dealt with in the preceding part, where both in the positive and in the negative dielectrophoresis regimes, particles were confined close to the grating surface. But the use of positive DEP regime is not trivial either: Particles not initially present in the channel tend to be collected in it, increasing the concentration. Furthermore, electrohydrodynamic flows develop on the tips of the electrodes generating recirculations of particles around the trapping zones.

In Figure 8.22, one can see a typical sequence of the migration of the particles (in that particular case, 0.5-μm latex), as the potential is switched from one pair of electrodes to the other.

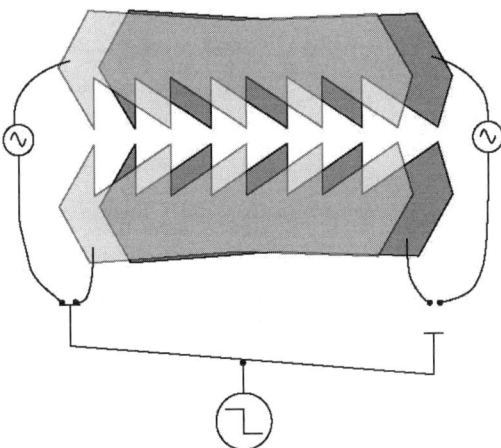

Figure 8.21 Stacked "Christmas-tree" asymmetric electrodes. These two pairs of electrodes are addressed successively.

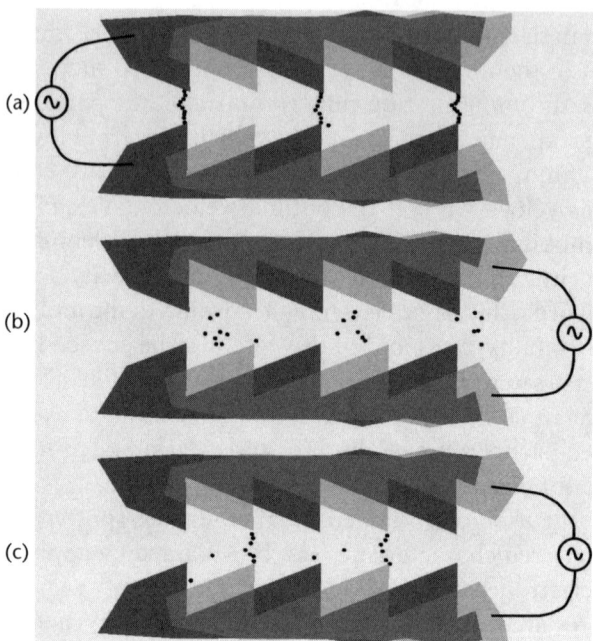

Figure 8.22 Behavior of 0.5-μm latex beads when the voltage is switched from (a) the lower set of electrodes to (b, c) the upper set; in (c), 7 seconds after switching. Note the pearl chaining between the tips of the electrodes.

The particles indeed move from one dielectrophoretic trap on one pair of electrodes to the next one on the other pair when switching the field from one plate to the other. The analysis of the experimental results is of the same nature as in the Brownian ratchet experiment. And the expected velocity regimes, $V = 0$ or $V = V_{opt}$, are indeed observed for two different sizes [62]. Although the dielectrophoretic behaviors are more complex in this geometry than they are in the Brownian ratchets case, the observed filter effect, along with the high velocities reached by the beads, make this realization highly promising in separations problems. For this particular device, as for many others presented here, an improvement in the performances will come from a better control of the microfluidics.

8.3 Electro-Osmosis

Let us consider a charged surface in an aqueous solution. The charge of the surface is balanced with the Debye layer of counter-ions extending over a distance κ^{-1} (8.5). An electric field E parallel to this surface acts on this excess of charges, which moves the fluid with them at a velocity $v(z)$. We can then write the Navier-Stokes equation (see Chapter 1)

$$\rho E + \eta \frac{\partial^2 v}{\partial z^2} = 0 \tag{8.27}$$

ρ and η are, respectively, the density and viscosity of the fluid. The other equation is the Poisson equation (8.1) that gives ρ as a function of the electric potential ψ.

$$\rho = -\varepsilon_l \nabla^2 \psi$$

where ε_l is the fluid permitivity. After substitution into (8.27), we obtain

$$E\varepsilon_l \frac{\partial^2 \psi}{\partial z^2} = \eta \frac{\partial^2 v}{\partial z^2} \tag{8.28}$$

On the surface, the usual boundary conditions apply: the velocity on the surface is 0 (no slippage) and the electric potential velocity is $\psi = \zeta$.

Far from the surface, $\dfrac{\partial \psi}{\partial z} = \dfrac{\partial v}{\partial z} = 0$.

Integrating twice yields

$$v(z) = \frac{\varepsilon_l E}{\eta}\left(\psi(z) - \zeta\right) \tag{8.29}$$

On a plane surface, ψ decreases exponentially within the double layer and vanishes far from it.

We can thus derive the electro-osmotic velocity V_{eo} by writing (8.29) far from the surface and we obtain

$$V_{eo} = -\frac{\varepsilon_l E \zeta}{\eta} \tag{8.30}$$

Depending on the application, these electro-osmotic flows are either a nuisance by interfering with the electrophoretic motilities or a powerful means to set liquids into motion by a controlled plug flow.

This calculation is also the basis of electrophoresis. We have considered the case of a fixed surface and have calculated the flows around it. Electrophoresis is the exact opposite situation: the solid moves under the application of a dc electric field in an immobile fluid (Figure 8.23).

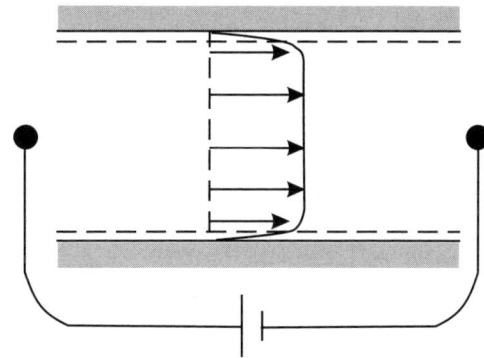

Figure 8.23 Velocity profile of an electro-osmotic plug flow in a negatively charged channel such as fused silica. The thickness of the Debye layer has been exaggerated.

References

[1] Tiselius, A., "A New Apparatus for Electrophoretic Analysis of Colloidal Mixtures," *Trans. Faraday Soc.*, Vol. 33, 1937, p. 524.

[2] Hunter, R. J., *Zeta Potential in Colloid Science*, New York: Academic Press, 1988.

[3] Smoluchowski, M. von, "Contribution à la théorie de l'endosmose électrique et de quelques phénomènes corrélatifs," *Bull. Int. Univ. Sci. Cracovie*, Vol. 8, 1903, pp. 182–200.

[4] Russel, W. B., D. A. Saville, and S. R. Schowalter, *Colloidal Dispersions*, Cambridge, England: Cambridge University Press, 1991.

[5] de Gennes, P. G., *Scaling Concepts in Polymer Physics*, Ithaca, NY: Cornell University Press, 1979.

[6] Long, D., J. L. Viovy, and A. Ajdari, "Simultaneous Action of Electric Fields and Nonelectric Forces on a Polyelectrolyte: Motion and Deformation" *Phys. Rev. Lett.*, Vol. 76, No. 20, 1996, pp. 3858–3861.

[7] Barrat, J. L., and J. F. Joanny, "Theory of Polyelectrolyte Solutions," *Adv. Chem Phys.*, Vol. 94, 1996, pp. 1–66.

[8] Ogston, A. G., "The Spaces in a Uniform Random Suspension of Fibres," *Trans. Faraday Soc.*, Vol. 54, 1958, 1754–1757.

[9] Rodbard, D., and A. Chrambac, "Unified Theory for Gel Electrophoresis and Gel Filtration," *Proc. Natl. Acad. Sci.*, Vol. 65, No. 4, 1970, pp. 970–977.

[10] Slater, G. W., and J. Noolandi, "New Biased Reptation Model for Charged Polymers," *Phys. Rev. Lett.*, Vol. 55, 1985, pp. 1579–1582.

[11] Schwartz, D. C., and C. R. Cantor, "Separation of Yeast Chromosome-Size DNAs by Pulsed Field Gradient-Gel Electrophoresis," *Cell*, Vol. 37, No. 1, 1984, pp. 67–75.

[12] Viovy, J. -L., "Electrophoresis of DNA and Other Polyelectrolytes: Physical Mechanisms," *Rev. Mod. Phys.*, Vol. 72, No. 3, 2000, pp. 813–872.

[13] Burmeister, M., and L. Ulanovsky, (eds.), "Pulsed-Field Gel Electrophoresis, Protocols, Methods,Theories," in *Methods in Molecular Biology No. 12*, Totowa, NJ: Humana Press, 1992.

[14] Gurrieri, S., et al., "Real-Time Imaging of the Reorientation Mechanisms of YOYO-Labelled DNA Molecules During 90° and 120° Pulsed Field Gel Electrophoresis," *Nucl. Acids Res.*, Vol. 24, No. 23, 1996, pp. 4759–4767.

[15] Auroux, P. -A., et al., "Micro Total Analysis Systems. 2. Analytical Standard Operations and Applications," *Anal. Chem.*, Vol. 74, No. 12, 2002, pp. 2637–2652.

[16] Isambert, H., et al., "Electrohydrodynamic Patterns in Charged Colloidal Solutions," *Phys. Rev. Lett.*, Vol. 78, No. 5, 1997, pp. 971–974.

[17] Volkmuth, W., and R. H. Austin, "DNA Electrophoresis in Microlithographic Arrays," *Nature*, Vol. 358, 1992, pp. 600–602.

[18] Bakajin, O., et al., "Separation of 100-Kilobase DNA Molecules in 10 Seconds," *Anal. Chem.*, Vol. 73, 2001, pp. 6053–6056.

[19] Doyle, P., et al., "Self-Assembled Magnetic Matrices for DNA Separation Chips," Science, Vol. 295, No. 5563, 2002, p. 2237.

[20] Branton, D., et al., "Characterization of Individual Polynucleotide Molecules Using a Membrane Channel," *Proc. Natl. Acad. Sci.*, Vol. 93, 1996, pp. 13770–13773.

[21] Tegenfeldt, J., et al. "The Dynamics of Genomic-Length DNA Molecules in 100 Nanometer Channels," *Proc. Nat. Acad. Sci.*, Vol. 101, 2004, p. 10979.

[22] Han, J., and H. G. Craighead, "Separation of Long DNA Molecules in a Microfabricated Entropic Trap Array," *Science*, Vol. 288, 2000, p. 1026.

[23] Pernodet, N., et al., "DNA Electrophoresis on a Flat Surface," *Phys. Rev. Lett.*, Vol. 85, 2000, pp. 5651–5654.

[24] Liebler, D., *Introduction to Proteomics: Tools for a New Biology*, Totowa, NJ: Humana Press, 2002.

[25] O'Farrell, P. H., "High Resolution Two-Dimensional Electrophoresis of Proteins," *J. Biol. Chem.*, Vol. 250, 1975, pp. 4007–4021.

[26] Toda, T., et al., "A Comparative Analysis of the Proteins Between the Fibroblasts from Werner's Syndrome Patients and Age-Matched Normal Individuals Using Two-Dimensional Gel Electrophoresis," *Mech. Ageing Dev.*, Vol. 100, 1998, pp. 133–143.

[27] Pohl, H. A., *Dielectrophoresis*, Cambridge, England: Cambridge University Press, 1978.

[28] Jones, T. B., *Electromechanics of Particles*, Cambridge, England: Cambridge University Press, 1995.

[29] Hughes, M. P., *Nanoelectromechanics in Engineering and Biology*, Boca Raton, FL: CRC Press, 2003.

[30] Dascalescu, L., et al., "Electrostatics of Conductive Particles in Contact with a Plate Electrode Affected by a Non-Uniform Electric Field," *J. Phys. D: Appl. Phys.*, Vol. 34, 2001, pp. 60–67.

[31] Gorre-Talini, L., S. Jeanjean, and P. Silberzan, "Sorting of Brownian Particles by the Pulsed Application of an Asymmetric Potential," *Physical Review E*, Vol. 56, 1997, pp. 2025–2034.

[32] Ballario, C., A. Bonincontro, and C. Cametti, "Dielectric Dispersions of Colloidal Particles in Aqueous Suspensions with Low Ionic Conductivity," *J. Coll. Interf. Sci.*, Vol. 54, No. 3, 1976, pp. 415–423.

[33] Foster, K. R., F. A. Sauer, and H. P. Schwan, "Electrorotation and Levitation of Cells and Colloidal Particles," *Biophysical J.*, Vol. 63, No. 1, 1992, pp. 180–190.

[34] Price, J. A. R., J. P. H. Burt, and R. Pethig, "Applications of a New Optical Technique for Measuring the Dielectrophoretic Behavior of Microorganisms," *Biochim. Biophys. Acta*, Vol. 964, No. 2, 1988, pp. 221–230.

[35] Schnelle, T., et al., "Three-Dimensional Electric Field Traps for Manipulation of Cells—Calculation and Experimental Verification," *Biochim. Biophys. Acta*, Vol. 1157, 1993, pp. 127–140.

[36] Chou, C. -F., et al., "Electrodeless Dielectrophoresis of Single- and Double-Stranded DNA" *Biophys J.*, Vol. 83, No. 4, 2002, pp. 2170–2179.

[37] Marquet, C., et al., "Rectified Motion of Colloids in Asymmetrically Structured Channels," *Physical Review Letters*, Vol. 88, 2002, pp. 168301–168304.

[38] Lapizco-Encinas, B. H., et al., "Dielectrophoretic Concentration and Separation of Live and Dead Bacteria in an Array of Insulators," *Anal. Chem.*, Vol. 76, No. 6, 2004, pp. 1571–1579.

[39] Fielder, S., et al., "Electrocasting—Formation and sStructuring of Suspended Microbodies Using A.C. Generated Field Cages," *Microsyst. Tech.*, Vol. 2, 1995, pp. 1–7.

[40] Washizu, M., and O. Kurosawa, "Electrostatic Manipulation of DNA in Microfabricated Structures," *IEEE Trans. Ind. App.*, Vol. 26, No. 6, 1990, pp. 1165–1172.

[41] Fuhr, G., U. Zimmermann, and S. G. Shirley, "Cell Motion in Time-Varying Fields: Principles and Potential," in *Electromanipulation of Cells*, U. Zimmermann and G. A. Neil, (eds.), Boca Raton, FL: CRC Press, 1996, pp. 259–328.

[42] Hughes, M. P., H. Morgan, and F. J. Rixon, "Measurements of the Properties of Herpes Simplex Virus Type 1 Virions with Dielectrophoresis," *Bioch. Biophys. Acta,* Vol. 1571, 2002, pp. 1–8.

[43] Kaler, K. V. I. S., and T. B. Jones, "Dielectrophoretic Spectra of Single Cells Determined by Feedback-Controlled Levitation," *Biophys. J.*, Vol. 57, No. 2, 1990, pp. 173–182.

[44] Hughes M. P, "Dielectrophoretic Behavior of Latex Nanospheres: Low-Frequency Dispersion," *J. Coll. Interf. Sci.*, Vol. 250, 2002, pp. 291–294.

[45] Hagedorn, R., et al., "Traveling-Wave Dielectrophoresis of Microparticles," *Electrophoresis*, Vol. 13, No. 1-2, 1992, pp. 49–54.

[46] Yang, J., et al., "Dielectric Properties of Human Leukocyte Subpopulations Determined by Electrorotation as a Cell Separation Criterion," *Biophys. J.*, Vol. 76, 1999, pp. 3307–3314.

[47] Fuhr, G., T. Schnelle, and B. Wagner, "Traveling Wave-Driven Microfabricated Electrohydrodynamic Pumps for Liquids," *J. Micromech. Microeng.*, Vol. 4, No. 4, 1994, pp. 217–226.

[48] Hu, Y., et al., "Observation and Simulation of Electrohydrodynamic Instabilities in Aqueous Colloidal Suspensions," *J. Chem. Phys.*, Vol. 100, No. 6, 1994, pp. 4674–4682.

[49] Schnelle, T., et al., "Trapping of Viruses in High-Frequency Electric Field Cages," *Naturwissenschaften*, Vol. 83, No. 4, 1996, pp. 173–176.

[50] Ramos, A., et al., "AC Electrokinetics: A Review of Forces in Microelectrodes Structures," *J. Phys. D*, Vol. 31, 1998, pp. 2338–2353.

[51] Suzuki, M., "Propagating Transitions of Electroconvection," *Phys. Rev. A*, Vol. 31, No. 4, 1985, pp. 2548–2555.

[52] Gascoyne, P. R. C., and J. Vykoukal, "Particle Separation by Dielectrophoresis," *Electrophoresis*, Vol. 23, No. 13, 2002, pp. 1973–1983.

[53] Markx, G. H., M. S. Talary, and R. Pethig, "Separation of Viable and Nonviable Yeast Using Dielectrophoresis," *J. Biotechnol.*, Vol. 32, No. 1, 1994, pp. 29–37.

[54] Morgan, H., M. P. Hughes, and N. G. Green, "Separation of Submicron Bioparticles by Dielectrophoresis," *Biophys. J.*, Vol. 77, No. 1, 1999, pp. 516–525.

[55] Markx, G. H., and R. Pethig, "Dielectrophoretic Separation of Cells—Continuous Separation," *Biotechnol. Bioeng.*, Vol. 45, 1995, p. 337.

[56] Huang, Y., et al., "Introducing Dielectrophoresis as a New Force Field for Field-Flow Fractionation," *Biophys. J.*, Vol. 73, No. 2, 1997, pp. 1118–1129.

[57] Ajdari A., and J. Prost, "Mouvement Induit par un Potentiel Périodique de Basse Symétrie: Diélectrophorèse Pulsée," *C. r. Acad. sci., Sér. 2*, Vol. 315, No. 13, 1992, pp. 1635–1639.

[58] Reimann, P., "Brownian Motors: Noisy Transport Far from Equilibrium," *Phys. Rep.*, Vol. 361, Nos. 2–4, 2002, pp. 257–265.

[59] Rousselet, J., et al., "Directional Motion of Brownian Particles Induced by a Periodic Asymmetric Potential," *Nature*, Vol. 370, 1994, pp. 446–448.

[60] Faucheux, L. P., and A. Libchaber, "Selection of Brownian Particles," *J. Chem. Soc. Faraday Trans.*, Vol. 91, 1995, pp. 3163–3166.

[61] Chauwin J. -F., A. Ajdari, and J. Prost, "Force-Free Motion in Asymmetric Structures—A Mechanism Without Diffusive Steps," *Europhys. Lett*, Vol. 27, No. 6, 1994, pp. 421–426.

[62] Gorre-Talini, L., J. Spatz, and P. Silberzan, "Dielectrophoretic Ratchets," *Chaos*, Vol. 8, No. 3, 1998, pp. 650–56.

CHAPTER 9

Conclusion

At the end of this book, let us recall the different steps that we have made toward the comprehension and prediction of the mechanical behavior of micro- and nano-particles and macromolecules.

Because these particles are immerged in a buffer liquid, we have focused first on carrier fluid behavior, by studying microfluidic flows in microsystems. Microfluidics may be considered a new science by itself. On many points it departs from the classic view of macroscopic fluid dynamics. We have presented and analyzed the different forms of microflows that can be found in microsystems for biology, from continuous microflows (when a single phase liquid flows continuously in microchannels), to plug flows (when plugs of immiscible liquids successively move in the microchannels), and to digital microfluidics (when separated microdrops are displaced step by step on a solid surface).

Second, starting from the observation that the biological molecules or the microparticles we want to track and guide do not have exactly the same behavior as the buffer fluid molecules due to their size and weight, we have focused on molecular diffusion and transport aspects. Because our approach aims for in vitro problems, as well as in vivo situations, we have given attention to the behavior of particles in confined volumes.

Next, as the main purpose of biochips is the study of biological targets like DNA, proteins, and cells, the principle of key-lock recognition has been presented; this principle is the basis of in vitro biochips as well as that for the in vivo interventions' addressing of specific targets—usually cells. Biorecognition is in reality a biochemical reaction that takes different forms. We have analyzed some fundamental reactions like DNA hybridization and protein enzymatic digestion. Their kinetics have been carefully detailed both from a theoretical standpoint and in the different forms they take in a biochip environment.

Because it has been observed that transport of particles by a buffer fluid is not always specific enough (i.e., the particles of interest cannot all be transported to the reactive surface), we have presented additional tools to transport and manipulate biological objects (e.g., the use of magnetic particles as transport vectors and the specific effects of electric fields on these biological objects).

Finally, as it was found that a more experimental point of view would complete the theoretical and modeling approach of the book, the last chapter has been dedicated to experimental manipulations of biological targets.

Through this entire book, we have shown that precise handling and manipulation of micro- and nanoparticles and macromolecules are at the heart of biotechnology. The theoretical background and the modeling approach presented here are and will likely remain the basis for the comprehension of the phenomena involved in biochips and bio-MEMS, even if biotechnology is rapidly evolving and new developments emerge constantly. Some new developments are already clearly foreseen, others are still unpredictable. Among the foreseen developments, we can list:

- *Digital microfluidics:* This is going to continue to gain momentum and become complementary to the more classic microfluidic solutions.
- *Downscaling,* with a continuous trend towards miniaturization of biochips and bio-MEMS (e.g., the use of nanotubes of carbon for improving the detection accuracy). This trend seems unavoidable since the ultimate goal is to work on single targets.
- *Integration:* This is because miniaturization means also designing biochips containing a maximum of functions. Such designs simplify the problems of microfluidic connection between the different functions.
- *Living cells:* Pharmaceutical industry involvement is going to contribute to the development of biochips for cells, where single living cells are isolated in microcusps. Testing of drugs directly on living cells is expected to boost the discovery of new drugs.
- *Cancerous protein markers:* In oncology, dramatic improvements are expected from the determination of marker proteins characterizing each specific type of cancer. In this field, new biotechnological tools for early detection of these markers are expected to totally change the prognostics of cancer.
- *Drug guidance:* Development of the techniques of drug guidance in the human body to directly address defective cells has become another major research topic. This is a domain where biotechnological developments could contribute to the efficiency of drug guidance.
- *Self-assembly:* Side applications of biotechnology have already appeared and are likely to increase with the techniques of self-assembly of nanoparticles and macromolecules.

An analysis of these listed trends shows that the theoretical background of this book will remain relevant in the future. On a modeling point of view, the progressive evolution toward smaller scales and smaller concentrations in particles is going to put forward discrete numerical approaches. Even at a very small scale, the buffer (carrier) liquid still satisfies the continuum assumption, and its behavior can be simulated by the classical Navier-Stokes formulation. On the other hand, discrete methods—like what we have presented in Chapters 4 and 5—are more appropriate to model the behavior of the transported particles when the number of these particles becomes small. Numerical methods using a relevant coupling of the two approaches will progressively be preferred to model such types of problems.

To conclude, the scientific domain of biotechnologies is complex and requires a wide and strong scientific background. However, it is rewarding because it addresses human conditions; it is also motivating because it is wide open to inventiveness and imagination.

About the Authors

Jean Berthier is a scientist at CEA/LETI (the Laboratory for Electronics, Technology and Information) in Grenoble, France, and he teaches mathematical methods in fluid mechanics at the University of Grenoble. He has also been working at the Sandia Laboratories and Los Alamos Laboratory on multiphase fluid flow computation. He received an engineering diploma from the Institut National Polytechnique, and an M.S in mathematics and an M.S. in fluid mechanics, both from Grenoble University.

Pascal Silberzan is a senior scientist at the Physico-Chimie Curie Laboratory, a joint laboratory between the Centre National de la Recherche Scientifique and the Institut Curie in Paris, France. He received his Ph.D. from the Collège de France and the University Paris 6, France.

Index

343